BEHAVIORAL PHARMACOLOGY
The Current Status

W0079716

FASEB MONOGRAPHS

General Editor: KARL F. HEUMANN

A Continuation Order Plan is available for this series. A continuation order will bring delivery of each new volume immediately upon publication. Volumes are billed only upon actual shipment. For further information please contact Plenum Press.

BEHAVIORAL PHARMACOLOGY
The Current Status

Edited by
Bernard Weiss and Victor G. Laties
University of Rochester School of Medicine and Dentistry

FASEB, Bethesda
PLENUM PRESS, New York and London

Library of Congress Cataloging in Publication Data

Main entry under title:

Behavioral pharmacology.

(FASEB monographs; v. 4)
Contains papers given at the 2-day symposium, which took place at
the Federation of American Societies for Experimental Biology meetings
in 1974.
"Originally appeared in Federation proceedings, vol. 34, No. 9, August
1975."
Includes bibliographical references and index.
1. Psychopharmacology—Congresses. 2. Behavior modification—Congresses. I. Weiss, Bernard. II. Laties, Victor G. III. Federation of American Societies for Experimental Biology, Federation proceedings. IV.
Series: Federation of American Societies for Experimental Biology.
FASEB monographs; v. 4.

RM315.B358 1976	615'.78	76-8408

ISBN 978-1-4684-2636-6   ISBN 978-1-4684-2634-2 (eBook)
DOI 10.1007/978-1-4684-2634-2

The material in this book originally appeared in *Federation Proceedings*
Vol. 34, No. 9, August 1975. First published in the present form
by Plenum Publishing Corporation in 1976.

Copyright ©1975, by the Federation of American Societies
Softcover reprint of the hardcover 1st edition 1975
for Experimental Biology

Plenum Press, New York is a Division of Plenum Publishing Corporation
227 West 17th Street, New York, N.Y. 10011

All rights reserved

No part of this book may be reproduced, stored in a retrieval system,
or transmitted, in any form or by any means, electronic, mechanical,
photocopying, microfilming, recording, or otherwise, without written
permission

Contents

SESSION III
BEHAVIORAL TOXICOLOGY

SESSION IV
CONTINGENCIES OF REINFORCEMENT AS
DETERMINANTS OF DRUG RESPONSE

Contents

The future of behavioral pharmacology

BERNARD WEISS

*Department of Radiation Biology and Biophysics, University of Rochester
School of Medicine and Dentistry, Rochester, New York 14642*

This issue of *Federation Proceedings* contains the papers given at the 2-day symposium, *Current Status of Behavioral Pharmacology*, which took place at the Federation meetings in 1974. We have here a survey of many of the most exciting fields of this discipline, from some of its most significant practitioners. The readers will also see that the word "future" in the title of this introductory essay is most suitable, since the issues and approaches that appear in the succeeding pages surely will dominate behavioral pharmacology for at least the next few years.

I take a special delight in seeing this project finally come to fruition in print. It began 2 years ago when, as president of the Division of Psychopharmacology of the American Psychological Association (Division 28), and also as a member of ASPET, I met with the ASPET council to discuss ways in which the two organizations might formalize some of the relationships that already had brought us together on a few issues. Many of us who had been involved with the affairs of Division 28 sought to develop a new, co-

herent and continuous liaison with ASPET.

The 1,600 members of Division 28 speak on drug issues for an organization of over 35,000 psychologists. ASPET, in many ways, is the drug-issue voice for the nearly 14,000 members of the Federation. My concern at that time was as much political as scientific; policy decisions were and are made, particularly in Government, in the absence of an adequate dialogue with the scientific community. The fault for this lies not only with those who shape Government policy but also with scientists themselves, who tend to exercise their influence in fragmentary ways.

Still, for both organizations, our main functions are scientific. The joint ASPET-Division 28 committee, formed to develop liaison mechanisms, confirmed these functions as the first priority of our collaboration. It is easy to see why. Although many pharmacologists and psychologists have learned to navigate one another's jargon, to accept one another's point of view, and even to acquire one another's skills, many

1

are still isolated within their own disciplines, a limitation of scope that is unhealthy. What better place to begin an ecumenical movement than the Federation Meetings; and, given the opportunity afforded by the apostasy of the biochemists in 1974, what better mechanism than a comprehensive survey of where we presently stand?

Symposia devoted at least in part to behavioral pharmacology have occurred regularly since 1958 (1–7). This one, which lasted 2 days, came to pass thanks to the eloquent persistence of pharmacologists such as Dr. Louis Harris, the sympathetic support of the ASPET council, the managerial expertise of the ASPET and FASEB central offices, the support of the Executive Board of Division 28, and, above all, the willingness of the participants.

I hope that you will find it as worthwhile as I did, and as exciting.

REFERENCES

1. Symposium on neuropharmacology. *Federation Proc.* 17: 1004–1043, 1958.
2. Effects of hallucinogenic drugs in man. *Federation Proc.* 20: 874–900, 1961.
3. Types of behavior on which drugs act. *Federation Proc.* 23: 799–850, 1964.
4. Central cholinergic transmission and its behavioral aspects. *Federation Proc.* 28: 89–159, 1969.
5. New concepts and approaches to the study of drug dependence and tolerance. *Federation Proc.* 29: 2–32, 1970.
6. Marihuana and its surrogates. *Pharmacol. Rev.* 23: 263–380, 1971.
7. Discriminable properties of drugs and state-dependent learning. *Federation Proc.* 33: 1785–1835, 1974.

Environmental influences affecting the voluntary intake of drugs: an overview

JAMES R. WEEKS

Experimental Biology Division, The Upjohn Company
Kalamazoo, Michigan 49002

ABSTRACT

Drug self-administration studies in animals have generally used drugs that are abused by man. The oral route by adding drug to the drinking water is simple, but the bitter taste of many drugs requires that the animals first be forced to consume treated water. The intravenous route, wherein relatively unrestrained rats or monkeys press a lever to obtain intravenous drugs, permits precise control over the dose and can be readily adapted to schedules and manipulations customarily used in the behavioral sciences. Environmental factors affecting drug intake include the dose, its schedule of administration, and conditioning of drug administration to secondary cues. There are differences in drug self-administration of stimulant drugs (as amphetamines) and depressants (as morphine and barbiturates). There is an inverse relation between the size of the dose and number of injections taken, but for stimulants daily intake will remain constant whereas for depressants smaller doses are only partially compensated for by increased numbers of injections. Likewise, drug intake of stimulants is better maintained on ratio schedules. Neutral stimuli, as lights or buzzers, paired with drug injections can be used to elicit conditioned responses. Such responses have been used to evaluate the reinforcing properties of drugs.—WEEKS, J. R. Environmental influences affecting the voluntary intake of drugs: an overview. *Federation Proc.* 34: 1755–1758, 1975.

The theme of this session is the behavior engendered by an experimental subject taking a drug, in contrast to giving a drug to the subject and then observing the drug's effect on some other behavior. The emphasis is on drugs of abuse, in animals mostly by the self-injection technique, and in humans by drinking alcohol under controlled conditions. My comments generally will ignore the extensive work in which alcohol has been added to the drinking water of animals, and be more concerned with self-administration of other drugs of abuse.

HISTORICAL INTRODUCTION

The first published work on self-injection by animals was by Headlee, Coppock and Nichols (17) in 1955. They clearly showed that in physically dependent rats morphine, injected intraperitoneally, was a reinforcing agent. These studies attracted little attention at the time, possibly because the rats were tightly · restrained and experiments were only of a 2-hour duration. A prolonged period of self-administration apparently was needed to qualify as a bona fide experimental addiction. Nichols and his colleagues, in the late 1950's, then concentrated on the use of morphine in the drinking water as a method of studying experimental addiction (24, 25).

In the early 1960's, techniques for intravenous self-administration in long-term experiments were developed. Methods for monkeys were developed at the University of Maryland by Schuster and Thompson (32) and at the University of Michigan by Deneau and Yanagita (9, 49), and I developed a method for rats at The Upjohn Company (36, 37). These methods involved placing a chronic indwelling catheter in the jugular vein, fitting the animal with a suitable harness and leash to allow relatively unrestricted movement, and connecting the catheter to a motor-driven syringe. Pressing a lever in the cage would then start the motor and infuse a fixed dose of drug. In terms of the behavioral sciences, pressing the lever would be the response and a subsequent injection the contingent reinforcement.

EXTERNAL ENVIRONMENTAL FACTORS

The dictionary defines environment as any outside influence which can affect life or behavior, and in addition Claude Bernard viewed living things as having both external and internal environments, the latter to be held constant by appropriate physiological and biochemical mechanisms. I shall now comment briefly on some environmental factors that may have bearing on the voluntary intake of drugs. First of all, we must realize that when drugs serve as reinforcements, the pharmacological actions of some drugs may affect response patterns beyond those usually generated by shock, noise, or food. Drugs of abuse can be divided into two classes: those that induce physical dependence and those that do not. Opiates and barbiturates, as morphine and pentobarbital, are typical of those that induce physical dependence, whereas central nervous system stimulants, as cocaine and amphetamines, do not. Once physical dependence is established, the reinforcing quality of a drug may be due to relief of the punishment of an impending withdrawal reaction, or a form of "escape training." The same drug, if offered to a naive subject, may either be without reinforcing properties or be reinforcing for a different reason. A drug which is self-administered by a naive animal, without pretreatment, is said to be a "primary reinforcing agent." Physical dependence may account for differences in the pattern of self-administration of opiates and stimulant drugs (35).

Three factors readily manipulated by the experimentor are 1) the dose (route, size and rate of administration), 2) the schedule of administration of the doses, and 3) conditioning of the drug doses to secondary cues.

Dose

Self-administration is usually by either the oral or intravenous route, although inhalation has been applied to monkeys (50). The intravenous route has the advantages of precise control over the dose, and the mechanics of injections can readily be adapted to manipulations customarily used in the behavioral sciences. Oral self-administration has the added complications of taste (often bitter and aversive), thirst, and the water needs of the animal. On the other hand, it has the obvious advantage of simplicity (see 31 for discussion).

By the intravenous route, there is an inverse relationship between the number of injections taken and the size of the dose. This is not surprising, of course. For the stimulant drugs, there was about a 10-fold range of doses over which animals held their total daily intake nearly constant by appropriate changes in the number of injections (46). However, for morphine and pentobarbital, the increased number of injections did not compensate for smaller doses, so there was a progressive decrease in the total amount of drug obtained (40, 47).

Recently, Yokel and Pickens (51) studied blood levels of *d*- and *l*-amphetamine over a fourfold range of self-administered doses. The rats maintained constant blood levels regardless of the size of the individual dose. When similar studies are done with opiates, we may gain some insight into the reasons for the differences in patterns of self-administration of stimulant and depressant drugs.

The rate at which an injection is given has not yet been studied thoroughly. If there is a positive reinforcing sensation in animals comparable to the "rush" described by human addicts, it would seem that a rapid injection, causing transient high blood levels in the brain, would have more reinforcing value than a slow infusion. In rats, Thompson and Pickens (35) found that dose-infusion times of cocaine ranging from 25 to 75 sec had no effect on the drug intake.

However, considering that recirculation time in the rat is at the most 3 sec (based on studies using thermal dilution cardiac output in unanesthetized rats (41)), injection times ranging from 0.5 to 3.0 sec may be necessary to reveal any such effect.

Schedules

Ratio schedules reward the animal after a set number of responses, while interval schedules reward the animal a set time after either the last injection or some auditory or visual cue. In general, behavior for drug reinforcements follows the same temporal patterns as other reinforcements (13, 39). For the stimulants, over a considerable range of fixed ratios total drug intake remains constant, the animal increasing his lever pressing activity to compensate for the requirements of the schedule (28, 29). In rats working for morphine, total daily drug intake at low fixed ratios is well maintained, but at higher fixed ratios intake decreases. The decrease was due not only to a longer working time required to get the injection, but also at high ratios the time from an injection until the rat started working for the next one increased (38). We attributed this longer interval as due to a partial loss of tolerance or dependence during the withdrawal enforced by the high ratio. Neither rats (Weeks and Collins, in preparation) nor monkeys (14) maintain self-injection of pentobarbital on fixed ratio schedules.

For all drugs the ratio can be increased to the point that no appreciable amount of drug will be taken. This is the basis of the "progressive ratio" test for the strength of the reinforcement: determining the highest ratio obtainable before behavior is extinguished (48). For cocaine in monkeys, this ratio has been as high as 6,400. Ratios achieved vary with the drug and the individual animal. This test may measure only the strength of primary reinforcements, since the smaller amount of drug obtained at higher ratios might, in physically dependent animals, secondarily lead to loss of some dependence and hence of the drive to continue responding.

Rats do not readily drink bitter solutions of morphine or pentobarbital, and so to induce voluntary oral drug intake they must in some way be forced to drink the solution. When no fluid other than water containing morphine is available to rats on a 7-hour daily schedule, the rats in time develop a preference for morphine solutions when offered a choice with plain water (19). However, if the morphine solutions were present continuously, and then rats offered a choice between morphine or water, water is preferred (18). Apparently the intermittent schedule, with recurrent development of withdrawal distress, is necessary for the reinforcing properties of the morphine to overcome the aversiveness of the bitter taste.

Although naive monkeys will initiate self-injection of pentobarbital (9), such behavior has not yet been reported for rats. However, when rats are given intermittent unavoidable shocks, they will take intravenous injections of barbiturates (6, 7). Nevertheless, drug intake is not maintained more than a few days in spite of continued shocks.

Another method to force rats to consume aversive solutions is schedule-induced polydipsia (11). This phenomenon is excessive drinking accompanying an intermittent schedule of food reinforcements. Rats have thus been induced to drink ethanol and pentobarbital solutions that they would otherwise reject (21, 22).

Conditioning

Neutral environmental stimuli can elicit conditioned changes in morphine self-administration in monkeys. While self-administering morphine, an injection of the morphine antagonist nalorphine elicits a series of rapid morphine injections, presumably to overcome the abstinence reaction precipitated by the nalorphine. When these nalorphine injections are repeatedly paired with a flashing red light, the light alone then elicits a series of morphine injections. Of course, this reaction quickly disappears on repeated presentation of the light (16). Nalorphine-conditioned abstinence changes persist as long as 3 months, even after the monkeys have been withdrawn from active morphine intake (15). Likewise, if rats are made dependent on morphine in their drinking water, and the treated water identified by a flavor, the flavor acts as a conditioned cue when the rats are later offered a choice between flavored and plain water (27, 43–45).

Crowder, Smith and associates (5) have applied conditioned responses to measure the reinforcing property of drugs. Rats are injected automatically with small doses of morphine and a buzzer is sounded during each injection. In a later session, the rats are retested in the same apparatus with a lever, which, when pressed by the rat, sounds the buzzer but now it is paired with a saline injection instead of morphine. The number of saline injections taken was greater than the spontaneous unreinforced rate without the buzzer. Furthermore, the number of these conditioned lever responses is directly related to the size of the morphine dose originally paired with the buzzer. The doses of morphine, ranging from 3.2 to 320 μg/kg, were so small that there was no evidence of physical dependence. The conditioned responses, then, are a consequence of the primary reinforcing property of morphine. It is also possible that the reinforcing property of such very small doses of morphine was enhanced by the brief infusion time of only 0.2 sec.

When α-methyl-p-tyrosine was given during the morphine-buzzer pairings, the buzzer did not acquire secondary reinforcing properties, indicating an inhibition of reinforcing property of the morphine (8). A compound given to a rat self-administering morphine might falsely decrease morphine intake if the compound caused motor impairment. However, by this method behavioral activity is not required during administration of the potential inhibitor. Haloperidol, for example, decreased morphine intake directly but had no effect on conditioned reinforcement (33).

Diurnal variation

This factor cannot readily be manipulated, but exists and must be recognized in planning and interpreting experiments. Monkeys tend to take less morphine or pentobarbital during the nighttime than daytime hours (9). On the contrary, we found that rats took more of these drugs during the nighttime, a pattern consistent with the nocturnal nature of this species (4, Weeks and Collins, unpublished).

INTERNAL ENVIRONMENTAL FACTORS

Differences in the animals themselves can influence self-administration experiments. Several laboratories have noted that in a given population of rats there are opiate "drinkers" and "nondrinkers," which may reflect inborn differences (1, 2, 24). Nichols and Hsiao (26), by selective breeding, produced two strains of rats which differed in their susceptibility to morphine addiction by the oral route. A hereditary factor in oral morphine

consumption in mice has also been demonstrated (10). If rats represent a mixed population regarding their susceptibility to drug self-administration, one is well advised to use adequate numbers of animals before generalizing too broadly.

The drugs themselves, once administration starts, may have effects which might affect subsequent behavior. State-dependent learning, wherein learning of a given task while in a drugged condition can be recalled later only in the drugged state, is an example. Pentobarbital has such an effect in rats (3). Stretch and Gerber (34) allowed monkeys to self-administer amphetamine and then extinguished the behavior by substituting saline. Now, when an intravenous injection of amphetamine is given, still with only saline available for self-injection, a pattern of responses follows as in the initial amphetamine sessions. Possibly state-dependent learning, triggered by the amphetamine injection, plays a role in this phenomenon.

Tolerance, rates of metabolism and excretion of drugs, blood levels and the penetration of brain tissue, all factors important for drug therapy, apply equally to self-administration studies (20).

CLOSING REMARKS

There are other possible applications of drug self-administration not yet explored. If we accept the concept that normal physiological processes maintain a constant internal environment, and that drugs can be used to restore a disturbed internal environment, perhaps then such drugs could be self-administered in a quantitative manner. Visceral and glandular responses can be learned, so such changes certainly can be sensed by the animal (23), implying that the drug control by self-injection is also a possibility. Weiss and Laties demonstrated that shaved rats in a cold environment could maintain their body temperature by operating heat lamps, an example of self-regulation of the internal environment (42).

Over 30 years ago Richter (30) showed that rats could self-select an extremely well-balanced diet from a wide variety of foods. Tastes paired with recovery from thiamin deficiency induce conditioned preferences (12, 52). Perhaps a diabetic rat could maintain an optimum insulin level, and thereby allow studies of dietary and pharmacologic factors affecting carbohydrate metabolism. Indeed, Headlee et al. (17) not only first described self-injection of morphine but also showed that hungry rats, offered intraperitoneal injections of either glucose or insulin, developed a preference (albeit of borderline statistical significance) for the glucose over insulin.

With imagination, the possibilities are manifold. On this note, hoping that I may stimulate another application of drug self-administration, I conclude my comments.

REFERENCES

1. CAPPELL, H., AND A. E. LeBLANC. Psychopharmacologia 21: 192, 1971.
2. CHERNOV, H. I., B. S. BARBAZ, R. L. BOSHEK AND M. M. FEIST. Arch. Int. Pharmacodyn. Ther. 195: 231, 1972.
3. CHUTE, D. L., AND D. C. WRIGHT. Science 180: 878, 1973.
4. COLLINS, R. J., AND J. R. WEEKS. Naunyn-Schmiedebergs Arch. Pharmakol. Exp. Pathol. 249: 509, 1965.
5. CROWDER, W. F., S. G. SMITH, W. M. DAVIS, J. T. NOEL AND W. R. COUSSENS. Psychol. Rec. 22: 441, 1972.
6. DAVIS, J. D., C. LULENSKI AND N. E. MILLER. Int. J. Addict. 3: 207, 1968.
7. DAVIS, J. D., AND N. E. MILLER. Science 141: 1286, 1963.
8. DAVIS, W. M., AND S. G. SMITH. Life Sci. (pt 1) 12: 185, 1973.
9. DENEAU, G., T. YANAGITA AND M. H. SEEVERS. Psychopharmacologia 16: 30, 1969.
10. ERIKSSON, K., AND K. KIIANMAA. Scand. J. Clin. Lab. Invest. 27: 43, 1971.
11. FALK, J. L. Science 133: 195, 1961.
12. GARCIA, J., F. R. ERVIN, C. H. YORKE AND R. A. KOELLING. Science 155: 716, 1967.
13. GOLDBERG, S. R. J. Pharmacol. Exp. Ther. 186: 18, 1973.
14. GOLDBERG, S. R., F. HOFFMEISTER, U. U. SCHLICHTING AND W. WUTTKE. J. Pharmacol. Exp. Ther. 179: 277, 1971.
15. GOLDBERG, S. R., AND C. R. SCHUSTER. J. Exp. Anal. Behav. 14: 33, 1970.
16. GOLDBERG, S. R., J. H. WOODS AND C. R. SCHUSTER. Science 166: 1306, 1969.
17. HEADLEE, C. P., H. W. COPPOCK AND J. R. NICHOLS. J. Am. Pharm. Assoc. 44: 229, 1955.
18. KHAVARI, K. A., AND M. E. RISNER. Psychopharmacologia 30: 291, 1973.
19. KUMAR, R., H. STEINBERG AND I. P. STOLERMAN. Nature (London) 218: 564, 1968.
20. KUMAR, R., I. P. STOLERMAN AND H. STEINBERG. Psychopharmacologia 21: 595, 1970.
21. MEISCH, R. A. Psychonomic Sci. 16: 16, 1969.
22. MEISCH, R. A., AND T. THOMPSON. Physiol. Behav. 8: 471, 1972.
23. MILLER, N. E. Science 163: 434, 1969.
24. NICHOLS, J. R., AND W. M. DAVIS. J. Am. Pharm. Assoc. 48: 259, 1959.
25. NICHOLS, J. R., C. P. HEADLEE AND H. W. COPPOCK. J. Am. Pharm. Assoc. 45: 788, 1956.
26. NICHOLS, J. R., AND S. HSIAO. Science 157: 561, 1967.
27. PARKER, L., A. FAILOR AND K. WEIDMAN. J. Comp. Physiol. Psychol. 82: 294, 1973.
28. PICKENS, R., AND W. C. HARRIS. Psychopharmacologia 12: 158, 1968.
29. PICKENS, R., AND T. THOMPSON. J. Pharmacol. Exp. Ther. 161: 122, 1968.
30. RICHTER, C. P. Harvey Lect. 38: 63, 1942/1943.
31. SCHUSTER, C. R., AND T. THOMPSON. Annu. Rev. Pharmacol. 9: 483, 1969.
32. SCHUSTER, C. R., AND T. I. THOMPSON. Lab. Psychopharmacol. Univ. Md. Tech. Rep. 62-29, 1962.
33. SMITH, S. G., AND W. M. DAVIS. Psychol. Rec. 23: 215, 1973.
34. STRETCH, R., AND G. J. GERBER. Can. J. Psychol. 27: 168, 1973.
35. THOMPSON, T., AND R. PICKENS. Federation Proc. 29: 6, 1970.

36. WEEKS, J. R. *Federation Proc.* 20: 397, 1961.
37. WEEKS, J. R. *Science* 138: 143, 1962.
38. WEEKS, J. R., AND R. J. COLLINS. *Psychopharmacologia* 6: 267, 1964.
39. WEEKS, J. R., AND R. J. COLLINS. In: *The Addictive States, Res. Publ. Assoc. Res. Nerv. Ment. Dis.* 46: 288, 1968.
40. WEEKS, J. R., AND R. J. COLLINS. *Proc. Int. Congr. Pharmacol. 5th, 1972,* abstracts p. 248.
41. WEEKS, J. R., AND E. CSORDAS. *Federation Proc.* 22: 400, 1963.
42. WEISS, B., AND V. G. LATIES. *Science* 133: 1338, 1961.
43. WIKLER, A., AND F. T. PESCOR. *Psychopharmacologia* 10: 255, 1967.
44. WIKLER, A., AND F. T. PESCOR. *Psychopharmacologia* 16: 375, 1970.
45. WILKER, A., F. T. PESCOR, D. MILLER, AND H. NORRELL. *Psychopharmacologia* 20, 103, 1971.
46. WILSON, M. C., M. HITOMI AND C. R. SCHUSTER. *Psychopharmacologia* 22: 271, 1971.
47. WOODS, J. H., AND C. R. SCHUSTER. *Int. J. Addict.* 3: 231, 1968.
48. YANAGITA, T. *Univ. Mich. Med. Cent. J.* 36: 216, 1970.
49. YANAGITA, T., G. A. DENEAU AND M. H. SEEVERS. *Int. Congr. Physiol. Sci. Lect. Symp. 23rd,* Tokyo, 1965, p. 453.
50. YANAGITA, T., S. TAKAHASHI, K. ISHIDA AND H. FUNAMOTO. *Jap. J. Clin. Pharmacol.* 1: 13, 1970.
51. YOKEL, R. A., AND R. PICKENS. *Psychopharmacologia* 34: 255, 1974.
52. ZAHORIK, D. M., AND S. F. MAIER. *Psychonomic Sci.* 17: 309, 1969.

An experimental analysis of behavioral factors in drug dependence

TRAVIS THOMPSON AND ROY PICKENS

University of Minnesota, Minneapolis, Minnesota 55455

Many biological scientists have viewed attempts to study the psychological or behavioral aspects of drug action as belonging near the questionable limits of scientific inquiry, if not floating well into the abyss of the objectively unknowable. Such scientific conservatism is not without reason, for as Sidman (62) noted, "A major factor contributing to the slowness of development of a science of Behavioral Pharmacology was the late recognition that behavior is a phenomenon amenable to study by the methods of Natural Science." The solution lies in the unequivocal demonstration of a viable laboratory approach to the scientific investigation of the behavioral actions of drugs. However, "The use of experimental animals in the analysis of behavioral effects of drugs has been hampered by a paucity of objective, quantitative methods of study" (19). Though an early study by Skinner and Heron (65) pointed the way toward such an objective analysis, it wasn't until the late 1950's that this promise began to be realized. The new scientific domain, behavioral pharmacology, grew principally out of an amalgamation of the concepts of an operational analysis of behavior propounded by Skinner (63, 64) and the more firmly established concepts of experimental pharmacology.

While many of the important contributions in the early years of behavioral pharmacology research were technological (e.g., the use of the Skinner box, the use of response rate as a dependent variable, the analysis of inter-response time distributions), a more important consequence was the emergence of a conceptual framework within which to formulate experimental questions (20, 75). Within this framework the actions of a drug were conceived of as a joint function of a given compound's physiological and biochemical effects on the organism *and* the environmental conditions under which the drug was administered.

Among the environmental variables with which an administered drug may interact are certain antecedent conditions (e.g., the organism's past history), the current stimulus conditions (e.g., the presence of relevant cues or discriminative stimuli), the type of response, and the nature and scheduling of behavioral consequences. Other papers in this symposium will discuss the former three classes of factors in depth. However, here we will deal with one subclass of behavioral consequences; indeed a rather unique set of events. The behavioral consequences we have in mind are the administration of drugs to organisms, contingent on occurrence of some prior response. Such an arrangement has been called drug self-administration experimental situation (59, 72). Within this frame of reference, any drug that, when self-administered by an organism, increases the rate of responding and/or maintains schedule control over the behavior leading to its administration, is said to serve as a reward or a *reinforcer*.

That drugs can serve as reinforcers for arbitrary learned responses was presaged by Shirley Spragg (66), who showed that physically dependent chimpanzees would learn to select one of two boxes concealing a syringe of morphine if the experimenter would then inject the animal with the drug. However, the demonstration that an animal will learn a response if that response produces access to a drug solution per se, is less important than the recognition that the *phenomenon* is an instance of the general class of events, reinforcement. The significance lies in the fact that, as such, drug reinforcement is a member of a larger class of other reinforcing events (e.g., presentation of food, water, brain stimulation); and a vast amount of information is available concerning the variables that influence the ability of other reinforcers to affect behavior. If generalizations based on research with other reinforcers hold for drug reinforcers, we are in a far better position to begin to understand basic mechanisms underlying the extensive, and at times devastating, control that drugs can exercise over human behavior (i.e., drug dependence). Furthermore, we may, as is indicated near the end of this paper, begin to apply these principles to reducing or eliminating the control that drugs can establish over human behavior.

Drug reinforcers initially gain control over behavior during a phase of exposure called *acquisition*. The nature of the drug, its dose and numerous other factors determine the rapidity and quality of acquisition. The greatest amount of research in drug self-administration has been devoted to variables influencing the *maintenance* of behavior by drug reinforcers. The type of drug, the drug dosage, the schedule of drug self-administration, and the presence of other drugs all dramatically influence the maintenance of behavior by drug reinforcers. Since much of this research has been reviewed elsewhere (cf. 54, 55, 59, 72, 73), only more recent research in factors maintaining drug self-administration will be discussed in depth here.

Finally, once a drug has gained control over behavior as a maintaining consequence, the problem of *reducing or eliminating* that control arises. Relatively limited attention has been paid to this area, largely because the basic background questions that were a prerequisite to this line of research remain only partially answered. Nonetheless, promising investigations have begun to emerge.

ACQUISITION

Acquisition of drug-reinforced behavior has been studied using both inducing and noninducing procedures. Inducing procedures are typically employed to establish self-administration of drugs having reinforcing effects but also having other effects that may be aversive to an organism, or to establish self-administration of drugs having only relatively weak reinforcing effects. Inducing procedures typically involve the use of nondrug reinforcers in establishing drug self-administration. Before the advent of the intravenous self-administration technology, drug self-administration in animals was studied almost entirely by the oral administration route, usually by allowing an animal to self-administer drugs by drinking from a water bottle containing a drug solution. Without inducement animals typically refused to do so, however, apparently because of the aversive taste of most drug solutions. Thus, to induce self-administration, the animals were water deprived and required to obtain their daily water supply by drinking from the drug solution. Nichols (48) first pretreated rats with morphine to produce physical dependence and then allowed them to drink morphine solutions when both thirsty and undergoing withdrawal. In later choice tests such animals drank significantly more morphine solution than controls. Using a similar procedure, Stolerman and Kumar (67) have found that pretreatment with morphine is not necessary for the development of morphine preference. Drug-naive rats developed a morphine preference after being given only morphine solutions to drink in order to relieve thirst. Moreover, it has recently been shown that water deprivation is also not necessary to establish ethanol as a reinforcer for rats. All that is necessary is previous training in lever pressing reinforced by water. Although responding is initially suppressed when a 4% ethanol solution is substituted for the water, over a 2-week period lever pressing maintained by ethanol becomes equal to or greater than that maintained by water in nondeprived rats (4).

A second procedure for inducing the oral self-administration of drugs with aversive tastes is based on Falk's (24) original observation that animals will consume large quantities of water when placed on intermittent schedules of food reinforcement. On such schedules, rats drink approximately 0.5 cc of water after each food reinforcement, consuming as much as one-half their body weight in water during a 3.17-hour experimental session (25). This phenomenon, termed schedule-induced polydipsia, was first used by Lester (37) and Senter and Sinclair (61) to induce drinking of large volumes of ethanol solution by previously drug-naive rats. In these studies, the animals were first made polydipsic to water as described above, and then ethanol substituted for water as the drinking solution. Using this procedure, large amounts of ethanol were consumed.

Meisch and Thompson (39, 40) used schedule-induced polydipsia to produce oral ethanol self-administration in rats, and compared the tendency of their subjects to drink various concentrations of ethanol before and after exposure to the inducing schedule. Following the schedule experience, responding for ethanol and intake volume increased significantly. The amount of ethanol consumed varied directly with the concentration, with all values being above water controls. On discontinuing the inducing schedule, following exposure to concentrations of from 4 to 32% (wt/vol), ethanol maintained lever pressing well above water control values, i.e., ethanol had come to serve as a reinforcer in its own right (39). Moreover, the rate at which ethanol can be established as a reinforcer using the schedule-induced polydipsia technique is surprisingly rapid. Rats were given the opportunity to self-administer ethanol or tap water for 2 hours. For the following 24 hours a food-reinforcement schedule was concurrently in effect, which induced consumption of a considerable amount of 8% ethanol. By the next session, ethanol had already been established as a potent reinforcer (44).

Recently Meisch, Henningfield and Thompson (45) have induced ethanol drinking by rhesus monkeys. The monkeys were exposed to 1 hour of liquid availability followed by 2 hours of concurrent food and liquid reinforcement. The food schedule was a Multiple Extinction 120-sec FR 1. During the first hour ethanol of varying concentrations or tap water was available, with at least 1 water day preceding and following each ethanol day. Figure 1 shows the number of liquid reinforcements earned on ethanol and water days at an initial 8%, 16% and 32% (wt/vol) and again at 8% retest, during which the inducing polydipsia schedule had been discontinued. Clearly, the monkey was consuming vastly greater quantities on ethanol days than on water days. Figure 2 shows the time course

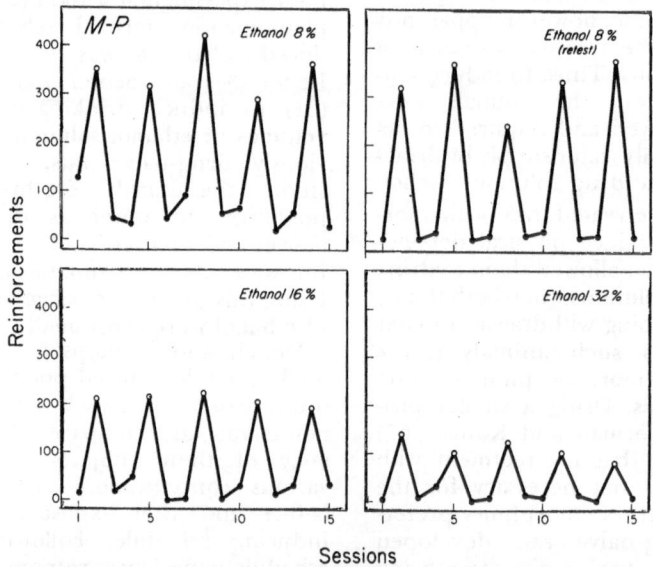

Figure 1. Reinforcements per 3-hr session as a function of ethanol concentration for monkey M-P (weight 4.9 kg). Ordinate: Reinforcements (or responses) per 3-hr session. Abscissa: Consecutive sessions. Unfilled circles: Ethanol sessions. Filled circles: Water sessions. Each reinforcement resulted in the delivery of 0.55 ml of liquid. The concentrations were presented in an ascending order of 8, 16 and 32% (wt/vol) and then back to 8% (retest) (45).

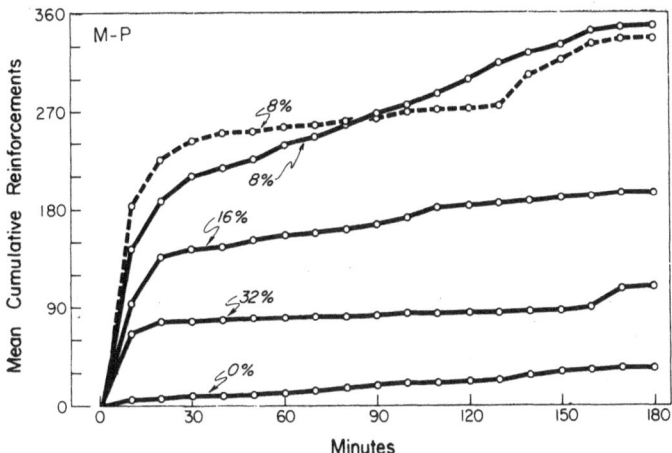

Figure 2. Cumulative reinforcements over 3-hr sessions as a function of concentration for monkey M-P. Ordinate: Mean cumulative reinforcements. Abscissa: Time within sessions. Each point on the zero percent curve is the mean of 40 observations. Each point on the ethanol curves is the mean of 5 observations. The dotted line connects the values for 8% (wt/vol) that were obtained when the monkey was returned to this concentration following the series of sessions at 32% (wt/vol). Note that the highest rate of intake occurred at the beginning of the session, and that ethanol values exceeded water control values (45).

of ethanol consumption subsequently when the drug was available over the entire 3 hour session. As can be seen, most of the drug is consumed in the first 30 min of the session.

Other techniques have been used to induce self-administration of drugs with either unknown or relatively weak reinforcing effects. Pickens, Thompson and Muchow (56) induced rhesus monkeys to smoke hashish by employing food reinforcement. The animals were food deprived and trained either to suck on a tube containing heated air or to lever-pull for food presentation. Number of responses required for reinforcement on each manipulandum was adjusted until no preference was shown. Hashish smoke was then made available through the sucking tube, and within 2–3 weeks a sucking-tube preference was clearly seen. The preference was reversed (week

7) when hashish was removed from the burning chamber, and reappeared (week 8) when hashish was again reintroduced (Fig. 3).

Findley, Robinson and Peregrino (26) periodically gave rhesus monkeys a forced choice between intravenously self-administering secobarbital and saline, chlordiazepoxide and saline, or secobarbital and chlordiazepoxide. If neither substance was self-administered within a specified time on each trial, the animal received an electric shock. Using this forced-choice procedure, a preference was shown for secobarbital over saline, chlordiazepoxide over saline, and secobarbital over chlordiazepoxide. Different injection doses might be expected to produce different results, however.

Hoffmeister et al. (32) studied the reinforcing effects of d-amphetamine, imipramine, morphine, and chloropromazine by use of a substitu-

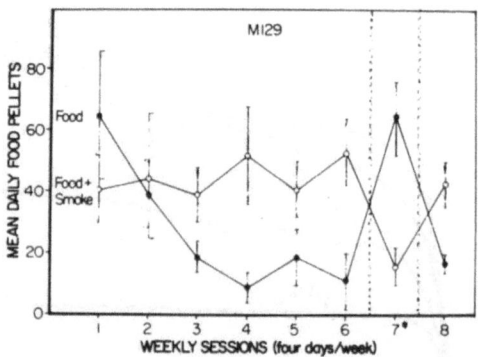

Figure 3. Hashish smoking by monkey M129. Food pellets could be obtained in either component of a two-component concurrent option by lever-pulling (closed circles) or sucking on a tube containing hashish (open circles). During week 7 heated air only was available through the smoking tube. Vertical lines are ± SE (56).

tion procedure. Rhesus monkeys were initially trained to respond for intravenous cocaine self-administration. After responding stabilized, saline or various dosages of the test compounds were substituted for cocaine as the injection solution. Response rate after test drug substitution was then compared to that following saline substitution. Stable responding was maintained by d-amphetamine and morphine, but imipramine produced response rates similar to that for saline, and chlorpromazine produced rates below that for saline. Pickens, Thompson and Muchow (56) obtained intravenous Δ^9-THC (tetrahydrocannabinol) self-administration in rhesus monkeys by initially establishing self-administration with phencyclidine and then substituting Δ^9-THC for phencyclidine in the self-administration procedure. Animals would not initiate Δ^9-THC self-administration without prior phencyclidine self-administration exposure or after cocaine self-administration exposure.

With noninducing procedures, acquisition of drug-reinforced respond-

ing is readily accomplished simply by making drugs available for self-administration. No further behavioral manipulations are necessary. The intravenous self-administration technique has been the most widely used noninducing procedure for studying drug reinforcement, perhaps because it is capable of rapidly and consistently delivering a relatively large amount of drug solution directly into the bloodstream and because it eliminates taste factors from influencing the results. Other routes of drug administration have also been used, including intraperitoneal (14), intragastric (29), inhalation (35), and intraventricular (Gustafson and Pickens, unpublished observations).

Noninducing procedures are most effective in establishing self-administration of the more powerful drug reinforcers. Drugs self-administered when made freely available to animals include members of the opiate, stimulant, and sedative classes. While early studies of opiate and sedative self-administration employed animals pretreated to produce physical dependence and tolerance (59, 76, 81), subsequent studies have found physical dependence to be unnecessary for the acquisition of intravenous self-administration of morphine (15, 60, 76), methohexital and ethanol (78).

Unlike inducing procedures, non-inducing procedures are relatively free of extraneous factors that may confound study of the acquisition process. There have been few attempts to study such processes, however. Dougherty and Pickens (21) have made the only systematic observations on the acquisition of the

uniformly-long interresponse times (IRTs) which characterize intravenous cocaine self-administration by rats (Fig. 4). Alternate-day drug sessions were limited to 100–120 reinforcements each. On the first drug session, responding (shown here in blocks of 20 IRTs) was highly variable with no discernible pattern of drug intake. On subsequent sessions, however, responding became increasingly less variable until typically by the third or fourth session the characteristic pattern of IRTs during the session was clearly seen.

To quantify the changes observed,

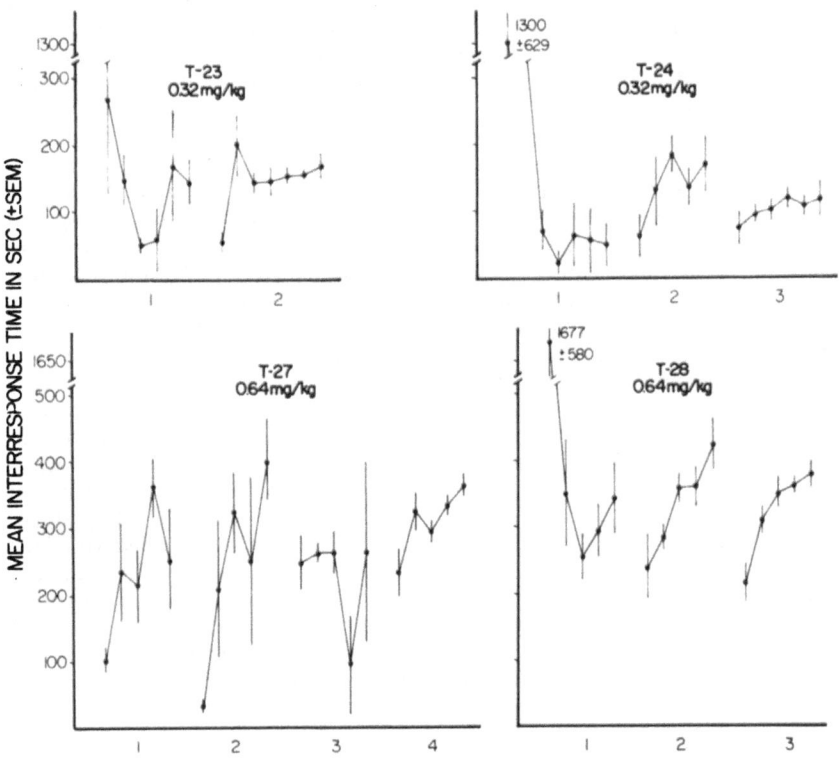

Figure 4. Development of stabilized patterns of responding during acquisition of intravenous cocaine self-administration of four naive rats (T-23, T-24, T-27 and T-28). Injection doses of either 0.32 or 0.64 mg/kg were employed. Each point indicates mean ± SE (vertical lines) of successive blocks of 20 interresponse times during each drug session (21).

successive interresponse times were divided into blocks of twenty and compared for degree of variability. A coefficient of variation was computed by dividing the standard deviation of the IRTs in each block by their mean. Coefficients below 1.0 indicate standard deviations smaller than the means. Coefficients of about 0.5 were correlated with stabilized response patterns as determined by gross visual inspection of the response records (21).

Davis and Smith (18) and Crowder et al. (13) showed acquisition of conditioned-reinforcing properties by a stimulus paired repeatedly with intravenous self-administration of reinforcing drugs. During self-administration, a buzzer was sounded contingent with amphetamine presentation. After extinction of drug-reinforced responding, response-contingent buzzer presentations alone increased response rate above operant levels in four rats tested. Pretreatment with the catecholamine-depleting agent, α-methyl-p-tyrosine, however, prevented the establishment of a buzzer as a conditioned reinforcer after 100 behaviorally non-contingent pairings of the buzzer with intravenous morphine injection (18). The same pretreatment also blocked the acquisition of morphine-reinforced behavior in nondependent, nontolerant rats (16).

MAINTENANCE

The two most frequently studied variables in the maintenance of drug self-administration have been injection dose and schedule of drug presentation. In general, increases in injection dose produce increases in total drug intake for most of the self-administered compounds. Meisch and Thompson (40) used schedule-induced polydipsia initially to obtain ethanol drinking in rats.

Subsequently, the inducing food schedule was discontinued and lever pressing was reinforced by ethanol alone. As the ethanol concentration was increased (2, 4, 8, 16, and 32% wt/vol), while the number of reinforcements increased and then decreased with increasing concentrations, the total intake of drug increased across all concentrations.

With intravenous self-administration techniques, Woods and Schuster (80) reported approximately the same number of morphine injections taken at each dose level over a 10-fold dose range by rhesus monkeys. The result was that drug intake increased proportionally with injection dose. Pentobarbital intake also increases with injection dose, but the increase was only moderate and unlike morphine was not proportional to injection dose (28).

The relationship between injection dose and drug intake has been studied most extensively with stimulants. Yokel and Pickens (82) compared self-administration of d- and l-amphetamine and methylamphetamine across a range of injection doses. Increases in injection dose produced decreases in response rate and only relatively slight increases in hourly drug intake. For d-amphetamine and l-amphetamine, drug intake increased about threefold and twofold, respectively, across a 10-fold range of injection dose (Fig. 5). The rate of drug intake was also reflective of the drugs' central potency. In maintaining responding, two to three times more l- than d-amphetamine and about four times more l- than d-methylamphetamine was self-administered per hour. At similar molar doses, about twice as much l-methylamphetamine as l-amphetamine was self-administered, but no difference in d-amphetamine and d-methylamphetamine was evident. Since response

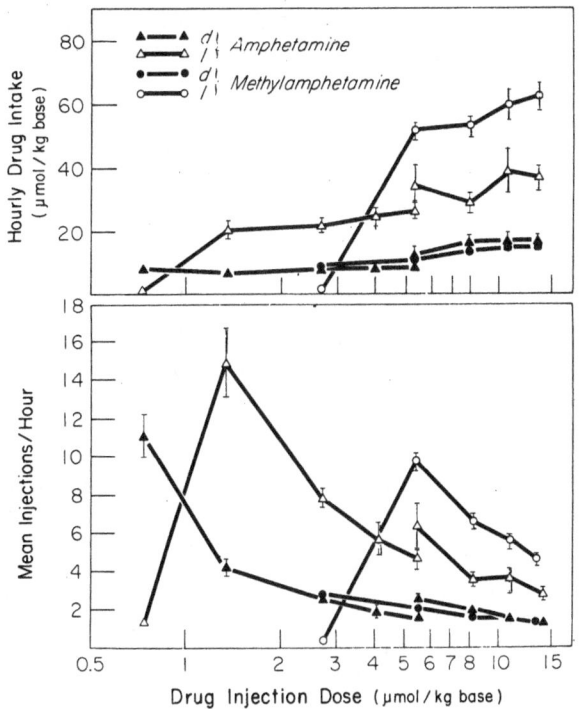

Figure 5. Mean drug intake and injections per hour for intravenous self-administration of amphetamine and methylamphetamine isomers as a function of injection dose (one response/injection). Each animal (no. = 13) was tested at least once at all doses for both isomers of either drug. All values are means ± SE (vertical lines) in micromoles per kilogram of base. In some cases, SE is less than point height (82).

rate in stimulant self-administration is inversely related to injection dose, the *l*-isomers appear to be functioning effectively as lower doses of the *d*-isomers for both drugs. This observation is supported by the fact that the *d*-isomers were also capable of maintaining responding at lower doses than the *l*-isomers.

While intake of stimulants appears to increase slightly with increases in injection dose, this relationship is obscured by differences in the amounts of drug metabolized at the various injection doses. With a constant percent of drug being metabolized per unit time, the duration of drug effect is proportionately less for the higher than for the lower injection doses. Consequently, drug intake would appear to increase with increases in injection dose, while the drug level at which responding occurs remains constant.

To study the relationship between drug level and responding for stimulant self-administration, Yokel and Pickens (83) computed theoretical whole body levels of *d*- and *l*-amphetamine at the time of each response during a period of drug self-administration. During the first 2 hours of each 6-hour session, results were highly variable with whole body levels initially sharply increasing and then decreasing. Thereafter, however, drug levels remained essentially constant across all injection doses for the remaining 4 hours of the drug session. While actual drug intake over the entire session increased with increases in injection dose (open circle, solid line), when theoretical whole body drug levels are calculated for hours 2–6 only (filled square, dashed line) no increasing trend is seen (Fig. 6). Approximately the same body level of drug is being maintained across all injection doses.

As part of the same experiment, blood was removed and [¹⁴C]amphetamine levels determined at the time of responding for self-administration (Fig. 7). Responding for drug was found to occur at about the same blood level across a fourfold range of injection doses (0.25–1.0 mg/kg). Blood levels were approximately three times higher for *l*-amphetamine than for *d*-amphetamine at the time of self-administration (83).

A second variable frequently studied in maintenance of drug self-administration is the schedule of drug presentation. In general, schedule-controlled behavior maintained by drug reinforcers is characteristically similar to that maintained by nondrug reinforcers, although differences do exist. Many of the apparent differences between drug and nondrug reinforcers can be resolved, however, when comparable experimental procedures are employed (51).

With simple schedules of drug presentation, Meisch and Thompson (43) studied fixed-ratio schedules of oral ethanol self-administration in rats. Response requirements for 8% (wt/vol) ethanol were geometrically increased until responding was no longer maintained. Number of responses per session increased with

the increase in response requirement up to fixed-ratio values of 16–64, and thereafter declined. Fixed-ratio responding could be maintained, however, by values up to 32, 64, 128, and 256 responses/reinforcement, in the four animals tested. The pattern of fixed-ratio responding was similar to that maintained by other reinforcers, except long pauses followed clusters of drug reinforcements.

Pickens and Thompson (53) studied fixed-ratio schedules of intravenous cocaine presentation in rats. Schedule values of 1 to 80 responses per injection were employed across the range of injection doses that maintained responding (0.5 to 3.0 mg/kg). Characteristic break-and-run fixed-ratio patterns were seen, but with relatively long pauses follow-

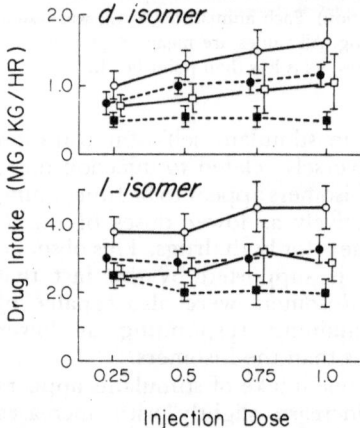

Figure 6. Mean drug intake ± SE (vertical line) of *d*- and *l*-amphetamine as a function of injection dose (no. = 5). Data shown are for actual (observed) drug intake and calculated (theoretical) whole body levels at the time of responding for each drug injection (83).

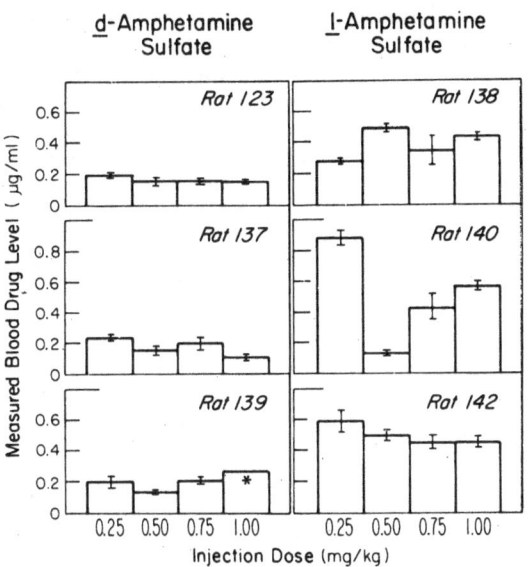

Figure 7. Mean ± SE of measured blood drug level (μg/ml) in rats at the time of responding for intravenous *d*- or *l*-¹⁴C-amphetamine self-administration. Blood samples were obtained via a carotid catheter at various times during 6-hr drug sessions. * = no replication (83).

ing each drug reinforcement. The duration of the pauses was dose dependent, with longer pauses being produced by the higher injection doses. As a result, approximately the same hourly amount of drug was taken at each injection dose across the entire range of fixed-ratio values. Responding could be maintained at higher fixed-ratio values by intermediate injection doses than by either higher or lower doses.

Goldberg and co-workers (28) studied the effects of fixed-ratio value and injection dose on intravenous cocaine and pentobarbital self-administration by rhesus monkeys. With cocaine, results essentially identical with those described above for rats were obtained. With 0.25 mg/kg pentobarbital, however, increasing response requirement from 1 to 10 responses/injection markedly decreased drug intake. With higher pentobarbital doses (2.0 mg/kg), similar amounts of drug were taken under both fixed-ratio 1 and 10 conditions.

With fixed-interval schedules of drug presentation, Dougherty and Pickens (22) studied the effects of interval duration and injection dose on fixed-interval schedules of intravenous cocaine self-administration in rats. Response patterns were seen that were characteristic of fixed-interval schedules involving other reinforcers. Following reinforcement, the duration of the pause before resumption of responding was directly related to injection dose and was about the same as that separating single responses on continuous schedules of cocaine presentation. Responding was unaffected by the fixed-interval contingency. Balster and Schuster (5) studied the effect of injection dose on a fixed-interval 9-min limited-hold 3-min schedule of intravenous cocaine presentation in rhesus monkeys. Unlike with rats, however, response rate increased with increases in injection dose. Whether these contradictory findings reflect species or procedural differences has not been determined.

With variable-interval and variable-ratio schedules of drug presentation, Pickens and Thompson (55) found intravenous cocaine to yield response rates and patterns similar to those produced by other reinforcers. Characteristics of simple schedules of drug reinforcement have been reviewed by Thompson and Pickens (73) and Pickens and Thompson (55).

With more complex schedules of drug presentation, Thompson and Schuster (74) employed a multiple schedule to study interactions among food-reinforced, shock-avoidance, and morphine-reinforced behaviors in rhesus monkeys. While all three behaviors were well maintained for several months, when the opportunity to self-administer morphine was eliminated, both shock-avoidance and food-reinforced behaviors deteriorated. Thompson, Bigelow and Pickens (69) used monkeys as subjects to study the influence of several variables on a complex chain of operants that were maintained by the opportunity to intravenously self-administer morphine. The chain consisted of as many as eight separate components, one of which involved social cooperation with a second animal. Removal of the opportunity to self-administer morphine resulted in deterioration of social as well as nonsocial behaviors.

Goldberg (27) has established second-order schedules of responding using intravenous cocaine reinforcement in squirrel monkeys. On such schedules, animals are initially trained to respond on fixed-ratio 20 schedules for cocaine injection. The drug schedule is then changed, with the com-

pletion of each fixed-ratio component producing only a brief-light presentation, and the first fixed-ratio completed after a 5-min interval producing both brief light and drug injection. Second-order schedule performance maintained by drug injection was then compared to that maintained by food presentation. Striking parallels were observed between the drug-maintained and food-maintained behaviors.

Iglauer and Woods (34) studied concurrent variable-interval schedules of intravenous cocaine self-administration by monkeys. When different injection doses were available on the two levers, monkeys preferred the higher dose.

ELIMINATION

Among the most viable procedures for reducing or eliminating the control a reinforcing drug may have over behavior are 1) weakening the reinforcing properties of the drug, 2) changing stimulus control of drug self-administration, and 3) increasing the probability of punishment associated with drug taking (70). Alteration of the reinforcing properties of drugs as a means of reducing drug-maintained responding has been studied within a number of infra-human settings. Yokel and Pickens (84) studied extinction of lever pressing following intravenous (i.v.) self-administration of amphetamine and methylamphetamine by rats. Following six hours of i.v. self-administration of various doses of d- and l-amphetamine, lever presses had no programmed consequence. They found that the number of responses to extinction was constant across drug and dose and between isomers, but *time* to extinction increased as a function of dose. Grove and Schuster (30) studied extinction of cocaine self-administration by monkeys. After training on a mult FR 1 FR 1 cocaine schedule, responses were extinguished in one schedule component. Extinction responses declined to a near zero level within three to four sessions. Response rates during nonextinction components increased for a short time, but eventually returned to their original levels. Meisch and Thompson (44) studied the extinction of lever-pressing reinforced by oral access to 8% ethanol in rats. When ethanol was replaced by water, the number of responses for liquid reinforcements decreased in a manner characteristic of extinction (Fig. 8). Over five 1-hour

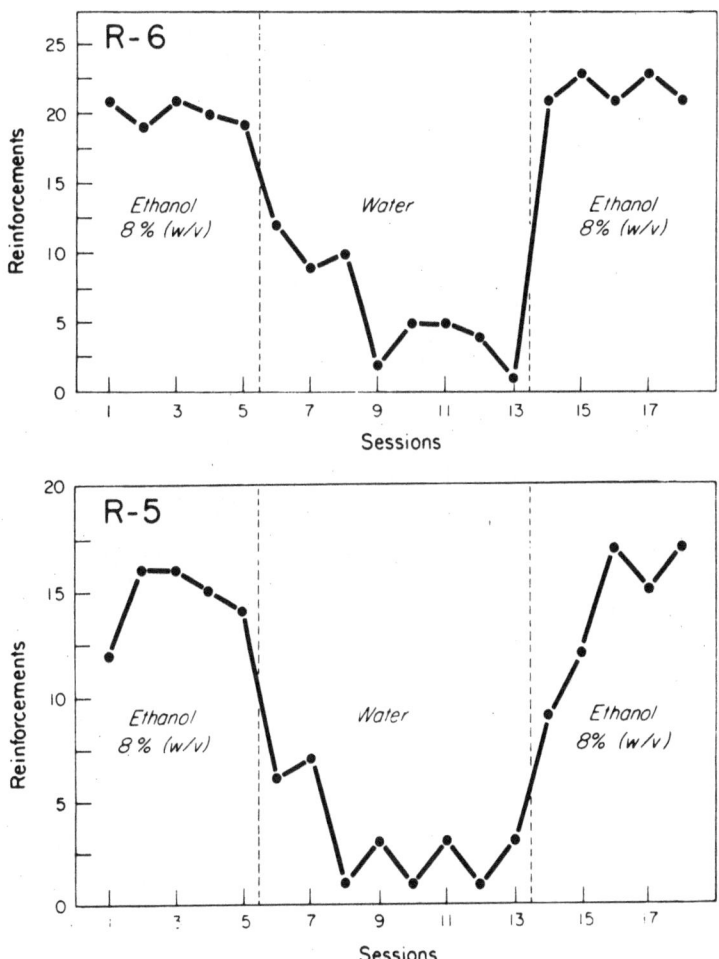

Figure 8. Reinforcements during consecutive daily 1-hr sessions for rats R-5 and R-6 when 8% (wt/vol) ethanol or water was present. Note that the introduction of water resulted in a decrease in the number of reinforcements, where-as the reintroduction of ethanol resulted in an increase in reinforcements back toward previous levels. The rats weighed 70% of their free-feeding weights: 403 g and 385 g respectively (44).

sessions, responding approached the original level. Pickens, Meisch and Dougherty (52) trained rats to self-administer methamphetamine. After establishing a stable baseline of responding for methamphetamine, various doses (5–80 mg/kg) of α-methyl-p-tyrosine (AMPT) were administered intraperitoneally. An increase in responding was observed, as normally would occur following a reduction in the injection dose of methamphetamine. At higher AMPT doses, responding was not maintained during periods of peak drug effect, with response rate immediately before and after such periods suggesting a marked reduction in the effective dose level of methamphetamine. Presumably, the effects of AMPT on methamphetamine self-administration were due to alteration of the reinforcing effects of the self-administered drug. This speculation is supported by the observation that AMPT eliminates the euphoria associated with intravenous injections of amphetamine by chronic amphetamine abusers (31).

With other compounds, Wilson and Schuster (77) reported chlorpromazine and trifluoperazine to produce dose-dependent increases and then decreases in responding for cocaine self-administration by monkeys. Dougherty and Pickens (23) found pretreatment with phenobarbital to increase and SKF-525A to decrease cocaine self-administration by rats. Haloperidol is also reported to antagonize morphine self-administration in rats and monkeys (17, 58). Similarly, Woods, Downs and Carney (79) reported a selective reduction of codeine self-administration by monkeys when pretreated with naloxone. Whether suppression or rate increments were obtained varied in a dose-dependent fashion. In a related study, conducted in our laboratory, rhesus

monkeys that were self-administering morphine intravenously received intramuscular injections of methadone. The rate of morphine self-administration was suppressed in direct proportion to the methadone dose, though the monkey's behavior maintained by other reinforcers was unaffected.

Altering the stimulus control over drug-maintained responding can be thought of in at least two ways. There are some stimuli that are cues or discriminative stimuli for drug-reinforced responding. For example, the light that indicates lever presses to produce an intravenous injection of cocaine is such a discriminative stimulus. There are other stimuli that, by virtue of their pairing with drug administration, come to be reinforcers in their own right. A red light which is illuminated during morphine infusion in monkeys comes to serve as a reinforcer. The implications of these two aspects of stimulus control are exemplified in several studies, as follows. Thompson and Ostlund (71) made rats physically dependent on morphine in one environment, detoxified them for 30 days, then readdicted half in a new environment and half in the original environment. Those readdicted in the original environment consumed morphine more readily than did those placed in a novel environment. Meisch and Thompson (43) studied ethanol self-administration on fixed ratio schedules by rats. On alternating days, rats had ethanol or tap water in a liquid reservoir concealed by the test panel. On ethanol days, the rats immediately responded and completed numerous ratios until intoxicated. On water days, when the odor of ethanol was absent, rats failed to complete even a single ratio. That is, the odor of ethanol had established powerful stimulus control over the drug-seeking behavior (42). In a related study Meisch and Thompson trained rats to orally self-administer ethanol on days alternating with water. Then lever pressing was extinguished, i.e., produced no programmed consequences. During the first 3 days of extinction tap water was in the reservoir, and for the next 3 days of extinction ethanol was present in the reservoir. The effect of the presence of ethanol markedly increased the previously ethanol-reinforced responding. It is important to bear in mind that in neither case did the animals *receive* the drug. The mere odor of ethanol obviously is a powerful cue. In one final study, Stretch et al. (68) trained monkeys to self-administer (i.v.) amphetamine. Then saline was substituted for amphetamine, and before the test session the animals were intraperitoneally administered amphetamine. They actively responded for the saline injections. That is, the internal stimulus state produced by the intraperitoneal amphetamine was a discriminative stimulus for lever pressing.

The conditioned reinforcing effects of stimuli paired with drug self-injection have been shown in several studies. Schuster and Woods (60) trained rhesus monkeys to intravenously self-administer morphine; during morphine injections a red light was presented. Then the animals were extinguished. On alternate days, lever presses produced either no programmed consequences, or a saline injection during which time the red light was illuminated. Response rates were considerably higher on days when the saline and red light were presented, presumably due to their conditioned reinforcing properties acquired during training. Meisch and Thompson (42) have shown a similar effect with oral ethanol self-administration by rats. Figure 9 shows the results obtained during extinction. The hatched bars show responses on sessions when every other response

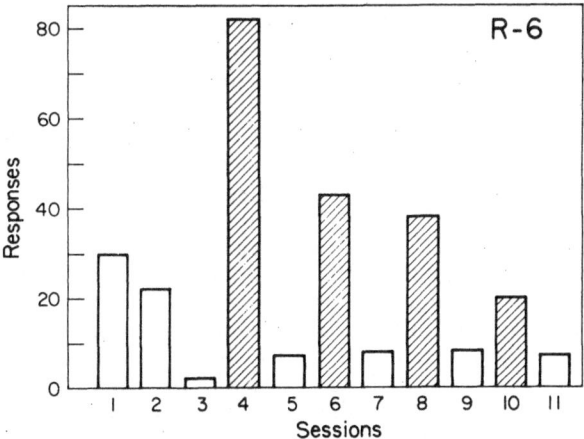

Figure 9. Number of responses emitted during extinction by rat 6 over consecutive daily 1-hr sessions. Hatched bars indicate results from sessions during which every second response (FR 2) produced onset of the second component stimuli that had been correlated with the opportunity for reinforcement. The first response in the presence of the second component stimuli produced their offset (42).

produced the onset of the stimulus paired with access to ethanol. The open bars show responding when no consequence was forthcoming following lever presses. Clearly, the light which had previously indicated ethanol availability had come to serve as a powerful reinforcer.

Relatively little attention has been paid to increasing the probability of punishment as a technique for reducing or eliminating infrahuman drug self-administration. Korman et al. (36) found that the relative amount of ethanol drunk per day was reduced when an intense sound followed 10 min of access to ethanol. Peacock and Watson (49) reported whole body gamma radiation paired with ethanol access temporarily suppressed ethanol drinking. Grove and Schuster (30) studied effects of shock on cocaine self-administration by monkeys. They used a multiple schedule in which each lever press produced a cocaine injection plus a painful shock during one component, and in the other component each press produced an infusion of the same cocaine dose but without shock. Infusions during the shock component decreased as the intensity increased while little systematic change in infusion rates was observed in nonshock periods.

HUMAN IMPLICATIONS

Seminal research in human drug self-administration was conducted by Mello and Mendelson (47) in which an operant analysis of human ethanol self-administration was provided. Recently, research emanating from the work of Mello, Mendelson and co-workers had been thoroughly reviewed (46). Allman et al. (2) have studied effects of stress and social isolation on human ethanol self-administration. They found that a combination of stress (i.e., telling subjects that their performance was inadequate to earn a monetary reinforcer at the end of the session) and social isolation suppressed alcohol drinking. In a related study, Bigelow, Liebson and Griffiths (7) studied the effect of isolation, contingent on ethanol self-administration, on total ethanol intake daily. They obtained a reduction of approximately 40% in amount drunk under contingent brief isolation.

Cohen and co-workers (10–12) provided for loss of privileges in alcoholics, contingent on ethanol self-administration. Subjects who consumed 5 oz of 95 proof ethanol or less per day could remain in an "enriched" environment; however, if they consumed more than 5 oz per day they were transferred to an "impoverished" living condition. These investigators found a marked reduction in ethanol intake, with many subjects successfully drinking less than the 5 oz limit.

Other procedures have been studied to suppress human drug self-administration based on operant principles.

In pilot investigations in our own laboratories, oral self-administration of drugs has been studied in human subjects. In one investigation, Pickens, Bigelow and Griffiths (50) allowed a chronic alcoholic free access to alcohol in a hospital setting. Drinks consisting of 1 oz 80 proof bourbon and 1 oz water plus ice were available upon request from the nursing staff. For 6 weeks, drug intake and behavioral factors associated with drinking were determined. During the period, an average of 30.7 oz of bourbon was consumed each day (range 24–40 oz), with essentially no abstinence periods occuring during the subject's waking day. Drinking was most frequent after awakening (8–9 AM) and in late afternoon and early evening (5–8 PM).

During the observation period, drinking was typically followed by socialization with other patients and staff. Therefore, in a subsequent treatment program, the relationship between drinking and socialization was reversed. While drinks remained freely available, at least 1 min of conversation with the staff was required to obtain each, and afterward the subject was confined to his room for ten min in social isolation. Over the course of the following 7 weeks,

drinking was gradually decreased to zero level (Fig. 10).

After drinking had been eliminated, an attempt was made to alter certain aspects of the discriminative control over drinking that is believed to be acquired by alcoholics from chronic exposure to alcohol under a variety of conditions. Since having a single drink was indicated by the subject to be the situation most frequently associated with a period of binge drinking, discriminative control by these stimuli was extinguished using a stimulus fading procedure.

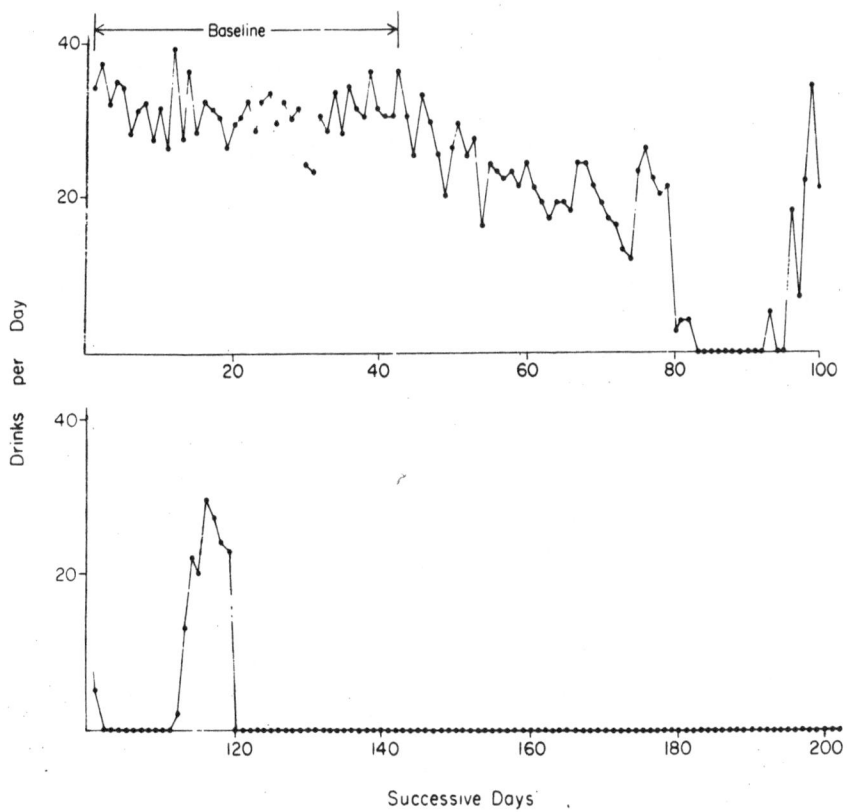

Figure 10. Number of drinks taken each day over a 6-week baseline period and following reversal of the drinking-socialization contingency. Each drink was 1 oz bourbon and 1 oz water plus ice (50).

Beginning with events remotely associated with drinking (e.g., sniffing
bourbon from bottle), stimuli more
and more closely associated with
drinking were added until single
drinks (1 oz bourbon with ice and
water) could be consumed without
initiating a period of further drinking. In another pilot investigation,
Pickens et al. (57) allowed a subject
with a 14-year history of sedative
abuse to orally self-administer barbiturates in a hospital setting. Drugs
were available from a multichannel
vending machine which was activated
by a special key issued to the research
subject. Programming equipment was
employed to place a minimum limit
on times between successive drug self-
administrations and a maximum limit
on total daily drug intake. All drug
self-administrations were verified by
the nursing staff.

On nontest days the subject was
permitted to self-administer 50 mg
pentobarbital capsules from one
channel of the vending machine. On
test days, a second drug was also
available for self-administration on a
second channel of the machine.
Selection could be made between the
standard drug (50 mg pentobarbital)
and a test drug (either 30, 50; or
100 mg pentobarbital or secobarbital).
All drugs were delivered in standard
unmarked capsules. On test days, the
subject was instructed to alternate between the standard and test compounds until a preference was established. For both pentobarbital and
secobarbital, 100 mg capsules were
preferred over the 50 mg pentobarbital standards, which in turn were
preferred over the 30 mg capsules
(Fig. 11).

After drug testing, the subject was
switched to 15 mg phenobarbital
capsules for self-administration and
allowed to self-control drug withdrawal. Successive reductions in the
number of capsules self-administered
each day were reinforced with extra
privileges, while increases in drug
intake or remaining at the same drug
level for more than 2 successive days
resulted in lost privileges. Using this
procedure, withdrawal was accomplished in approximately 1 week.
Altman, Myer and co-workers (3)
have attempted to extinguish intravenous heroin self-administration by
physically dependent heroin addicts.
In addition, they have studied the
effect of manipulating the response
requirement per heroin reinforcement. Patients were treated with
naloxone or naltrexone, and allowed
to self-administer heroin. Whereas
all subjects would self-administer the
maximum dosage permissible under
baseline conditions, when pretreated
with these narcotic antagonists, responding dropped to near zero levels
for heroin. They also found that the
amount of heroin consumed was
inversely related to the response cost
—i.e., if more work was required,
less heroin was consumed. In two
related studies, Liebson et al. (38)
and Bigelow and Liebson (6) studied
the effect of fixed-ratio requirement
per ethanol reinforcement in alcoholics. As in the Altman study,
high fixed-ratio requirements were
associated with a suppression of
drug self-administration.

More recent efforts have been directed toward devising practicable
therapeutic procedures based on an
operant interpretation of drug dependence. Hunt and Azrin (33) have
devised a community-based treatment program for chronic alcoholics.
In their program, a variety of material
and social reinforcers (e.g., access to
money, job, wife, friends, home, T.V.,
telephone, and the like) are con-
tigent on not drinking. Drinking
alcohol leads to immediate loss of one
or more of these reinforcers for a

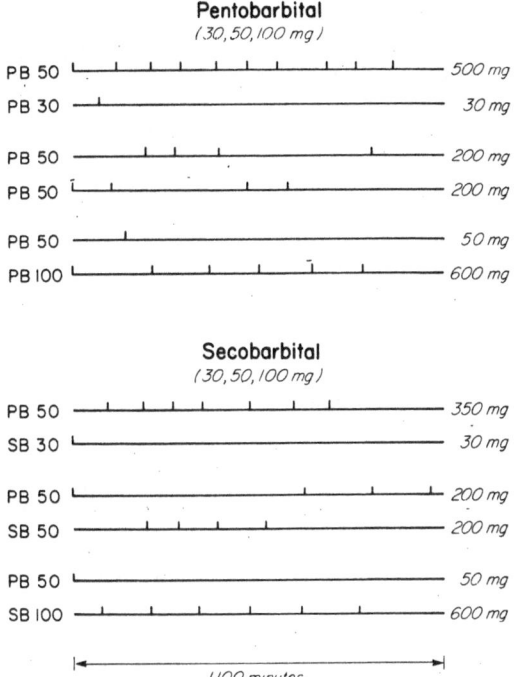

Figure 11. Oral self-administration of barbiturates by human sedative abusers. Vertical lines indicate relative distribution of individual drug self-administrations during preference studies on test days. On left side of figure, PB = pentobarbital and SB = secobarbital, followed by capsule dosage in milligrams. On right side of figure, total daily intake of each dosage. All drugs were administered in standard unmarked capsules (57).

specific period (e.g., 1 day). This procedure has proven effective, thus far, in preventing relapse to uncontrollable drinking. Boudin (9) has developed a broad spectrum community-based behavior therapy program for a wide array of drug abusers (e.g., opiates, stimulants, barbiturates), involving many of the same concepts as the Hunt-Azrin procedure. In addition, Boudin teaches clients responses which are incompatible with drug use through biofeedback and relaxation training. Among the procedures employed in Boudin's Drug Project is requiring the client to make a sizable monetary or other material deposit (e.g., $500, the client's motorcycle). Each week that

the drug user successfully carries out the terms of his contract, a portion of the deposit is returned. If the client violates the terms of the contract during that period, the portion that would have been returned is sent to some organization the views of which are repugnant to the client, or the matieral deposit is disassembled and sold. Using this procedure, Boudin has completed treatment on nine clients. On follow-up (ranging from 2 to 22 months) the rate of total abstinence has been 100%.

Finally, Bigelow, Liebson and Lawrence (8) have attempted to reinforce behavior incompatible with ethanol consumption in alcoholics. They have used three procedures with different alcoholic populations. In combined alcoholic-heroin addicts, they have made methadone availability contingent on consuming disulfuram daily. In another out-patient population, the opportunity to go to work (i.e., to hold a job) was contingent on taking disulfuram each morning. In the third group, a monetary deposit was earned back by taking disulfuram regularly. In all instances, the amount of ethanol consumed was dramatically reduced when the patients were reinforced for engaging in the incompatible behavior of taking disulfuram, a drug that induces vomiting on consuming alcohol.

CONCLUSIONS

In the early 1960's when an operant interpretation of drug self-administration and dependence was first explicitly suggested, many viewed this approach as radical if not downright simple-minded. In the decade and a half that has followed, sufficient data have accumulated to support the operant approach. Moreover, the laboratory findings and principles that have emerged appear to possess great potential when extrapolated to human drug abuse in natural settings.

While we are clearly not suggesting that all of the answers are in or that anything approximating a complete understanding of the complex set of processes involved in drug dependence is at hand, we wish to point out that a behavioral problem as complex and as subtle as drug dependence *has* proven amenable to scientific analysis. What has appeared to some to be beyond the pale of scientific inquiry was awaiting the systematic application of a set of principles that takes into consideration the interaction of drug variables with the environmental conditions under which they are administered.

REFERENCES

1. ALLMAN, L. Group drinking during stress: effect on alcohol intake and group processes. *Int. J. Addict.* 8: 475–488, 1973.
2. ALLMAN, L. R., H. A. TAYLOR, AND P. E. NATHAN. Group drinking during stress: effects on drinking behavior, affect, and psychopathology. *Am. J. Psychiat.* 129: 669–678, 1972.
3. ALTMAN, J. L., R. E. MEYER, S. M. MIRIN AND H. B. MCNAMEE. Opiate antagonists and the modification of heroin self-administration behavior in man: an experimental study. American Psychiatric Association Meeting, May 1974.
4. ANDERSON, W. W., AND T. THOMPSON. Ethanol self-administration in water-satiated rats. *Pharmacol., Biochem. Behav.* 2: 447–454, 1974.
5. BALSTER, R. L., AND C. R. SCHUSTER. Fixed-interval schedule of cocaine reinforcement: effect of dose and infusion duration. *J. Exp. Anal. Behav.* 20: 119–129, 1973.
6. BIGELOW, G., AND I. LIEBSON. Cost factors controlling alcoholic drinking. *Psychol. Rec.* 22: 305–314, 1972.
7. BIGELOW, G., I. LIEBSON AND R. GRIFFITHS. Alcoholic drinking: suppression by a brief time-out procedure. *Behav. Res. Ther.* 12: 107–115, 1974.
8. BIGELOW, G., I. LIEBSON AND C. LAWRENCE. Prevention of alcohol abuse by reinforcement of incompatible behavior. Presented at the meeting of the Association for the Advancement of Behavior Therapy, Miami, December 1973.
9. BOUDIN, H. Environmental intervention with young drug offenders. *Environmental Approaches to Emotional Disturbances.* Columbia, Mo.: Univ. Extension Div., Univ. of Missouri, 1973, p. 19–43.
10. COHEN, M., I. LIEBSON AND L. FAILLACE. Controlled drinking by chronic alcoholics over extended periods of free access. *Psychol. Rep.* 32: 1107–1110, 1973.
11. COHEN, M., I. A. LIEBSON, L. A. FAILLACE AND R. P. ALLEN. Moderate drinking by chronic alcoholics: a schedule-dependent phenomenon. *J. Nerv. Ment. Dis.* 153: 434–444, 1971.
12. COHEN, M., I. A. LIEBSON, L. A. FAILLACE AND W. SPEERS. Alcoholism: controlled drinking and incentives for abstinence. *Psychol. Rep.* 28: 575–580, 1971.

13. CROWDER, W. F., S. G. SMITH, W. M. DAVIS, J. T. NOEL AND W. R. COUSSENS. Effect of morphine dose size on the conditioned reinforcing potency of stimuli paired with morphine. *Psychol. Rec.* 22: 441–448, 1972.

14. DAVIS, W. M., AND J. R. NICHOLS. A technique for self-injection of drugs in the study of reinforcement. *J. Exp. Anal. Behav.* 6: 233–235, 1963.

15. DAVIS, W. M., AND S. G. SMITH. Alpha-methyltyrosine to prevent self-administration of morphine and amphetamine. *Curr. Ther. Res.* 14: 814–819, 1972.

16. DAVIS, W. M., AND S. G. SMITH. Blocking of morphine based reinforcement by alpha-methyltyrosine. *Life Sci.* 12: 185–191, 1973.

17. DAVIS, W. M., AND S. G. SMITH. Haloperidol effects on morphine self-administration: testing for pharmacological modification of the primary reinforcement mechanism. *Psychol. Rec.* 23: 215–221, 1973.

18. DAVIS, W. M., AND S. G. SMITH. Behavioral control exerted by an amphetamine based conditioned reinforcer. In: *Drug Addiction, Vol. III: Neurobiology, Behavior and its Influences*, edited by J. M. Singh and H. Lal. New York: Intercontinental Medical Book Corp., 1974.

19. DEWS, P. B. Differential sensitivity to pentobarbital of pecking performance in pigeons depending on the schedule of reward. *J. Pharmacol. Exp. Ther.* 113: 393–401, 1955.

20. DEWS, P. B., AND W. MORSE. Behavioral pharmacology. *Annu. Rev. Pharmacol.* 1: 145, 1961.

21. DOUGHERTY, J. A., AND R. PICKENS. Development of temporal patterns of cocaine self-administration. *Proceedings, 81st Annual Convention of the American Psychological Association*, Montreal, 1973, p. 1003–1004.

22. DOUGHERTY, J. A., AND R. PICKENS. Fixed-interval schedules of intravenous cocaine presentation in rats. *J. Exp. Anal. Behavior.* 20: 111–118, 1973.

23. DOUGHERTY, J. A., AND R. PICKENS. Effects of phenobarbital and SKF 525A on cocaine self-administration in rats. In: *Drug Addiction, Vol. III: Neurobiology, Behavior and its Influences*, edited by J. M. Singh and H. Lal. New York: Intercontinental Medical Book Corp., 1974.

24. FALK, J. L. Production of polydipsia in normal rats by an intermittent food schedule. *Science* 133: 195–196, 1961.

25. FALK, J. L. Studies on schedule-induced polydipsia. In: *Thirst*, edited by M. J. Weyner. New York: Macmillan, 1964, p. 95–116.

26. FINDLEY, J. D., W. W. ROBINSON AND L. PEREGRINO. Addiction to secobarbital and chlordiazepoxide in the rhesus monkey by means of a self-infusion preference procedure. *Psychopharmacologia* 26: 93–114, 1972.

27. GOLDBERG, S. R. Comparable behavior maintained under fixed-ratio and second-order schedules of food presentation, cocaine injection or *d*-amphetamine injection in the squirrel monkey. *J. Pharmacol. Exp. Ther.* 186: 18–30, 1973.

28. GOLDBERG, S. R., F. HOFFMEISTER, U. U. SCHLICHTING AND W. WUTTKE. A comparison of pentobarbital and cocaine self-administration in rhesus monkeys: effects of dose and fixed-ratio parameter. *J. Pharmacol. Exp. Ther.* 179: 227–283, 1971.

29. GOTESTAM, K. G. Intragastric self-administration of medazapam in rats. *Psychopharmacologia* 28: 87–94, 1973.

30. GROVE, R., AND C. R. SCHUSTER. Effects of response-contingent shock and extinction on cocaine self-administration by monkeys. In press.

31. GUNNE, L. M., E. ANGGARD AND L. E. JONSSON. Blockade of amphetamine effects in human subjects. Presented at the International Institute on the Prevention and Treatment of Drug Dependence. Lausanne: I.C.A.A., 1970.

32. HOFFMEISTER, F., S. R. GOLDBERG, U. SCHLICHTING AND W. WUTTKE. Self-administration of *d*-amphetamine, morphine and chlorpromazine by cocaine "dependent" rhesus monkeys. *Naunyn-Schmiedebergs Arch. Pharmakol.* 266: 359–360, 1970.

33. HUNT, G. M., AND N. H. AZRIN. Community reinforcement approach to alcoholism. Presented at the Meeting of the American Psychological Association, 1972.

34. IGLAUER, C., AND J. WOODS. Concurrent performances: reinforcement by different doses of cocaine in rhesus monkeys. *J. Exp. Anal. Behav.* 22: 179–196, 1974.

35. JARVIK, M. E. Tobacco smoking in monkeys. *Ann. N.Y. Acad. Sci.* 142: 280, 1967.

36. KORMAN, M., I. J. KNOFF AND R. L. LEAR. Alcohol as a discriminative stimulus: a preliminary report. *Tex. Rep. Biol. Med.* 20: 61–63, 1962.

37. LESTER, D. Self-maintenance of intoxication in the rats. *Q. J. Stud. Alcohol* 22: 223–231, 1961.

38. LIEBSON, I. A., M. COHEN, L. A. FAILLACE AND R. F. WARD. The token economy as a research method in alcoholism. *Psychiatr. Q.* 45: 574–581, 1971.

39. MEISCH, R. A., AND T. THOMPSON. Ethanol intake in the absence of concurrent food reinforcement. *Psychopharmacologia* 22: 72–79, 1971.

40. MEISCH, R. A., AND T. THOMPSON. Ethanol intake during schedule-induced polydipsia. *Physiology and Behavior* 8: 471–475, 1972.

41. MEISCH, R. A., AND T. THOMPSON. Ethanol reinforcement: effects of concentration during food deprivation. *Finn. Found. Alc. Stud.* 20: 71–75, 1972.

42. MEISCH, R. A., AND T. THOMPSON. Ethanol as a reinforcer: an operant analysis of ethanol dependence. In: *Drug Addiction, Vol. III: Neurobiology and Influences on Behavior,* edited by J. M. Singh and H. Lal. New York: Intercontinental Medical Book Corp., 1974.

43. MEISCH, R. A., AND T. THOMPSON. Ethanol as a reinforcer: effects of fixed-ratio size and food deprivation. *Psychopharmacologia* 28: 171–183, 1973.

44. MEISCH, R. A., AND T. THOMPSON. Rapid acquisition of ethanol as a reinforcer for rats. *Psychopharmacologia* 37: 311–321, 1974.

45. MEISCH, R.A., J.E. HENNINGFIELD, AND T. THOMPSON. Establishment of ethanol as a reinforcer for rhesus monkeys via the oral route: initial results. In: *Advances in Experimental Medicine and Biology,* Vol. 59, edited by M. M. Gross. New York: Plenum, 1975.

46. MELLO, N. Behavioral studies of alcoholism. In: *The Biology of Alcoholism,* edited by B. Kissin and H. Begleitter. New York: Plenum, 1972.

47. MELLO, N. K., AND J. H. MENDELSON. Operant analysis of drinking patterns of chronic alcoholics. *Nature* 206: 43–46, 1965.

48. NICHOLS, J. R. A procedure which produces sustained opiate-directed behavior (morphine addiction) in the rat. *Psychol. Rep.* 13: 895–904, 1963.

49. PEACOCK, L. J., AND J. A. WATSON. Radiation-induced aversion to alcohol. *Science* 143: 1462–1463, 1964.

50. PICKENS, R., G. BIGELOW AND R. GRIFFITHS. An experimental approach to treating chronic alcoholism: a case study and one-year follow-up. *Behav. Res. Ther.* 11: 321–325, 1973.

51. PICKENS, R., W. C. BLOOM AND T. THOMPSON. Effects of reinforcement magnitude and session length on response rate of monkeys. *Proceedings, 77th Annual Convention of the American Psychological Association,* 1969, p. 809–810.

52. PICKENS, R., R. A. MEISCH AND J. A. DOUGHERTY. Chemical interactions in methamphetamine reinforcement. *Psychol. Rep.* 23: 1267–1270, 1968.

53. PICKENS, R., AND T. THOMPSON. Cocaine reinforcement behavior in rats: effects of reinforcement magnitude and fixed-ratio size. *J. Pharmacol. Exp. Ther.* 161: 122–129, 1968.

54. PICKENS, R., AND T. THOMPSON. Characteristics of stimulant drug reinforcement. In: *Stimulus Properties of Drugs,* edited by T. Thompson and R. Pickens. New York: Appleton-Century-Crofts, 1971.

55. PICKENS, R., AND T. THOMPSON. Simple schedules of drug self-administration in animals. In: *Drug Addiction, Vol. I: Experimental Pharmacology,* edited by J. M. Singh, L. H. Miller and H. Lal. New York: Futura, 1972, p. 107–120.

56. PICKENS, R., T. THOMPSON AND D. MUCHOW. Cannabis and phencyclidine self-administration by animals. In: *Psychic Dependence,* edited by L. Goldberg and F. Hoffmeister. Berlin: Springer-Verlag, 1973, p. 78–86.

57. PICKENS R., L. GUSTAFSON, M. CUNNINGHAM AND L. HESTON. Barbiturate self-administration by human sedative abusers. *Rep. Res. Lab. Dept. of Psychiatry, Univ. of Minnesota,* No. PR-75-1, January, 1975.

58. POZUELO, J., AND F. W. KERR. Suppression of craving and other signs of dependence in morphine-addicted monkeys by administration of alpha-methyl-para-tyrosine. *Mayo Clin. Proc.* 47: 621–628, 1972.

59. SCHUSTER, C. R., AND T. THOMPSON. Self administration of and behavioral dependence on drugs. *Annu. Rev. Pharmacol.* 9: 483–502, 1969.

60. SCHUSTER, C. R., AND J. WOODS. The conditioned reinforcing effects of stimuli associated with morphine reinforcement. *Int. J. Addict.* 3: 223–230, 1968.

61. SENTER, R. J., AND J. D. SINCLAIR. Self-maintenance of intoxication in the rat: a modified replication. *Psychonomic Sci.* 9: 291–292, 1967.

62. SIDMAN, M. Behavioral pharmacology. *Psychopharmacologia* 1: 1–19, 1959.

63. SKINNER, B. F. *The Behavior of Organisms*. New York: Appleton-Century-Crofts, 1938.

64. SKINNER, B. F. *Science and Human Behavior*. New York: Macmillan, 1953.

65. SKINNER, B. F., AND W. T. HERON. Effects of caffeine and benzedrine upon conditioning and extinction. *Psychol. Rec.* 1: 340–346, 1937.

66. SPRAGG, S. D. S. *Comparative Psychology Monographs*. 15: No. 7, 1940.

67. STOLERMAN, I. P., AND R. KUMAR. Preferences for morphine in rats: validation of an experiment model of dependence. *Psychopharmacologia* 17: 137–150, 1970.

68. STRETCH, R., G. J. GERBER AND J. M. WOODS. Factors affecting behavior maintained by response-contingent intravenous infusions of amphetamine in squirrel monkeys. *Can. J. Physiol. Pharmacol.* 48: 581–589, 1971.

69. THOMPSON, T., G. BIGELOW AND R. PICKENS. Environmental variables influencing drug self-administration. In: *Stimulus Properties of Drugs*, edited by T. Thompson and R. Pickens. New York: Appleton-Century-Crofts, 1971.

70. THOMPSON, T., R. GRIFFITHS AND R. PICKENS. Behavioral variables influencing drug self-administration by animals: implications for controlling human drug abuse. In: *Psychic Dependence*, edited by L. Goldberg and F. Hoffmeister. Berlin: Springer-Verlag, 1973, p. 88–103.

71. THOMPSON, T., AND W. OSTLUND. Susceptibility to re-addiction as a function of the addiction and withdrawal environment. *J. Comp. Physiol. Psychol.* 59: 388–392, 1965.

72. THOMPSON, T., AND R. PICKENS. Drug self-administration and conditioning. In: *Scientific Basis of Drug Dependence*, edited by H. Steinberg. London: Churchill, 1969.

73. THOMPSON, T., AND R. PICKENS. Drugs as reinforcers: schedule considerations.

In: *Schedule Effects: Drugs, Drinking and Aggression*, edited by R. Gilbert and J. D. Keehn, Toronto: Univ. of Toronto Press, 1972.

74. THOMPSON, T., AND C. R. SCHUSTER. Morphine self-administration, food-reinforced and avoidance behaviors in rhesus monkeys. *Psychopharmacologia* 5: 87–94, 1964.

75. THOMPSON, T., AND C. R. SCHUSTER. *Behavioral Pharmacology*. New Jersey: Prentice Hall, 1968.

76. WEEKS, J. Experimental morphine addiction: method for automatic intravenous injection in unrestrained rats. *Science* 138: 143–144, 1962.

77. WILSON, M., AND C. R. SCHUSTER. Pharmacological modification of the self-administration of cocaine and SPA in the rhesus monkey. NAS-NRC Committee on Problems of Drug Dependence, Indianapolis, Indiana, 1968.

78. WINGER, G., AND J. H. WOODS. The reinforcing property of ethanol in the rhesus monkey: I. Initiation, maintenance and termination of intravenous ethanol-reinforced responding. *Ann. N. Y. Acad. Sci.* 215: 162–175, 1973.

79. WOODS, J. H., D. A. DOWNS AND J. M. CARNEY. Behavioral functions of narcotic antagonists: response-drug contingencies. *Federation Proc.* 34: 1777–1784, 1975.

80. WOODS, J. H., AND C. R. SCHUSTER. Reinforcement properties of morphine, cocaine, and SPA as a function of unit dose. *Int. J. Addict.* 3: 231–237, 1968.

81. YANAGITA, T., S. TAKAHASHI, K. ISHIDA AND H. FINAMOTO. Voluntary inhalation of volatile anesthetics and organic solvents by monkeys. *Japan. J. Clin. Pharmacol.* 1: 13–16, 1970.

82. YOKEL, R. A., AND R. PICKENS. Self-administration of optical isomers of amphetamine and methylamphetamine by rats. *J. Pharmacol. Exp. Ther.* 187: 27–33, 1973.

83. YOKEL, R. A., AND R. PICKENS. Drug level of *d-* and *l*-amphetamine during intravenous self-administration. *Psychopharmacologia* 34: 255–264, 1974.

84. YOKEL, R. A., AND R. PICKENS. Extinction following amphetamine and methylamphetamine self-administration by rats. In press.

Second-order schedules of drug injection[1]

S. R. GOLDBERG, R. T. KELLEHER AND W. H. MORSE

Laboratory of Psychobiology, Department of Psychiatry
Harvard Medical School, Boston, Massachusetts, 02115
and New England Regional Primate Research Center
Southborough, Massachusetts 01772

ABSTRACT

Key-press responding of squirrel monkeys produced intravenous injections of cocaine under two simple types of schedule. Under a fixed-ratio schedule, every 30th response produced an injection; steady responding at high rates of over one per second were maintained during each fixed-ratio component. Under a fixed-interval schedule, the first response occurring after a fixed time of 5 min produced an injection; there was a pause at the start of each interval and then progressively increasing responding until cocaine was injected at the end of the interval. Both squirrel monkeys and rhesus monkeys also were studied under second-order schedules of drug injection. Under one type of second-order schedule, studied only in squirrel monkeys, completion of each fixed-interval component produced only a 2-sec light; completion of the 10th fixed-interval component produced the brief light and an intravenous injection of cocaine. Under a second type of second-order schedule, each fixed-ratio component completed during a fixed time interval (5 or 60 min) produced only a 2-sec light; the first fixed-ratio component completed after the interval of time elapsed produced the brief light and an intravenous (squirrel monkeys) or intramuscular (rhesus monkeys) injection of cocaine. Under both types of second-order schedules, repeated sequences of responding were maintained during each session and characteristic fixed-interval or fixed-ratio patterns of responding were controlled by the brief visual stimuli. — GOLDBERG, S. R., R. T. KELLEHER AND W. H. MORSE. Second-order schedules of drug injection. *Federation Proc.* 34: 1771–1776, 1975.

[1] Preparation of this manuscript was supported by Public Health Service Research Grants DA-00499, MH-02094, MH-07658 and Research Career Program Award 1-K5-MH22589 (R.T.K.) with facilities and services furnished by the New England Regional Primate Research Center, Harvard Medical School, Southborough, Massachusetts (Public Health Service Grant RR-00168, Division of Research Resources, National Institutes of Health).

During the past decade it has been repeatedly demonstrated that experimental animals will engage in behavior that leads to the intravenous injection of certain drugs from several pharmacologic classes (1, 2, 6, 7, 9, 15, 16). The injection of drug thus functions as a reinforcer to maintain operant behavior in these experimental animals. As with any environmental event, the suitability of a drug injection to maintain operant behavior depends, in large part, on the sequence of events leading to the injection. With human addicts, injection of drug is usually preceded by a complex sequence of behavior that involves obtaining money, purchasing the drug, and then preparing to inject the drug. The environmental stimuli which are repeatedly associated with this complex behavioral sequence come to play a very important role in the control of human drug-seeking behavior (13, 14). This paper will review some recent experiments which demonstrate that complex sequences of behavior can also be maintained by injection of certain drugs in non-human primates such as squirrel monkeys and rhesus monkeys. The experiments reviewed will further demonstrate that environmental stimuli play an important role in the control of these complex sequences of drug-seeking behavior in monkeys.

The techniques used to study behavior maintained by drug injection in monkeys were modifications of operant techniques developed by B. F. Skinner and his colleagues (3). Usually, the animal was surgically prepared with a chronic venous catheter (2, 6) which could be connected to an injection pump during experimental sessions. A readily repeatable behavioral response, such as pressing a small switch or key, was then controlled by the scheduled presentation of a stimulus complex consisting of an intravenous injection of drug accompanied by presentation of a brief light. This approach emphasizes the temporal and sequential relations between stimuli presented to the animal, responses of the animal, and further stimuli consequent on these responses; these interrelationships are called schedules of reinforcement.

SIMPLE SCHEDULES OF DRUG INJECTION

Although there are many possible schedules of reinforcement, we need consider only two representative types, one based on number of responses (ratio) and the other based on time (interval). If key-press responding is maintained by the intravenous injection of cocaine, then under a fixed-ratio schedule, cocaine injection follows the occurrence of a constant number of key-press responses. Under a fixed-interval schedule, cocaine injection follows the first key-press response to occur after a constant interval of time has elapsed; responses during this interval of time have no specified consequences. Each of these two types of schedules engenders its own characteristic pattern of behavior.

Figure 1 shows cumulative-response records of characteristic patterns of responding maintained by each type of schedule. In these records the recording pen steps vertically with each response and the recording paper moves horizontally at a constant speed with time. The short diagonal strokes indicate the injection of cocaine. The positive slope of the record is directly related to the rate of responding.

The records at the left of Fig. 1 show representative performances of two squirrel monkeys under a 30-response fixed-ratio schedule. Under this schedule, every 30th key-press

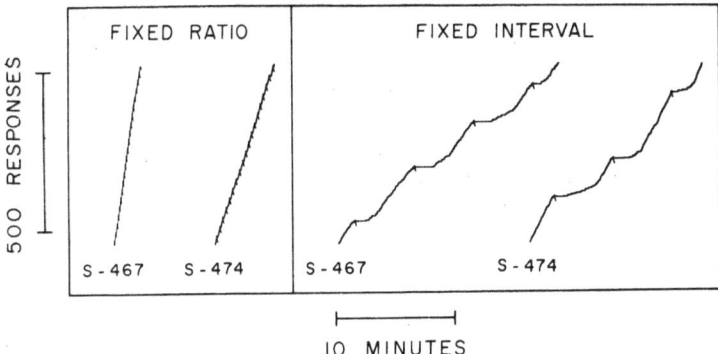

Figure 1. Characteristic performances under fixed-ratio or fixed-interval schedules of intravenous cocaine injection in the squirrel monkey (monkeys S-467 and S-474). *Ordinate*: cumulative number of responses; *abscissa*; time. *Left frame*: sample of records in which a 30-response fixed-ratio schedule was in effect in the presence of a green light. *Right frame*: sample of records in which a 5-min fixed-interval schedule was in effect in the presence of a red light. Each short diagonal stroke on the record indicates an intravenous injection of 25 µg/kg of cocaine hydrochloride and a 2-sec presentation of a yellow light. The recorder was stopped during the 1-min time-out periods after each injection. Patterns of responding under these fixed-ratio and fixed-interval schedules of cocaine injection are similar to those observed when other events are presented under comparable schedules.

response in the presence of a green light produced an intravenous injection of 25 µg/kg cocaine; the injection was accompanied by a brief, 2-sec yellow light. Each injection was followed by a 1-min time-out period, not shown in these records, during which the experimental chamber was dark and responses of the monkey had no specified consequences. Under optimal conditions, this type of schedule engenders a steady high rate of responding during each ratio component that is often (monkey S-474), but not always (monkey S-467), preceded by an initial brief period of no responding often referred to as a pause.

The records at the right of Fig. 1 show representative performance of the same two squirrel monkeys several months after the schedule of cocaine injection was changed to a 5-min fixed interval schedule. Under this schedule, the first key-press response after 5 min elapsed in the pres-

ence of a red light produced an intravenous injection of 25 µg/kg cocaine; the injection was accompanied by a brief, 2-sec yellow light. Again, each injection was followed by a 1-min time-out period, not shown on the records, during which the experimental chamber was dark and responses of the monkey had no specified consequences. Under this schedule, there is an initial period of no responding at the start of each interval and then progressively increasing responding up to the point where drug is injected at the end of the interval.

These characteristic schedule-controlled patterns of responding are of particular interest since they are highly reproducible under diverse conditions and can be maintained by a variety of other events. Figure 2 shows representative fixed-interval performance maintained under diverse conditions by food or water presentation. The upper frame shows

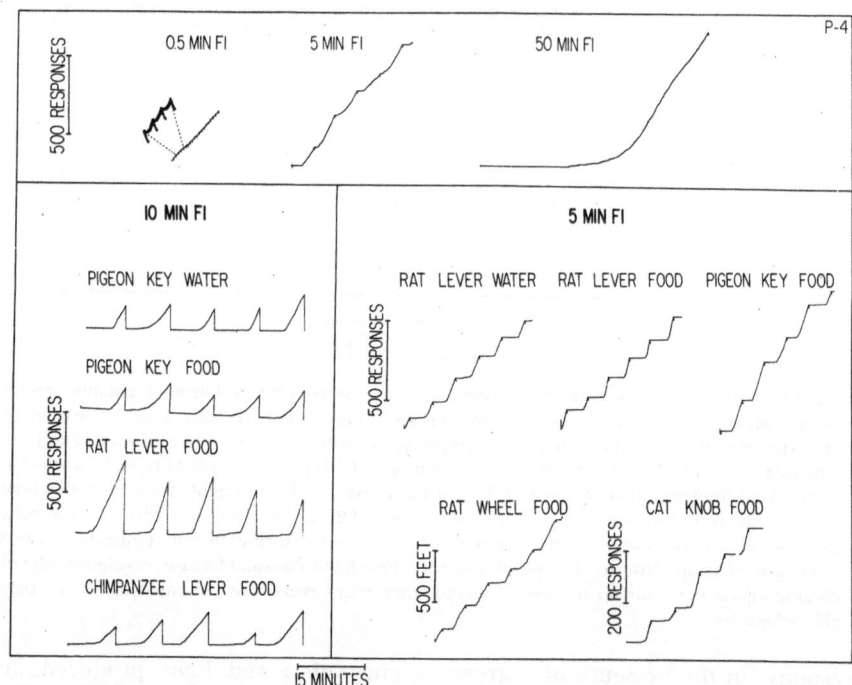

Figure 2. Characteristic fixed-interval performance under diverse conditions. *Ordinate*: cumulative number of responses; *abscissa*: time. A fixed-interval (FI) schedule of food or water presentation was in effect in all examples shown in this figure. *Upper frame*: pigeon (P-4) pecking plastic key (food). The general pattern persists despite the 100-fold change in the duration of the fixed-interval. Food presentations are marked by short diagonal strokes. *Lower left frame*: performances under a 10-min fixed-interval schedule. Food or water presentations are marked by the resetting of the recording pen to the baseline. *Lower right frame*: performances under a 5-min fixed-interval schedule. The species, the type of switch recording the response, and the maintenance event are indicated above the record. The pigeon pecks a plastic key with its beak; the rat and chimpanzee press a horizontal lever with their paws; the cat depresses a rounded knob with its paw. The rat turns a wheel by running; only a turn of 180° is reinforced, but the cumulative distance the wheel turns is recorded directly. (From Kelleher and Morse (12) with permission.)

cumulative-response records from a pigeon pecking a plastic key under a schedule of food presentation at different schedule values. The species, the type of switch recording the response, and the event used to maintain the behavior are indicated above the other records. It can be seen that the characteristic fixed-interval pattern of responding occurs over a wide range of parameter values

and that it can be engendered in many different species. In the squirrel monkey, fixed-interval performance comparable to that maintained by intravenous cocaine injection can be maintained by presentation of food or by termination of a stimulus associated with occasional electric shocks (12). In addition, comparable fixed-ratio performance can be maintained in the squirrel monkey both by

presentation of food and by intravenous injection of cocaine when
they are studied under comparable
behavioral schedules (6). In contrast,
the same event (cocaine injection, Fig.
1), scheduled in different ways, engenders different rates and patterns
of responding. Such findings indicate
that the schedule of reinforcement
can be a more important determinant
of rates and temporal patterning of
behavioral responses than the type of
event maintaining the behavior.

SECOND-ORDER SCHEDULES OF DRUG INJECTION

It is possible to use schedules, such
as the fixed-ratio or fixed-interval
schedules just described, as building
blocks for more complex schedules.
For example, behavior can be maintained under schedules in which food
presentation occurs only after the
completion of several consecutive
schedule components; each component schedule terminates with the
brief presentation of a stimulus that
has been associated with food presentation (6, 10, 11). Complex sequences
of this type have been termed
second-order schedules. We have
been particularly interested in apply-
ing this same high-order scheduling
technique to the study of behavior
maintained by injection of drugs.

In one series of experiments (4–6),
key-press responding of squirrel
monkeys was initially developed and
maintained under the simple 30-
response fixed-ratio schedule described earlier. When behavior
stabilized, the schedule of intravenous cocaine injection was changed
to a second-order schedule with fixed-
ratio components. Stable performance of three squirrel monkeys under
this schedule is shown in Fig. 3. Under
this schedule, every 30-response fixed-
ratio component completed by the

monkey during a 5-min interval of
time produced only a 2-sec yellow
light; each presentation of the yellow light is indicated by a short
diagonal stroke on the cumulative
record. The first fixed-ratio component completed by the monkey
after the 5-min interval elapsed
produced both the brief light and an
intravenous injection of 100 μg/kg
cocaine. Injection of cocaine is indicated by the cumulative response
pen resetting to the bottom of the
record and also by a mark on the
lower event line. Each cocaine injection was followed by a 1-min
time-out period, not shown on the
records. Each daily session lasted
until the monkey received 15 injections. Although injections of cocaine were far more intermittent
under this second-order schedule
than under the simple fixed-ratio
schedules previously described, repeated sequences of rapid responding
were maintained throughout each
daily session. From 400 to 1,000 key-
press responses preceded each injection of cocaine under this second-
order schedule.

In another series of experiments,
squirrel monkeys, whose behavior
had been previously maintained
under the simple 5-min fixed-interval
schedule described earlier, were
studied under a second-order schedule with fixed-interval components.
Stable performance of a monkey
under this schedule is shown in Fig. 4.
Completion of each 5-min fixed-
interval component produced a 2-sec
yellow light, indicated by a short
diagonal stroke on the cumulative
record. The completion of every
tenth fixed-interval component, however, produced both the brief light
and an intravenous injection of 100
μg/kg cocaine. Each cocaine injection was followed by a 100-sec
time-out period. The downstroke of

the event pen indicates the injection of cocaine and the event pen then remained down during the time-out period. The recording pen reset to the bottom of the record whenever 1,100 responses had cumulated and also at the end of time-out periods. Each daily session lasted until the monkey received three injections. A full session is shown here from top to bottom for monkey S-474. Although the maximum frequency of

cocaine injections was less than one per 50 min under this second-order schedule, repeated sequences of positively-accelerated responding, characteristic of fixed-interval schedules, were maintained throughout each daily session.

Figure 5 shows in more detail the characteristic patterns of responding maintained during individual fixed-ratio or fixed-interval components under each type of second-order

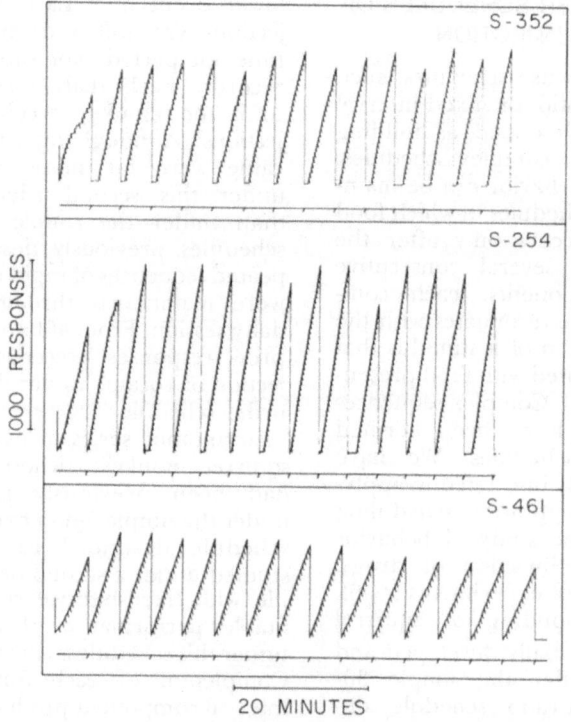

20 MINUTES

Figure 3. Representative performances of three squirrel monkeys (S-352, S-254 and S-461) under a second-order schedule in which the first completion of a 30-response fixed-ratio component after 5 min produced an intravenous injection of cocaine. The completion of each fixed-ratio produced a 2-sec yellow light but had no other programmed consequences until the 5-min interval of time elapsed. Short diagonal strokes on the cumulative records indicate presentations of the yellow light; diagonal strokes on the event record and the resetting of the cumulative record indicate intravenous injections of 100 μg/kg of cocaine hydrochloride. After each injection, there was a 1-min time-out period. The recorder was stopped during 2-sec presentations of the yellow light and during time-out periods. Note the high rates of responding in the fixed-ratio components. (From Goldberg (5) with permission.)

schedule shown in Figs. 3 and 4. The records at the left show representative performances of two squirrel monkeys under the second-order schedule with fixed-ratio components. This type of schedule engendered a steady high rate of responding. Responding during each ratio component was usually (monkey S-60), but not always (monkey S-254), characterized by an initial brief pause followed by an abrupt change to a steady high rate of responding until the ratio was completed and the brief yellow light was presented. The rates and patterns of responding controlled by the brief lights under this second-order schedule were almost identical to those controlled by cocaine injections under the simple fixed-ratio schedules described earlier.

The records at the right of Fig. 5 show representative performances of

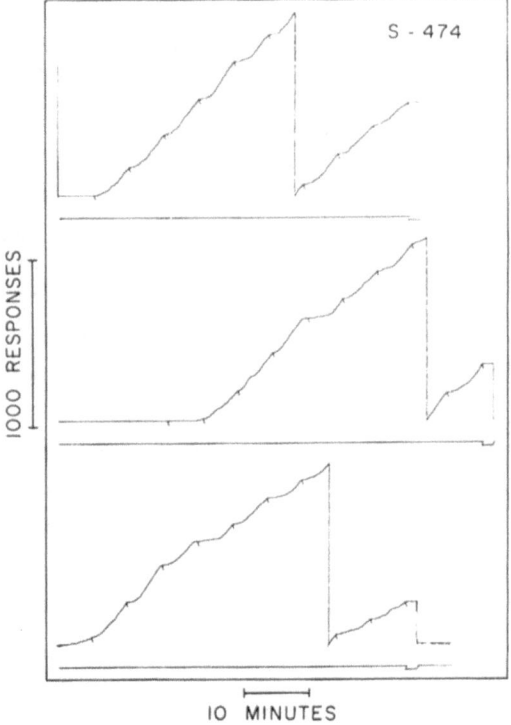

Figure 4. Representative performance of squirrel monkey S-474 under a second-order schedule in which every tenth completion of a 5-min fixed-interval produced an intravenous cocaine injection. The completion of each fixed-interval produced a 2-sec yellow light. Short diagonal strokes on the cumulative record indicate presentations of the yellow light. The downstrokes on the event record indicate intravenous injections of 100 μg/kg of cocaine hydrochloride and the beginning of a 100-sec time-out period during which the event pen remained down. This figure shows a complete experimental session comprising three sequences (from top to bottom), each terminating in a cocaine injection and a time-out period. The recording pen reset to the bottom of the record whenever 1,100 responses had cumulated and at the end of each sequence. Note the characteristic fixed-interval pattern of responding in most components.

two squirrel monkeys under the second-order schedule with fixed-interval components. There was an initial period of no responding at the start of each interval and then responding progressively increased until the brief light was presented at the end of the interval. Again, the rates and patterns of responding controlled by the brief lights under this second-order schedule were almost identical to those controlled by cocaine injections under the simple fixed-interval schedules described earlier.

Under second-order schedules, drug injections occur only when the monkey has completed a sequence of schedule components. These schedules extend the range of conditions and the parameters under which drug-maintained behavior can be studied. With the high degree of intermittency possible under second-order schedules, long and orderly sequences of behavior can be maintained and powerful moment-to-moment control exerted over rates

and patterns of responding at times when the direct effects of the drug are minimal or absent. For example, it is possible to arrange the parameter values of the second-order schedules so that drug injection occurs only at the end of an experimental session.

A recent experiment with rhesus monkeys illustrates one way that second-order schedules which restrict drug injection to the end of each session can be applied (5,8). Three rhesus monkeys lived in individual primate cages provided with response keys and enclosed in isolation chambers. During experimental sessions, which were conducted on Monday, Wednesday and Friday, the chamber door was closed and every 10-response fixed-ratio component completed during a 60-min interval of time produced a 2-sec red light. The first 10-response fixed-ratio component completed by the monkey after the 60-min interval elapsed produced a red light which remained on for 2 min while the chamber door was opened; the monkey, who had

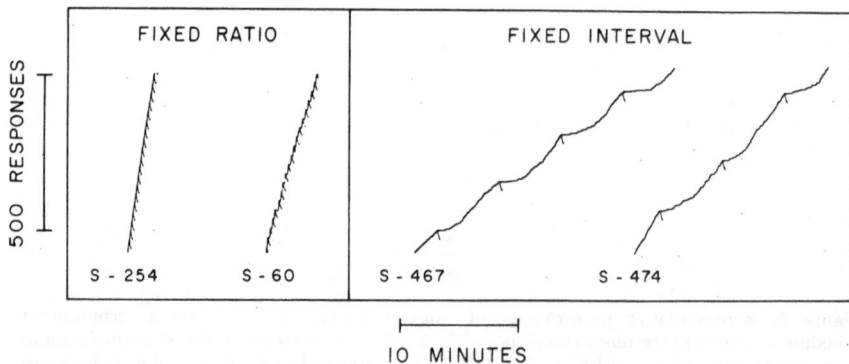

Figure 5. Sample records of performances under fixed-ratio or fixed-interval components of second-order schedules of cocaine injection (monkeys S-254, S-60, S-467, and S-474). *Ordinate*: cumulative number of responses; *abscissa*: time. Short diagonal strokes on the records indicate presentations of the yellow light. The types of second-order schedules from which these components were taken have been described in Figs. 3 and 4. Note the similarity of the patterns of responding maintained under the components of the second-order schedules to those maintained under the simple schedules of cocaine injection shown in Fig. 1.

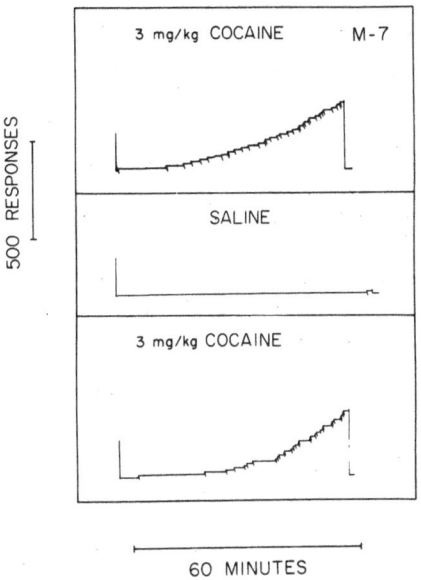

Figure 6. Performance of rhesus monkey M-7 showing the effect of substituting saline for cocaine under the second-order schedule of intramuscular cocaine injection. *Abscissa*, time; *ordinate*, cumulative number of responses. Short diagonal strokes on the cumulative records indicate presentations of a 2-sec red light. Each session ended with an intramuscular injection of 3 mg/kg of cocaine hydrochloride or saline accompanied by a red light, which is indicated by the recording pen resetting to the bottom of the cumulative record. The fourth saline-substitution session (*middle panel*) is compared to representative performance before (*top panel*) and after (*bottom panel*) saline substitution. (From Goldberg (5) with permission.)

been trained to extend his arm out through a hole in the side of the cage, was then given an intramuscular injection of cocaine by a technician.

Performance of a rhesus monkey under this second-order schedule of intramuscular drug injection is shown in Fig. 6. Repeated sequences of rapid responding were maintained during each session by intramuscular injection of cocaine. The brief visual stimuli controlled characteristic fixed-ratio patterns of responding; pauses in responding occurred after each red light and were followed by steady high rates of responding until the light was produced again. When intramuscular saline injections were substituted for cocaine injections, responding markedly decreased. Responding was restored when cocaine injections were reinstated. The average rate of fixed-ratio responding maintained under the second-order schedule of intramuscular cocaine injection was lower than under the second-order schedule of intravenous cocaine injection, but the repeated sequences of characteristic fixed-ratio patterns of responding were similar. It is clear that stimuli associated with drug administration powerfully modulate the control of drug-maintained behavior.

SUMMARY

Under second-order schedules, drug injections occur only when the monkey has completed a sequence of schedule components. This is a useful technique for analyzing sequences of drug-seeking behavior because it is possible to investigate the ways in which scheduled stimulus presentations affect responding in the sequence, without changing the frequency of drug injection. The results reviewed demonstrate that brief environmental stimuli that have been associated with drug injection can come to control rates and patterns of behavioral responding that are strikingly similar to those initially maintained by the cocaine injections themselves. Second-order schedules of the type described in this paper provide a valuable technique for the experimental study of environmental factors involved in drug dependence.

REFERENCES

1. BALSTER, R. L., C. E. JOHANSON, R. T. HARRIS AND C. R. SCHUSTER. Phencyclidine self-administration in the rhesus monkey. *Pharmacol. Biochem. Behav.* 1: 167–172, 1973.
2. DENEAU, G. A., T.YANAGITA AND M. H. SEEVERS. Self-administration of psychoactive substances by the monkey. A measure of psychological dependence. *Psychopharmacologia* 16: 30–48, 1969.
3. FERSTER, C. B., AND B. F. SKINNER. *Schedules of Reinforcement.* New York: Appleton-Century-Crofts, 1957.
4. GOLDBERG, S. R. Sequences of rapid responding maintained by cocaine self-injection in squirrel monkeys. *Pharmacologist* 13: 28, 1971.
5. GOLDBERG, S. R. Control of behavior by stimuli associated with drug injections. In: *Psychic Dependence: Definition, Assessment in Animal and Man,* edited by L. Goldberg and F. Hoffmeister. Berlin: Springer-Verlag, 1973, p. 106–109.
6. GOLDBERG, S. R. Comparable behavior maintained under fixed-ratio and second-order schedules of food pre-
7. GOLDBERG, S. R., F. HOFFMEISTER, U. U. SCHLICHTING AND W. WUTTKE. A comparison of pentobarbital and cocaine self-administration in rhesus monkeys: effects of dose and fixed-ratio parameter. *J. Pharmacol. Exp. Ther.* 179: 277–283, 1971.
8. GOLDBERG, S. R., AND W. H. MORSE. Behavior maintained by intramuscular injections of morphine or cocaine in the rhesus monkey. *Pharmacologist* 15: 236, 1973.
9. HOFFMEISTER, F., AND U. U. SCHLICHTING. Reinforcing properties of some opiates and opioids in rhesus monkeys with histories of cocaine and codeine self-administration. *Psychopharmacologia* 23: 55–74, 1972.
10. KELLEHER, R. T. Chaining and conditioned reinforcement. In: *Operant Behavior: Areas of Research and Application,* edited by W. K. Honig. New York: Appleton-Century-Crofts, 1966, p. 160–212.
11. KELLEHER, R. T. Conditioned reinforcement in second-order schedules. *J. Exp. Anal. Behav.* 9: 475–485, 1966.
12. KELLEHER, R. T., AND W. H. MORSE. Determinants of the specificity of behavioral effects of drugs. *Ergeb. Physiol. Biol. Chem. Exp. Pharmakol.* 60: 1–56, 1968.
13. VAILLANT, G. E. The natural history of urban narcotic drug addition—some determinants. In: *Scientific Basis of Drug Dependence,* edited by H. Steinberg. London: Churchill, 1969, p. 341–361.
14. WIKLER, A. Conditioning factors in opiate addiction and relapse. In: *Narcotics,* edited by D. M. Wilner and G. G. Kassebaum. New York: McGraw-Hill, 1965, p. 85–100.
15. WILSON, M. C., M. HITOMI AND C. R. SCHUSTER. Psychomotor stimulant self-administration as a function of dosage per injection in the rhesus monkey. *Psychopharmacologia* 22: 271–281, 1971.
16. WOODS, J. H., F. IKOMI AND G. WINGER. The reinforcing property of ethanol. In: *Biological Aspects of Alcohol,* edited by J. P. Creaven and M. K. Roach. Austin: Univ. of Texas Press, 1971, p. 371–388.

Behavioral functions of narcotic antagonists: response-drug contingencies[1]

JAMES H. WOODS, DAVID A. DOWNS AND JOHN CARNEY

Department of Pharmacology, University of Michigan
Ann Arbor, Michigan 48104

ABSTRACT

Behavioral effects of the narcotic antagonist naloxone are discussed in terms of stimulus functions. As an eliciting stimulus, the effects of naloxone depend on prior administration of narcotic. Administered independently of responding, naloxone can increase or decrease rates of narcotic-reinforced responding depending on the dose of naloxone. When naloxone is administered as a consequence of narcotic self-injection, the future probability of that behavior is reduced; thus, naloxone can function as a punishing stimulus. As a negatively-reinforcing stimulus, naloxone can maintain behavior which terminates or prevents delivery in morphine-dependent monkeys. In animals with previous naloxone avoidance-escape experience, unavoidable-inescapable injections of naloxone produce increases in avoidance-escape response rates. In these animals, responding subsequently can be maintained, at least temporarily, when naloxone is administered only as the consequence of responding.—WOODS, J. H., D. A. DOWNS AND J. CARNEY. Behavioral functions of narcotic antagonists: response-drug contingencies. *Federation Proc.* 34: 1777–1784, 1975.

In the experimental analysis of behavior, stimuli may be functionally classified as eliciting, reinforcing, punishing, and discriminative depending on the changes in behavior observed when a given stimulus is presented (28). Changes in behavior produced by narcotic antagonists also may be classified in terms of these stimulus functions. The purpose of this paper is to discuss some of the stimulus functions of narcotic antagonists and to provide examples of situations in which these behavioral effects are modulated by pharmacological and behavioral conditions.

[1] Supported by Public Health Service Grants DA 00154, DA 00254, and GM 00198.

45

RESPONSE-ELICITING FUNCTIONS OF NARCOTIC ANTAGONISTS

In morphine-dependent rhesus monkeys, narcotic antagonists, e.g., naloxone, can elicit a number of responses. The set of responses known as the morphine abstinence syndrome includes piloerection, emesis, diarrhea, rhinorrhea, miosis, and hyperreflexia (27). The magnitudes of some of the antagonist-elicited responses are related to the magnitude (dose) of the drug stimulus (31). Such elicited responses occur upon the initial administration of naloxone in morphine-dependent monkeys and usually are similar to the abstinence syndrome produced by an appropriate period of deprivation from morphine (34). This eliciting-stimulus function appears to be largely independent of the relationship between antagonist administration and ongoing behavior. Such responses typically are observed regardless of the behavior the animal may have been emitting prior to antagonist administration. Because of these criteria, the characterization of some effects of narcotic antagonists as eliciting is consistent with the same characterization for other stimuli (28).

The major difference between narcotic antagonists and other eliciting stimuli is that for antagonists we are able to specify a necessary historical condition for obtaining many of the elicited responses, i.e., prior administration of narcotic. Moreover, the magnitude of an antagonist-elicited response is often accepted as a measure of the degree of physical dependence on narcotics (32).

Naloxone and other narcotic antagonists may also elicit various responses in nondependent rhesus monkeys, e.g., convulsive or preconvulsive phenomena. In general, however, such responses are obtained at doses of antagonist much greater than doses which can elicit abstinence responses in dependent animals. Thus, antagonist-elicited responses depend primarily on prior pharmacological conditions; namely, the presence or absence of a history of narcotic administration.

ANTAGONISM OF BEHAVIORAL EFFECTS OF NARCOTICS

Narcotic antagonists can also produce changes in behavior by antagonizing acute effects of narcotics. There are several ways in which the antagonism of behavioral effects of narcotics may be demonstrated. In one example, administration of narcotic can produce a decrease in the rate of schedule-controlled responding; if an appropriate dose of a narcotic antagonist, e.g., naloxone, is then administered, it is often possible to reinstate prenarcotic levels of performance. Figure 1 shows the effects of naloxone on morphine-induced decreases in food-reinforced responding in the pigeon. Under a requirement of 30 responses to produce food delivery (fixed-ratio 30 schedule), morphine produced dose-related decreases in responding. In this case, naloxone's blockade of morphine's effects represents a shift to the right in the dose-response curve. This effect can be demonstrated with naloxone doses that by themselves have no response rate increasing actions. At relatively high doses, naloxone alone may cause decreases in response rates in similar behavioral situations (23).

Narcotics not only can decrease operant responding, but also can function to increase and maintain responding as reinforcing stimuli (35). Under fixed-ratio schedules of narcotic reinforcement, responding is characteristically distributed in a

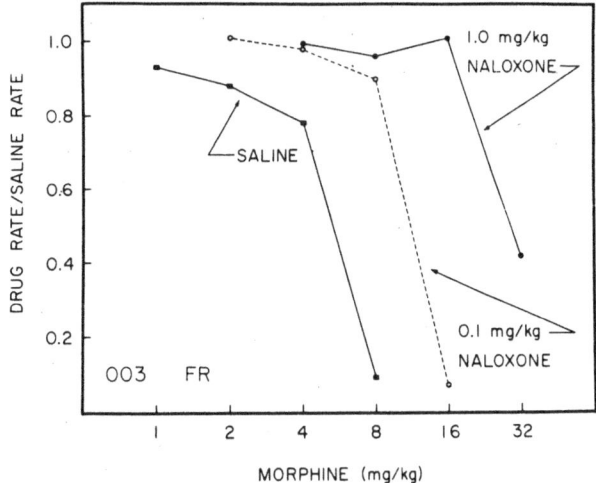

Figure 1. Effects of intramuscular injection of morphine plus saline, or morphine plus naloxone on responding under a 30-response fixed-ratio schedule in one pigeon. *Ordinate*: ratio of response rate following drug injection divided by response rate following injection of saline alone. Saline alone values were obtained on sessions immediately preceding those when morphine plus saline or morphine plus naloxone was given. *Abscissa*: morphine dose. Each data point represents a single observation.

negatively accelerated pattern within sessions (6). We have been interested in determining whether this decline in response rate over successive narcotic self-injections represented a general depression of schedule-controlled responding. A procedure was implemented to evaluate this possibility by training rhesus monkeys to respond for food pellets and codeine injections during alternating components within daily sessions (Fig. 2a). Each food component or drug component terminated after 15 reinforcements or about 15 min. Food pellets and codeine injections were delivered following 30 lever-press responses (FR 30). Food and drug components were signaled by different colored stimulus lights.

Under control conditions, response rates were comparable in the first food and drug components of each session (Fig. 2a). Over successive codeine components, however, there was a gradual decline in response rate and in the number of codeine injections. In contrast, the rate of food-reinforced responding remained relatively constant across components, thus demonstrating that the typical negative acceleration of codeine-reinforced responding is not simply a general depression of performance.

The administration of a low dose of naloxone (0.01 mg/kg) at the start of the session produced an obvious change in the pattern of codeine-reinforced responding without changing food-reinforced responding (Fig. 2b). The high rate of responding for codeine in the initial drug component was unchanged. The progressive decline in response rate across subsequent drug components, however, was markedly attenuated. Thus, the rate and pattern of responding main-

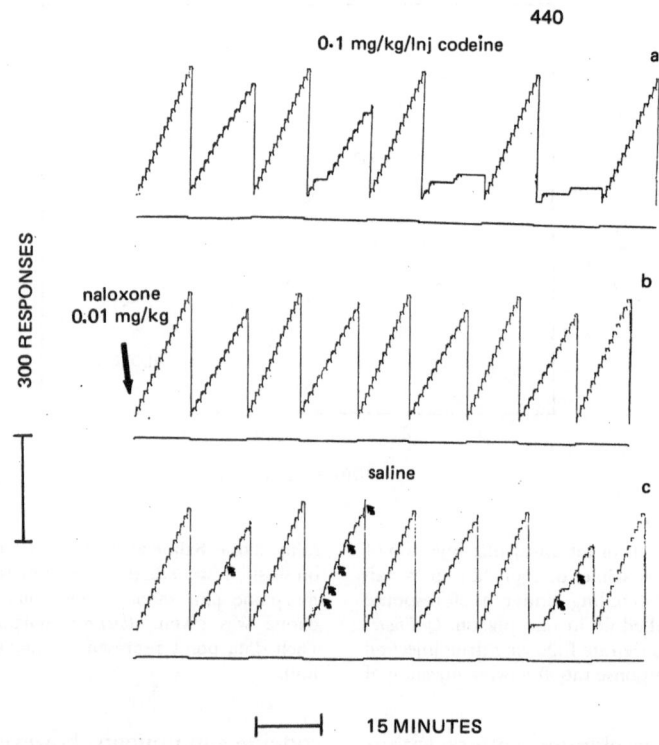

Figure 2. Cumulative records of responding in alternating food-reinforcement (event pen up) and drug-reinforcement (event pen down) components. In all cases, 30 responses in the presence of a white stimulus light produced a 300-mg banana-flavored food pellet in food-reinforcement components, while 30 responses in the presence of a green stimulus light produced an intravenous injection of codeine (0.1 mg/kg per injection) in drug-reinforcement components. A 30-sec timeout followed each codeine injection. Following each food pellet delivery, there was a time-out period in which each response reset the time-out clock for an additional 30 sec (DRO 30 sec). Time-outs and DRO periods were differentially associated with particular stimulus lights. Record *a* shows typical rates and patterns of responding over successive components. Record *b* shows a session following intravenous pretreatment with 0.01 mg/kg of naloxone. Record *c* shows a session in which saline was substituted for codeine. The arrows in record 2*c* indicate examples of responding during saline injections.

tained by codeine became relatively constant across drug components. Larger pretreatment doses of naloxone result in depression of narcotic-reinforced responding in procedures similar to the above without changing the rate of food-reinforced responding (5).

It is possible that the effects observed after a low dose of naloxone are the result of complete cancellation of the behavioral effects of codeine. A partial test of this hypothesis was conducted in two monkeys by substituting saline for codeine in the drug components. When saline was substituted for codeine there were marked increases in the rate of responding. This was similar to the effect of pretreatment with naloxone

(Fig. 2b). However, the pattern of responding was quite different from that obtained with codeine either with or without naloxone pretreatments (see arrows, Fig. 2c). In both monkeys tested, there was a disruption of the usual pattern of responding such that a greater number of responses were emitted during injections. Responses during injections had no programmed consequence, and relatively few responses were normally emitted. Thus, naloxone at the 0.01 mg/kg dose does not appear to produce the change in behavior by making codeine equivalent to saline, since the patterns of responding observed were quite different when saline was substituted for codeine. However, it is possible that naloxone partially antagonized codeine and resulted in a functionally lower dose per injection of codeine. Lower doses of codeine have not been assessed in this situation.

Another possible explanation for the rate-increasing effect of naloxone on codeine-reinforced responding is that naloxone simply functions to increase behavior that is emitted at low rates. Therefore it would be expected that all comparably low rates of responding should be increased by naloxone to a similar extent; however, the low rates of responding obtained during drug injections and time-out periods were unchanged at the dose of naloxone tested. Thus, naloxone appears to alter codeine-reinforced responding by interacting selectively with the drug reinforcer and not exclusively the ongoing rate of responding.

Differential effects of antagonists on narcotic-reinforced and cocaine-reinforced responding have been reported previously (13, 35). These experiments have shown that naloxone can increase or decrease narcotic-reinforced responding selectively depending on dose of antagonist

and previous experience with narcotic antagonist. They thus serve to emphasize that the behavioral effects of narcotic antagonists, like other drugs, depend on the environmental conditions and behavioral history of the subject. The following sections of the paper illustrate some of the ways that the behavioral effects of naloxone can be modulated by the nature of its contingent relation to behavior.

PUNISHMENT OF CODEINE-REINFORCED RESPONDING BY INTRAVENOUS NALOXONE DELIVERY

Punishment is said to have occurred when the future probability of behavior that precedes the delivery of the punishing stimulus is reduced (1). In this section, we shall describe the behavioral effects of naloxone in terms of the stimulus function of punishment.

Rhesus monkeys, following adaptation to restraint and catheterization, were exposed to conditions in which a lever press was followed by an intravenous injection of codeine (0.1 mg/kg per injection). Codeine availability was limited to 1-hr periods separated by 11 hr. Under these conditions, responding was maintained at an average rate of 53 to 64 inj/hr (Fig. 3). Response-contingent naloxone delivery was accomplished by adding naloxone to the codeine solution. Naloxone–codeine combinations were examined in single sessions followed by at least three codeine-reinforced sessions at baseline levels. With each of the monkeys, the dose of naloxone was increased until responding fell to or below 10 inj/hr and then the sequence of doses was reversed. A naloxone dose of 0.03 mg/kg per injection was sufficient to produce partial suppression in all animals in the ascending series; greater suppression was observed with higher doses. There was a more profound suppression of responding with the various doses in the descending series.

Figure 4 shows the recovery of codeine-reinforced responding following sessions in which 0.01 and 0.1 mg/kg per injection naloxone were added to the codeine solution. Codeine-reinforced responding did not return in the sessions immediately following the naloxone–codeine sessions; rather, there was a dose-related persistence of the suppression, lasting for as long as three sessions in the case of the 0.1 mg/kg per injection naloxone–codeine combination.

Another set of observations was obtained with three monkeys in which naloxone was administered just prior to the experimental session not contingent on codeine-reinforced responding. Each dose was administered on three or more occasions before the next dose in the sequence was begun. Doses were increased, then decreased and were given approximately every 2 days. In the ascending sequence of doses, there was a dose-related decrease in codeine-reinforced responding. Again, the descending series of doses produced larger rate-decreasing effects than the ascending series (Fig. 5). Com-

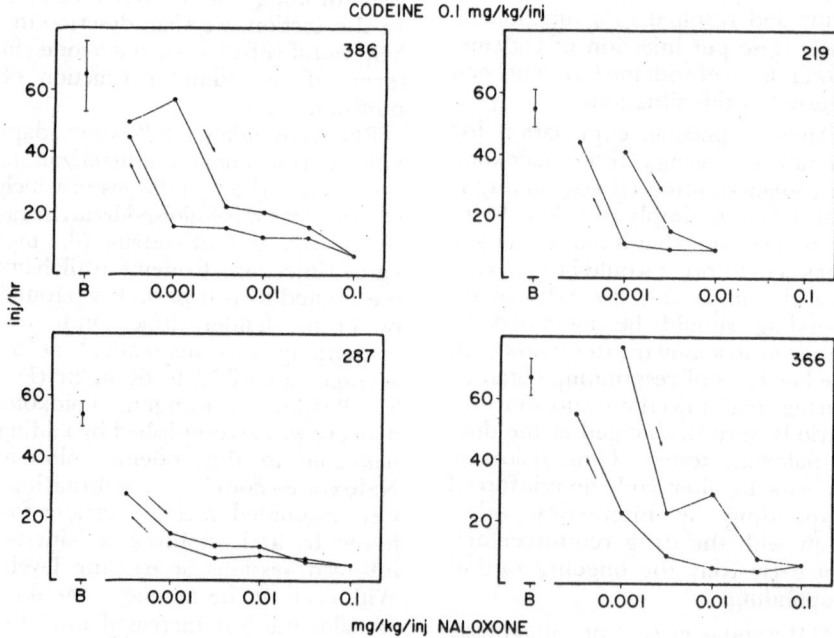

Figure 3. Effects of acute response-contingent naloxone injections on rate of 0.1 mg/kg per injection codeine-reinforced responding. Each graph represents the performance of an individual monkey. *Ordinate*: injections/hour. *Abscissa*: response-contingent dose of naloxone, log scale. B: Baseline rates of codeine self-injection. The solid circle at B is the average of each session that preceded a session in which dose of naloxone was manipulated. The bracket includes ±2 standard errors. Naloxone dose was increased to a maximum dose and then decreased, and the arrows indicate the sequence of naloxone administration.

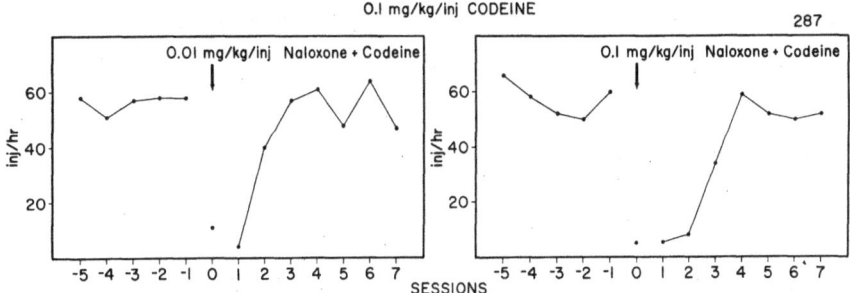

Figure 4. Recovery of codeine self-administration responding following sessions in which 0.01 or 0.1 mg/kg per injection naloxone was administered along with 0.1 mg/kg per injection codeine. *Ordinate*: injections/hour. *Abscissa*: five 1-hr sessions preceding naloxone administration, the 1-hr session in which naloxone was self-administered with codeine (0), and the succeeding seven sessions in which codeine alone was available. Each graph represents the performance of one individual monkey (287) in the ascending series of naloxone doses (see Fig. 3). (From (34) with permission.)

parable effects of antagonist dose sequence have been reported previously (13) and these effects are similar to those obtained with response-contingent antagonist delivery.

In order to compare the amount of response-contingent naloxone delivered to the amount of naloxone delivered noncontingently, i.e., injected before a session, it is necessary to multiply the number of injections times the naloxone dose. The response-contingent delivery of naloxone produced only slightly greater suppression of responding than comparable amounts of noncontingently

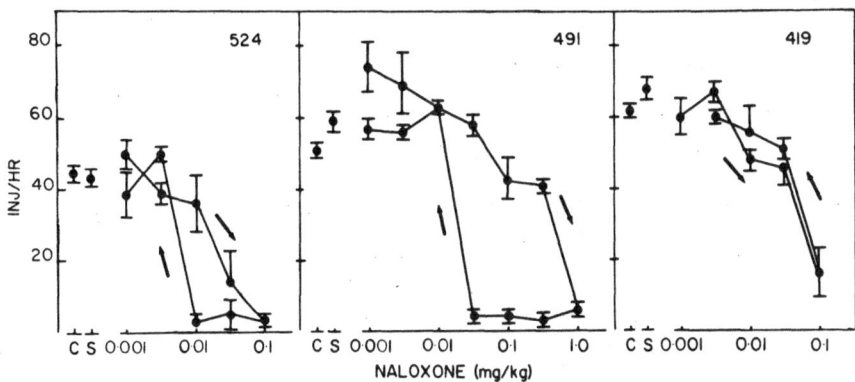

Figure 5. Effects of acute noncontingent naloxone administration on the rate of 0.1 mg/kg per injection of codeine-reinforced responding. Each graph represents the performance of an individual monkey. *Ordinate*: injections/hour. *Abscissa*: naloxone dose, mg/kg, log scale. C: Injections/hour, average of 9 sessions, brackets indicate ±1 standard error. S: Saline presession administration, average number of injections/hour. Other points represent the average (brackets indicate range) of three or more observations at a particular naloxone dose. Arrows indicate direction of dose sequence (ascending or descending).

delivered naloxone. One should also note, of course, that the temporal pattern of antagonist delivery is quite different for the two methods of administration. Although both the contingent and noncontingent procedures suppressed responding during the test sessions, the response-contingent delivery of naloxone clearly produced greater suppression in sessions subsequent to those in which the naloxone was delivered. Figure 6 shows the largest doses of naloxone given noncontingently that suppressed codeine-reinforced responding by 80% or more and a smaller dose from the descending series that produced comparable suppression. Neither of these doses was effective in suppressing codeine-reinforced responding in subsequent sessions in marked contrast to response-contingent delivery of naloxone (Fig. 4) at comparable total amounts of drug. The differentiation in favor of greater suppression with the response-contingent procedure was applicable to each of the monkeys described above.

The experimental analysis of punished responding typically has utilized procedures that involve the maintenance of the punishment condition for longer than a single session. In protocols involving multiple sessions, certain transition effects have been noted for both the introduction and cessation of punishment (1). We were therefore interested in examining chronic effects of response-contingent naloxone and, secondarily, in the effects of codeine dose as a determinant of the suppressing effects of naloxone. Rhesus monkeys with prior experience at various doses of acute response-contingent naloxone were exposed to 10 consecutive sessions at each codeine–naloxone dose combination, starting with the lowest dose of each drug

for 10 sessions. Then, response-contingent naloxone was discontinued until responding recovered to the previous codeine level. Each of two higher doses of naloxone was then examined, followed by an increase in codeine dose. Then, each of the three doses of naloxone was replicated with a still higher dose of codeine. Figure 7 shows the results of these manipulations in three monkeys. The highest dose of codeine alone (0.3 mg/kg per injection) maintained a slightly lower rate of responding than 0.1 mg/kg per injection. This is consistent with other findings regarding codeine dose and response rate (6, 14). Each of the doses of naloxone clearly suppressed responding in a dose-related fashion. At both codeine doses, the highest dose of naloxone suppressed responding in all cases, whereas the effects of the lower doses of naloxone were more pronounced in reducing codeine-reinforced responding maintained by 0.1 mg/kg per injection. This finding suggests that punishment of codeine-reinforced responding is a joint function of naloxone and codeine dose. A major additional finding is that in the chronic punish-

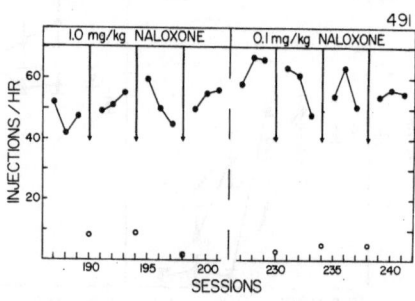

Figure 6. Effects of a series of acute naloxone administrations (1.0 mg/kg, left panel; 0.1 mg/kg, right panel) on the number of 0.1 mg/kg per injection codeine self-injections in a single monkey (491). Those sessions designated by arrows were preceded by intravenous naloxone administration.

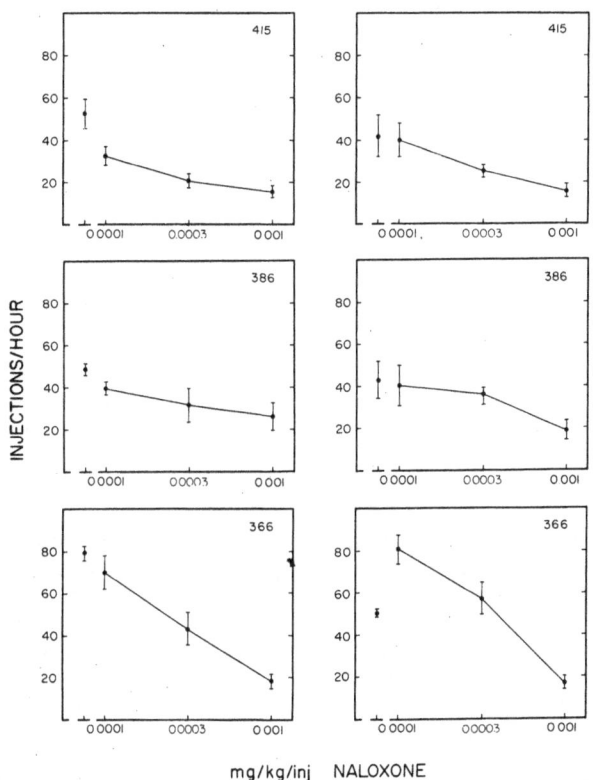

mg/kg/inj NALOXONE

Figure 7. Effects of chronic response-contingent naloxone injections on codeine-reinforced responding in three monkeys (415, 386, 366). In the left panels, 0.1 mg/kg per injection codeine was the reinforcer; in the right panels, 0.3 mg/kg per injection codeine was delivered as the reinforcer; indicated at the far left are averages of at least 10 sessions prior to the introduction of naloxone. Brackets indicate ±2 standard errors.

ment situation, suppression of codeine-reinforced responding was remarkably stable. Moreover, it should be noted that greater suppression was obtained with the chronic procedure than with acute exposures at the same doses of naloxone.

AVOIDANCE OF AND ESCAPE FROM NARCOTIC ANTAGONIST

Narcotic antagonists also can function as negative reinforcers (7, 11, 15) in that responding may be maintained by response-contingent termination (escape) or nonoccurrence (avoidance) of antagonist administration. Figure 8 shows cumulative records of responding maintained by time-out from a continuous intravenous infusion of naloxone. In this escape procedure, morphine-dependent (10 mg/kg per day) rhesus monkeys received a continuous infusion of naloxone in the presence of a blue stimulus light. Lever-press responding during the infusion resulted in a 1-min interruption (time-out ac-

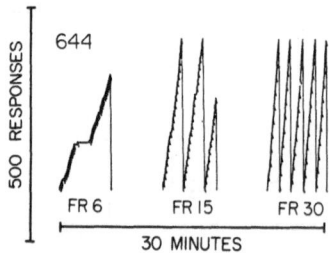

Figure 8. Cumulative response records of naloxone (0.001 mg/kg per minute) escape performance under different fixed-ratio values in monkey 644. Records should be read from left to right. Each response stepped the pen upwards while the paper moved at a constant speed. Oblique deflections of response pen indicate completion of each fixed ratio and presentation of a 1-min time-out from the otherwise continuous infusion of naloxone. The paper did not advance and responses were not recorded during time-outs. Each session lasted for about 1 hr.

companied by houselight) of the otherwise continuous delivery of naloxone. The cumulative records show that rates and patterns of responding typical of fixed-ratio performance (8) were obtained when up to 30 responses were required to produce each timeout (FR 30 escape). In the animal shown, response rates approaching 4.0/sec were emitted under the FR 30 escape schedule at a naloxone dose of 0.001 mg/kg per minute. Dose-response curves (Fig. 9) in two other monkeys under an FR 20 escape schedule indicate that naloxone escape response rate is an inverted U-shaped function of naloxone dose. As dose is increased, response rate increases to a maximum and then declines as dose is further increased. Similar inverted U-shaped functions have been reported for nondrug negative reinforcers such as shock (33), intense light (16), and sound (3). Moreover, comparable functions have been reported for various drug positive reinforcers (10, 36).

To account for such dose-response relationships with drug positive reinforcers, it has been suggested that drugs may have concomitant effects

which, at higher doses, are incompatible with high response rates (2, 25). A similar suggestion has been proposed for negative reinforcement using escape from intense light (16). The present naloxone escape experiment suggested such a disruptive

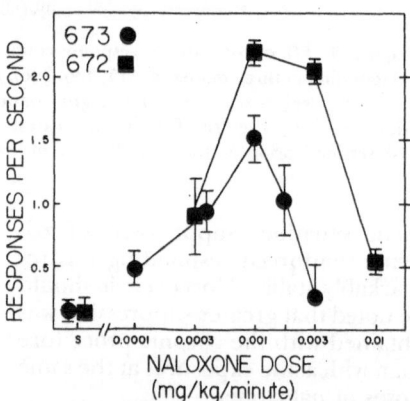

Figure 9. Response rate as a function of naloxone dose (mg/kg per minute) or saline (S) in each of two monkeys under the FR 20 naloxone escape procedure. Each point represents the mean of the last three out of up to seven sessions at each dose averaged over from one to three dose replications in each monkey. Vertical lines indicate one standard error above or below each point. (Adapted from (7).)

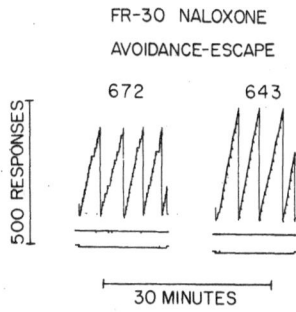

FR-30 NALOXONE

AVOIDANCE-ESCAPE

Figure 10. Cumulative response records of individual sessions under the FR 30 naloxone avoidance-escape procedure (0.002 mg/kg per injection) in each of two monkeys. Oblique deflections of the response pen (upper tracing) indicate completion of 30 responses and the presentation of a 1-min time-out. Downward deflections of the first event pen (center tracing) indicate completion of 30 responses after onset of naloxone injection (escape). Downward deflection of the second event pen (lower tracing) indicates onset of the session; upward deflections indicate unavoided and inescaped injections. The paper drive did not operate during time-outs. (Adapted from (7).)

process in that the highest naloxone doses maintained high response rates at the beginning of each session; as each session progressed and more naloxone was delivered, however, responding showed irregular patterning and reduced rates (7).

In a slightly different experimental procedure (avoidance-escape), naloxone injections could be avoided by responding within 30 sec after the onset of a blue light. If a naloxone injection was not avoided, it could nevertheless be escaped by responding during the 10-sec injection. Figure 10 shows cumulative records of FR 30 avoidance-escape sessions in two morphine-dependent monkeys when the naloxone dose was 0.002 mg/kg per injection. In both animals, response rates were sufficiently high to avoid or escape from nearly all naloxone injections. Again, response patterns were typical of fixed-ratio schedules of reinforcement. Figure 11 shows naloxone dose-response

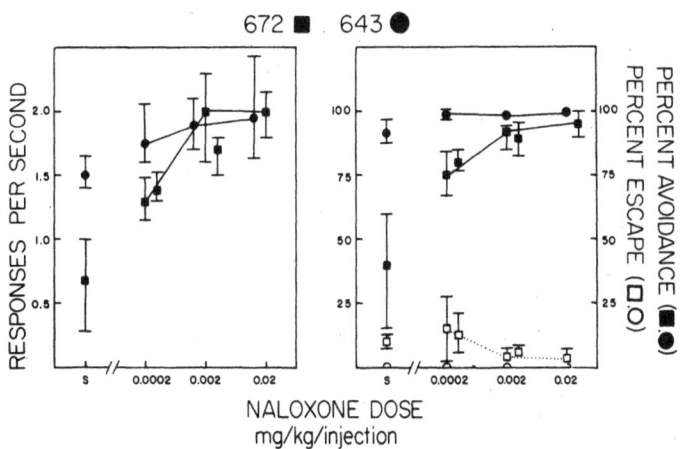

Figure 11. *Left panel*: response rate as a function of naloxone dose or saline in the FR 30 avoidance-escape procedure in each of two monkeys. Each point represents the mean of the last five out of 20 sessions at each dose. One exception is the point at S for 672, which represents the mean of five sessions only. Vertical lines indicate the range of each mean. Replications are offset from original determinations. *Right panel*: percent avoidance (solid symbols) and percent escape (open symbols) means and ranges corresponding to response rate data shown in left panel. (From (7).)

curves for the avoidance-escape procedure. Compared to the escape procedure, naloxone dose variation had small effects on avoidance-escape response rates. In addition, response rates were more directly related to naloxone dose in avoidance-escape than in the escape procedure. Similar relationships have been obtained with drug positive reinforcers when there are relatively long time periods between successive injections (2, 12). In such cases, the response-disrupting effects of the drug reinforcers presumably are minimized.

EFFECTS OF UNAVOIDABLE-INESCAPABLE NALOXONE ON AVOIDANCE-ESCAPE RESPONDING

There are additional functional similarities between naloxone and other negatively reinforcing or punishing stimuli. Under some schedules of negative reinforcement with electric shock, the delivery of unavoidable or inescapable additional shocks can result in increases in response rates (e.g., 4), and similar rate increases have been reported under some conditions of punishment of shock-avoidance responding (26). Naloxone (0.002 mg/kg per injection) avoidance-escape response rates also were increased markedly when unavoidable-inescapable naloxone (0.002 mg/kg per injection) injections were superimposed. Table 1 shows mean response rates and percent avoidance values during various conditions of superimposed unavoidable-inescapable naloxone. In the series of manipulations represented in Table 1, an unavoidable-inescapable naloxone injection first was delivered after every fifth completed FR 30 while the time-out after each FR 30 was one min. The time-out following each fixed ratio

then was reduced to 30 sec and finally 15 sec, while unavoidable-inescapable naloxone was delivered after every 10th completed ratio. Each of these procedural changes resulted in higher mean response rates than under the original condition (1-min time-out after every FR 30 unit; no unavoidable-inescapable naloxone). Thus, naloxone appears comparable to electric shock in that negatively reinforced responding may be increased in rate by the delivery of additional naloxone injections.

EFFECTS OF RESPONSE-CONTINGENT NALOXONE IN MORPHINE-DEPENDENT MONKEYS WITH HISTORIES OF NALOXONE AVOIDANCE-ESCAPE

Several investigators have shown that electric shocks also can maintain responding when shocks are delivered as the consequence of responding (18, 21, 29). In order to determine if naloxone would also maintain responding which led to its delivery in monkey 643, the avoidance-escape requirements were completely eliminated, the time-out following each completed FR 30 unit was reduced to 1.5 sec (a brief flash of the houselight), and every tenth completed FR 30 produced an injection of naloxone (0.002 mg/kg per injection) plus a 1-min time-out. These changes effected a second-order schedule (17) of naloxone-maintained responding, designated as FR 10 (FR 30:S). Under this schedule, responding was maintained in monkey 643 for eight sessions. The naloxone pump was then disconnected so that no naloxone was delivered. All other aspects of the procedure remained the same. When the naloxone pump was disconnected, responding declined to extremely low levels within three sessions. The nal-

TABLE 1. Effects of superimposed naloxone on responding
maintained by naloxone avoidance-escape

	Monkey 643		Monkey 672	
	Responses/ sec	% Avoidance	Responses/ sec	% Avoidance
Control, 0.002 mg/kg/per injection)	1.93	99.4	1.54	90.2
Naloxone every fifth ratio, 60-sec time out	2.40	100.0	2.24	98.0
Naloxone every tenth ratio, 30-sec time out	2.58	99.4	2.20	93.8
Naloxone every tenth ratio, 15-sec time out	2.13	96.0	2.75	98.5

oxone pump was then reconnected and the schedule requirement was reduced to FR 5 (FR 30:S) (naloxone was delivered after every fifth completed FR 30 unit). When the naloxone pump was reconnected, there was an immediate reinstatement of responding (Fig. 12). This performance was maintained for 10 additional sessions, whereupon the pump was again disconnected. Response rates again declined to very low levels within four sessions. The naloxone pump was reconnected and responding increased immediately for one session. In subsequent sessions, however, naloxone-reinforced responding was maintained only at very low rates. High response rates nevertheless could be generated and maintained by the delivery of occasional noncontingent naloxone injections.

A similar protocol was followed for monkey 672. In this animal, responding was maintained at high rates over 15 consecutive sessions under the FR 10 (FR 30:S) schedule of response-contingent naloxone. In the sixteenth session, however, the monkey responded at a low overall rate and received only two naloxone injections. After two additional sessions with low response rates, the schedule was reduced to FR 5 (FR 30:S). In the first session under the reduced schedule requirement, the monkey was

given two noncontingent ("priming") injections of naloxone (0.002 mg/kg per injection) at the start of the session. This produced a reinstatement of responding which continued throughout the session. Over subsequent sessions, high response rates could be maintained when "priming" injections of naloxone were administered before or during sessions, but low rates or no responses were emitted when such injections were not given. Kelleher and Morse (19) have similarly reported that noncontingent shocks could increase low rates of shock-maintained responding. The effectiveness of "priming" injections in reinstating responding, moreover, ·resembles effects obtained with positively reinforcing electrical brain stimulation to the extent that noncontingent brain shocks are sometimes necessary to initiate or maintain performance (9, 24, 30).

These results indicate that responding may be maintained under some conditions by response-contingent naloxone in morphine-dependent rhesus monkeys in a manner similar to the maintenance of responding by electric shock. The decline of responding when no "priming" injections were given, however, suggests that the schedules employed in the present study may be less than optimal for prolonged maintenance of

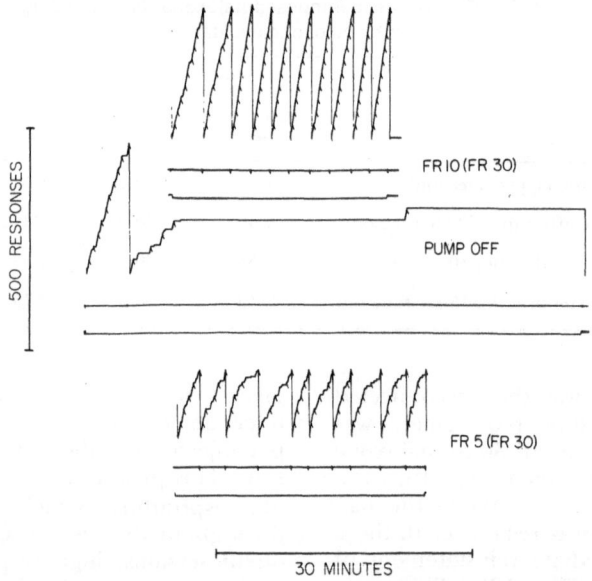

Figure 12. Cumulative records of responding maintained by naloxone. The upper record shows the eighth session of responding when every completed FR 30 unit produced a 1.5-sec flash of the houselight and every tenth completed FR 30 unit produced an injection of naloxone (0.002 mg/kg per injection) plus a 1-min time-out accompanied by houselight illumination. The center record shows performance in the third session with the naloxone infusion pump disconnected. All other aspects of the procedure were as described above. The lower record shows reinstatement of naloxone-maintained responding in the first session with the naloxone pump reconnected. In this and subsequent sessions, the schedule value was reduced to FR 5 (FR 30). Injections of naloxone or saline are indicated by downward deflections of the center event pen. Each session was terminated after 10 injections or about 1 hr.

such responding. Some fixed-ratio schedules of response-contingent shock, for example, have been found to result in decrements in response rates over sessions (22) and the eventual suppression of responding (19). On the other hand, fixed-interval schedules of electric shock presentations have been found to maintain responding consistently over sessions (4, 20, 29). Thus it remains to be determined whether schedules that have been shown to effectively maintain responding for electric shock over longer periods will function similarly with naloxone.

In conclusion, these data demonstrate that depending on the experi-

mental situation, a narcotic antagonist may elicit the narcotic abstinence syndrome, antagonize acute effects of narcotics, punish narcotic-reinforced responding, maintain escape and avoidance responding, increase naloxone avoidance-escape response rates, and finally, maintain responding, at least temporarily, when delivered as the consequence of responding.

Each of these cases has been illustrated within a single species. In all cases, there are common interactions between narcotic and narcotic antagonist which presumably occur at molecular levels. These common interactions, however, cannot simply account for the variety of behavioral effects reported herein. Rather, the comparability of a variety of stimulus functions of naloxone and various nondrug stimuli provides additional support for the generalization that such stimulus functions depend on schedules of reinforcement, properties of ongoing behavior, and experimental history.

These experiments are of interest behaviorally in that they extend the generality of effects obtained with nondrug stimuli to stimulus functions of a narcotic antagonist. On the other hand, pharmacological interactions that occur with narcotics and narcotic antagonists are also important determinants of behavioral functions of such drugs. Naloxone, for example, may be a negative reinforcer in monkeys with histories of morphine administration but may function much less potently, if at all, in monkeys without such histories (15). Thus, a general characterization of stimulus functions of narcotic antagonists must take into account the dynamic interaction of behavioral and pharmacological events.

REFERENCES

1. AZRIN, N. H., AND W. C. HOLZ. In: *Operant Behavior: Areas of Research and Application*, edited by W. K. Honig. New York: Appleton-Century-Crofts, 1966, p. 380.
2. BALSTER, R. L., AND C. R. SCHUSTER. *J. Exp. Anal. Behav.* 20: 119, 1973.
3. BARRY, J. J., JR., AND J. M. HARRISON. *Psychol. Rep.* 3: 3, 1957.
4. BYRD, L. D. *J. Exp. Anal. Behav.* 12: 1, 1969.
5. CARNEY, J. *Federation Proc.* 32: 726, 1973.
6. DOWNS, D. A. AND J. H. WOODS. *J. Pharmacol. Exp. Ther.* 191: 179, 1974.
7. DOWNS, D. A. AND J. H. WOODS. *J. Exp. Anal. Behav.* 23: 415, 1975.
8. FERSTER, C. B., AND B. F. SKINNER. *Schedules of Reinforcement*. New York: Appleton-Century-Crofts, 1957.
9. GALLISTEL, C. R. *Psychol. Bull.* 61: 23, 1964.
10. GOLDBERG, S. R. *J. Pharmacol. Exp. Ther.* 186: 18, 1973.
11. GOLDBERG, S. R., F. HOFFMEISTER, U. SCHLICHTING AND W. WUTTKE. *J. Pharmacol. Exp. Ther.* 179: 268, 1971.
12. GOLDBERG, S. R., AND W. H. MORSE. *Pharmacologist* 15: 236, 1973.
13. GOLDBERG, S. R., J. H. WOODS AND C. R. SCHUSTER. *J. Pharmacol. Exp. Ther.* 176: 464, 1971.
14. HOFFMEISTER, F., AND U. U. SCHLICHTING. *Psychopharmacologia* 23: 220, 1972.
15. HOFFMEISTER, F., AND W. WUTTKE. *Psychopharmacologia* 33: 247, 1973.
16. KAPLAN, M., B. JACKSON AND R. SPARER. *J. Exp. Anal. Behav.* 8: 321, 1965.
17. KELLEHER, R. T. In: *Operant Behavior: Areas of Research and Application*, edited by W. K. Honig. New York: Appleton-Century-Crofts, 1966, p. 160.
18. KELLEHER, R. T., AND W. H. MORSE. *J. Exp. Anal. Behav.* 11: 819, 1968.
19. KELLEHER, R. T., AND W. H. MORSE. *J. Exp. Anal. Behav.* 12: 1063, 1969.
20. MCKEARNEY, J. W. *J. Exp. Anal. Behav.* 12: 301, 1969.
21. MCKEARNEY, J. W. *Science* 160: 1249, 1968.
22. MCKEARNEY, J. W. *J. Exp. Anal. Behav.* 14: 1, 1970.
23. MCMILLAN, D. E., P. S. WOLF AND R. A. CARCHMAN. *J. Pharmacol. Exp. Ther.* 175: 443, 1970.
24. OLDS, J. *Science* 127: 315, 1958.
25. PICKENS, R., AND T. THOMPSON. *J. Pharmacol. Exp. Ther.* 161: 122, 1968.
26. POWELL, R. W., AND G. MORRIS. *J. Exp. Anal. Behav.* 12: 149, 1969.
27. SEEVERS, M. H., AND G. A. DENEAU. In: *Physiological Pharmacology*, edited by W. S. Root and F. G. Hoffman. New York: Academic, 1963, p. 565.
28. SKINNER, B. F. *The Behavior of Organisms: An Experimental Analysis*. New York: Appleton-Century-Crofts, 1938.
29. STRETCH, R., E. R. ORLOFF AND G. J. GERBER. *Can. J. Psychol.* 24: 117, 1970.
30. TROWILL, J. A., J. PANKSEPP AND R. GANDELMAN. *Psychol. Rev.* 76: 264, 1969.
31. VILLARREAL, J. E., AND M. G. KARBOWSKI. In: *Narcotic Antagonists*, edited by M. C. Braude, L. S. Harris, E. L. May, J. P. Smith and J. E. Villarreal. New York: Raven, 1973, p. 273.
32. WAY, E. L., H. H. LOH AND F. H. SHEN. *J. Pharmacol. Exp. Ther.* 167: 1, 1969.
33. WINOGRAD, E. *J. Exp. Anal. Behav.* 8: 117, 1965.
34. WOODS, J. H., D. A. DOWNS AND J. E. VILLARREAL. In: *Psychic Dependence. Definition, Assessment in Animals and*

Man, edited by L. Goldberg and F. Hoffmeister. Berlin: Springer-Verlag, 1973, p. 114.

35. WOODS, J. H., AND C. R. SCHUSTER. In: *Stimulus Properties of Drugs*, edited by T. Thompson and R. Pickens. New York: Appleton-Century-Crofts, 1971, p. 163.

36. WOODS, J. H., AND C. R. SCHUSTER. *Int. J. Addict.* 3: 231, 1968.

Experimental models for the modification of human drug self-administration

Methodological developments in the study of ethanol self-administration by alcoholics[1]

GEORGE BIGELOW, ROLAND GRIFFITHS AND IRA LIEBSON

Department of Psychiatry and Behavioral Sciences
The Johns Hopkins University School of Medicine, Baltimore 21205
and the Department of Psychiatry, Baltimore City Hospitals
Baltimore, Maryland 21224

ABSTRACT

Experimental studies of human ethanol self-administration are reviewed, and a description is provided of the procedural evolution that has occurred in the experimental study of the determinants of human ethanol self-administration. Human experimental models of alcoholism have been established within residential laboratories which permit chronic availability of ethanol to volunteer alcoholic subjects. Experimentation within such environments has progressed from observational and descriptive studies of experimental intoxication to studies that manipulate experimental variables so as to modify (reduce) ethanol self-administration by alcoholic subjects. To observe systematic effects of manipulated variables it has been necessary to develop sensitive baselines of ethanol self-administration. When ethanol intake has been relatively unrestricted, wide spontaneous fluctuations have made difficult the evaluation of manipulated variables. When a variety of restrictions on ethanol availability have been imposed, sensitive self-administration baselines have been established which have permitted the direct experimental assessment of some of the determinants of ethanol self-administration. Six methodological principles are suggested for enhancing the information yield of future research on the

[1] Supported by grant No. AA-00179 from the National Institute on Alcohol Abuse and Alcoholism.

determinants of ethanol self-administration. The same general methodology is suggested for research with other varieties of drug self-administration.—BIGELOW, G., R. GRIFFITHS AND I. LIEBSON. Experimental models for the modification of human drug self-administration: methodological developments in the study of ethanol self-administration by alcoholics. *Federation Proc.* 34: 1785–1792, 1975.

Animal models of drug abuse and drug dependence have been established by arranging for animals to self-administer drugs within controlled experimental settings (25). Similarly, human models of drug abuse can be established by arranging for humans to self-administer drugs within controlled experimental settings. Such experimental models permit the direct observation of the course, the correlates, and the consequences of drug self-administration. Perhaps most importantly, such experimental models can provide the opportunity to attempt to modify drug self-administration. To the extent that such modification is possible, we then have an experimental model of drug abuse treatment.

Research in human drug self-administration is certainly not yet common. However, initial experimental programs involving both opiates and barbiturates are beginning in several different laboratories. These efforts have been preceded by a decade of development and refinement of procedures for the experimental study of ethanol self-administration by alcoholics. Examination of this body of ethanol research can, of course, provide valuable guidance for future ethanol research, and may also suggest promising procedures for application to the experimental study of other varieties of human drug self-administration.

Within its brief history, experimental study of human ethanol self-administration has advanced from the phase of descriptive studies to the phase of experimental modifica-

tion of ethanol self-administration. The present report describes this transition, describes some of the procedures shown to influence ethanol self-administration, and summarizes some of the methodological principles that have evolved from this research.

All of the research described here has been conducted within residential hospital research settings. Detoxified volunteer chronic alcoholics have been admitted to these research environments and given the opportunity to self-administer considerable quantities of ethanol. The nature of these research environments has permitted the establishing of experimental control over the conditions of ethanol availability and over major aspects of subject's behavior.

This experimental approach derives from the pioneering experimental study of ethanol consumption by human alcoholics reported a decade ago by Mendelson (18) and colleagues. This study represented a major effort to observe some of the physiological, psychological, and behavioral concomitants of sustained ethanol intoxication in alcoholics, as well as to examine aspects of the appearance of ethanol withdrawal manifestations. In this initial study ethanol intake was not optional; rather, ethanol was administered on a temporally programmed schedule.

OPTIONAL ETHANOL SELF-ADMINISTRATION

Programmed schedules of ethanol administration were soon supplanted by experiments involving optional ethanol self-administration, in which the pattern and volume of ethanol intake came under the alcoholic subject's own control. The first such report was that of Mello and Mendelson (16), who examined the drinking of two alcoholic subjects given access to an unrestricted quantity of alcohol for a 14-day experimental drinking period. This study represented a first major step toward experimental modification of ethanol self-administration — self-administration achieved the status of a dependent variable that was free to fluctuate.

An excellent review of the large body of research conducted by Mello, Mendelson and colleagues involving optional ethanol self-administration is provided by Mello (14). In these studies subjects were given access to virtually unlimited quantities of ethanol. Ethanol was either earned by performance of some simple operant task or made freely available. Observation of subjects throughout the course of drinking episodes has provided suggestive information concerning the determinants of ethanol self-administration.

Surprisingly, it has been noted repeatedly that subjects' anxiety increases rather than decreases after drinking begins (13, 20). This same result was noted by Mendelson, LaDou and Solomon (19) following programmed ethanol administration. This finding is an apparent contradiction of the common view that ethanol self-administration is maintained via anxiety reduction.

Frequent note has also been made of the apparent controlling influence upon drinking exerted by social factors within the experimental environment (20, 21, 27, 28). However these impressions have been based on subjective and anecdotal observations.

Mello, McNamee and Mendelson (15) demonstrated experimentally that the effort required to obtain ethanol will influence the rate of ethanol self-administration. Two groups of subjects earned ethanol by working on a simple operant vigilance task. The response requirement per 10 ml of bourbon was either 16 or 32 consecutive correct responses. Over the course of the 7-day drinking period, the blood alcohol level of the high response requirement group averaged approximately one-half that of the low response requirement group.

MANIPULATIONS WITHIN EXPERIMENTS

The research reviewed above by Mendelson and Mello and colleagues did not manipulate variables during the course of experimental drinking periods. These investigators were primarily interested in assessing the effects and correlates of drinking and the determinants of physical addiction to ethanol (14). Consequently, they permitted drinking to proceed in a relatively unrestrained fashion and without experimental manipulations during the drinking periods.

For the study of the determinants of ethanol self-administration to proceed it was necessary that variables be manipulated to determine how they would influence ethanol consumption. The procedure of within-subject manipulation of variables during the course of an experiment is one of the most powerful of research techniques in the behavioral sciences. Since within-subject variation is generally less than between-subject variation, the probability of finding significant differences between experimental and control conditions is generally greater when both conditions are observed within the same subject.

Nathan and colleagues (22–24)

have pioneered the use of within-subject experimental manipulations superimposed on the relatively free-access optional drinking baselines introduced to the laboratory by Mello and Mendelson. Nathan's typical experimental design has involved a predrinking period, a drinking period, and a postdrinking period with 2 or 3 day subperiods of socialization and isolation alternating throughout. Subjects have access to all ward areas during socialization and are restricted to their private bedrooms during isolation. Thus, in all of these studies there has been an effort to assess the relevance of socialization versus isolation to drinking behavior, mood, and psychological status. This group's initial study (23) reported a tendency for alcoholic subjects to drink less during periods of isolation. However, they noted that this tendency appeared in only half of the subjects, and subsequent reports from this research group have not suggested any strong effect of the isolation condition upon drinking, although the effect continues to appear in the data of occasional subjects.

Nathan and O'Brien (22) have compared the behavior of a group of alcoholics to the behavior of a group of demographically-matched non-alcoholics when each was exposed to the same baseline conditions of ethanol availability and alternating periods of socialization and isolation. They found that alcoholics drank more than nonalcoholics and engaged in less social behavior than nonalcoholics. The within-subject socialization versus isolation manipulation did not appear to have a consistent effect on either group. The authors interpret the low level of social interaction shown by alcoholics to mean that social interac-

tions are not highly relevant to the control of alcoholics' drinking. This interpretation is, however, quite speculative.

Nathan et al. (24) used the same alternating-periods baseline to evaluate the effect upon drinking of the type of beverage served. Higher congener content beverage (bourbon) was compared to low congener content beverage (vodka or ethanol). The type of beverage was changed at the midpoint of the drinking period. No differential effect of the type of beverage was observed.

Allman, Taylor and Nathan (2) and Allman (1) reported a study in which a stress manipulation was added to the usual experimental design. The result was the three usual phases of predrinking, drinking, and post-drinking, all subdivided into alternating 4-day periods of stress versus no stress, with further subdivision into alternating 2-day periods of socialization versus isolation. Stress was manipulated by telling subjects that their rate of operant task performance was not sufficient to earn a financial bonus at the end of the experiment. During nonstress periods subjects were told that their performance rate was sufficient. The authors suggest that the combination of stress and isolation acted to suppress drinking. However extreme variability in the data, combined with the fact that the three subjects all participated simultaneously, suggests the need for further replication.

Goldman et al. (10) have reported use of a similar within-subject design to assess the effect upon drinking of group versus individual decision-making, socialization versus isolation, and token economy reinforcement for reduced ethanol consumption versus no reinforcement. The results suggest that the token economic contingencies were effective in reducing ethanol self-administration. However, wide within-subject variability makes the assessment of experimental effects difficult.

Cohen et al. (9) have manipulated incentives for abstinence under conditions of intermittent ethanol availability. Alcoholic subjects were given access every third day to a maximum of 720 ml (24 fluid ounces) of 95-proof (47.5%) ethanol. A financial incentive was offered for abstinence, and the amount was increased over successive trials if the subject refused it and chose to drink. It was possible to obtain abstinence from all four subjects, with incentives ranging between $7.00 and $20.00 per day. Further studies revealed that: a) abstinence could still be obtained, though at a slightly higher price, following consumption of a priming dose of 300 ml (10 fluid ounces) of 95-proof ethanol, and b) if payment of the financial incentive was to be delayed by several days the price of abstinence increased. Unfortunately, these investigators failed to obtain a baseline determination of whether subjects would uniformly fail to abstain in the absence of incentives.

STABLE AND SENSITIVE BASELINES

Sidman (26) describes the appropriate sequencing of experimental methodologies for the analysis of behavior. Initial studies within a problem area should properly be within the descriptive and correlational area. That is, a set of conditions should be established and behavior observed under that set of conditions for a period of time without further intervention. The aim in this initial strategy is to determine the form of behavioral baseline that results. Only later should experimental efforts move to the second phase, that of conducting manipulative experiments — manipulating variables within the experimental setting to assess their effect on the baseline performance measure. Properly, such manipulative experiments should be undertaken only after conditions have been established that maintain a stable baseline performance. Excessive variability of the baseline performance measure could result in serious misinterpretation of the effects of manipulated variables — either obscuring the legitimate effects of variables, or resulting in the erroneous appearance of an effect.

When drinking has been relatively unrestricted large fluctuations in alcohol consumption have been observed — with periodic spontaneous abstinence (17, 22). Such wide variability has made it difficult to determine the effects of experimentally manipulated variables under such drinking conditions.

Within our laboratory at Baltimore City Hospitals baselines of ethanol self-administration by alcoholic subjects have been developed which are sufficiently stable and sensitive to permit the gathering of systematic data concerning the controlling variables of ethanol self-administration. Since the procedures relevant to maintaining such baselines have not been systematically evaluated, it seems appropriate to describe baseline procedures as they have been established in actual studies of controlling variables. Therefore, in addition to describing procedures and results, the experimental summaries below will attempt to specify the baseline conditions of ethanol availability under which effects have been observed.

Contingent loss of privileges

A series of studies (5, 6, 8) has investigated the controllability of alcholics' drinking by contingent loss of privileges. Subjects were given access to a maximum of 720 ml (300 ml (10 fluid ounces) in the 1972 study) of 95-proof ethanol daily, 5 days per week. Abstinence was enforced on weekends. On alternate weeks a contingency was in effect such that if a subject consumed more than 150 ml (5 fluid ounces) of the available ethanol on any one day he immediately entered on "impoverished" condition which imposed a major loss of privileges: he was restricted to his bedroom, opportunities for activities and socialization were severely restricted, and he received pureed food rather than regular hospital meals. Impoverishment remained in effect for 24–48 hours. The alternate weeks when the contingency was not in effect served as a control condition. During these times subjects were assigned either to "impoverishment" or to the usual "enriched" ward environment independently of their drinking behavior. Drinking during these noncontingent baseline weeks was generally excessive and relatively stable regardless of whether the "impoverished" or the "enriched" condition prevailed. The results showed that in comparison to these control baselines, the experimental manipulation had a clear, consistent effect—when loss of privileges was contingent on excessive drinking, the drinking was suppressed.

These studies involved a rather complex set of conditions governing access to ethanol. The alternating contingent and noncontingent weeks made it possible for subjects to drink excessively without consequence on alternate weeks.

The 5-day-week experimental schedule, with enforced abstinence on weekends, resulted in enforced sobriety always preceding each change in experimental conditions. Also, it may be important to note that excessive drinking during the contingent period resulted in a loss of access to ethanol on the next day. Therefore during the contingent period intoxication was possible for only one day at a time. This further guaranteed that during the contingent period every day of exposure to the possibility of excessive drinking began as a sober day.

A subsequent study (7) has successfully utilized the loss-of-privileges contingency to prevent excessive drinking while the opportunity for a long continuous period of heavy drinking was present. This experiment began with the contingency in effect; thus the contingency condition constituted the baseline condition, and the experimental manipulation consisted of removing the contingency. None of the three subjects ever drank excessively during the contingency; all did so after the contingency was removed.

Fixed ratio cost of drinks

Studies by Liebson et al. (12) and Bigelow and Liebson (3) report that ethanol self-administration is reduced when greater effort is required to obtain each drink. Both studies varied the number of responses (fixed ratio cost) required to earn each drink containing 30 ml (1 fluid ounce) of 86-proof or 95-proof ethanol. The fixed ratio values were varied in a mixed order over a small number of days. Both studies imposed a limit to the rate of alcohol consumption—12 drinks per 8 hours (12) or 3 drinks per hour (3). Data are shown in Fig. 1 for a representa-

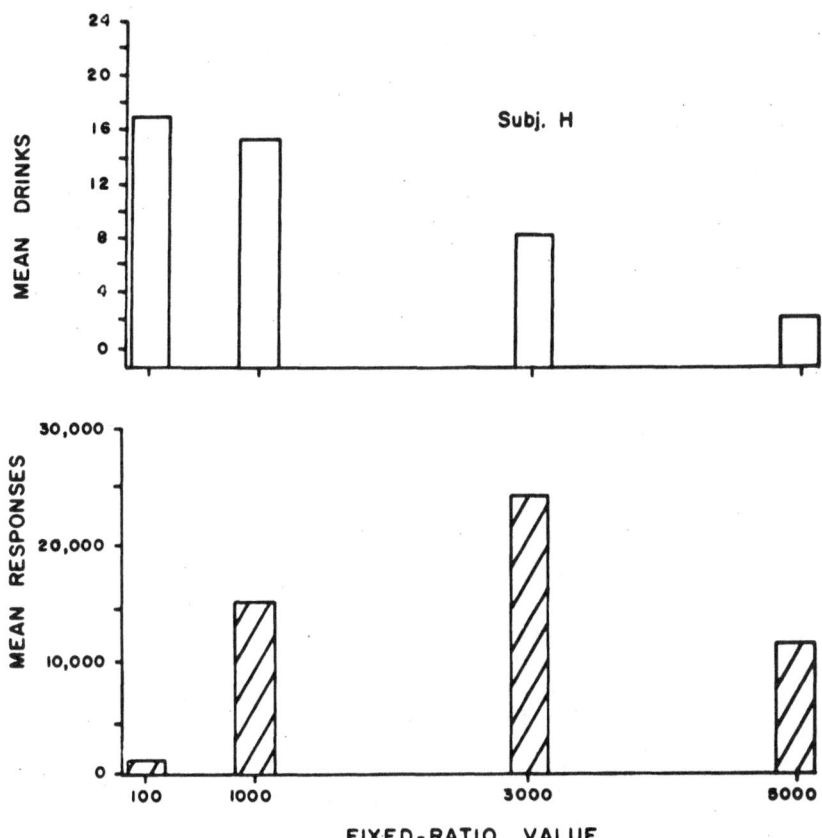

Figure 1. Mean daily ethanol consumption (*above*) and mean daily responding to acquire ethanol (*below*) are shown as a function of the number of operant responses required to obtain each drink (fixed ratio value).

tive subject from the Bigelow and Liebson (3) report. Ethanol self-administration decreased as fixed ratio value increased. Total response output increased at moderate costs and decreased at the highest fixed ratio value.

Brief contingent isolation

A recent study from our laboratory (4) examined the effect of contingent isolation on stable baselines of ethanol self-administration. A total

of 12 to 24 drinks (number varied between individual subjects) were made available daily between 7:00 AM and 11:00 PM. Aside from these restrictions, this baseline procedure permitted continuous daily drinking. During the experimental condition, isolation for either 10 or 15 min within a small 3-sided booth adjacent to the main ward dayroom was an immediate consequence to receipt of each alcoholic drink (containing 30 ml (1 fluid ounce) of 95-proof ethanol). Control periods requiring no isolation both preceded and followed the contingency period. Data from a representative subject are shown in Fig. 2. Typically, the baselines were sufficiently stable and the effect sufficiently strong so as to be readily apparent by visual inspection of data from any single subject. It is clear that contingent isolation was effective in suppressing an on-going drinking episode. A summary of the results for 10 subjects is shown in Fig. 3. During the contingent isolation phase only 52% of available alcoholic drinks were consumed, while 95% and 92% were consumed during the preceding and following control periods.

Required temporal spacing

Four subjects have participated in a study in which the minimum required interval between receipt of successive drinks was manipulated during an experimental drinking baseline. Drinks containing 30 ml (1 fluid once) of 95-proof ethanol were available upon request during the 16-hour period between 7:00 AM and 11:00 PM daily. A maximum of 18 drinks was available each day. Across days, the minimum interval required between receipt of successive drinks was randomly varied between 0, 30, 45, 60, and 90 min. Each subject was exposed to each interval on two separate occasions. On a given day no two subjects were exposed to the same required interval. The 60- and 90-min spacing conditions necessarily resulted in a reduction of the total daily alcohol to 16 and 11 drinks respectively. Presented in Fig. 4 is the mean percent of available drinks consumed as a function of the required minimum interdrink interval for

each of four subjects. When drinks were available without temporal constraint (0 min interdrink interval required) 100% of available drinks were consumed. As the minimum interdrink interval increased the percent of available drinks consumed declined. When 90 min were required between successive drinks an average of only 52.1% of the available alcoholic beverage drinks were consumed. Again, the experiment illustrates that stable baseline conditions help reveal orderly controlling relationships between environmental variables and ethanol self-administration.

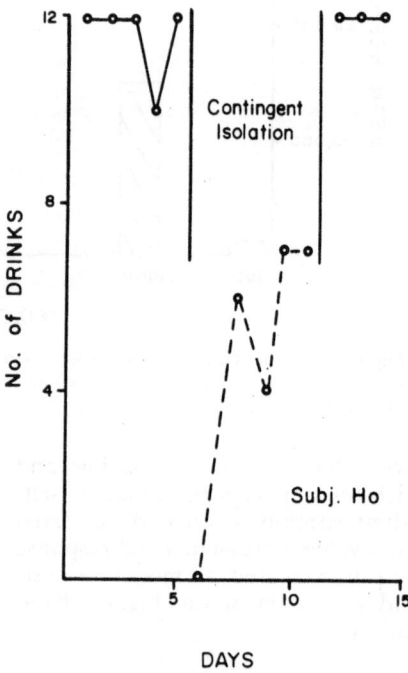

Figure 2. The effect on a stable daily baseline of ethanol self-administration is shown of introducing a requirement that a 10-min period of physical and social isolation be imposed immediately contingent on receipt of each alcoholic drink.

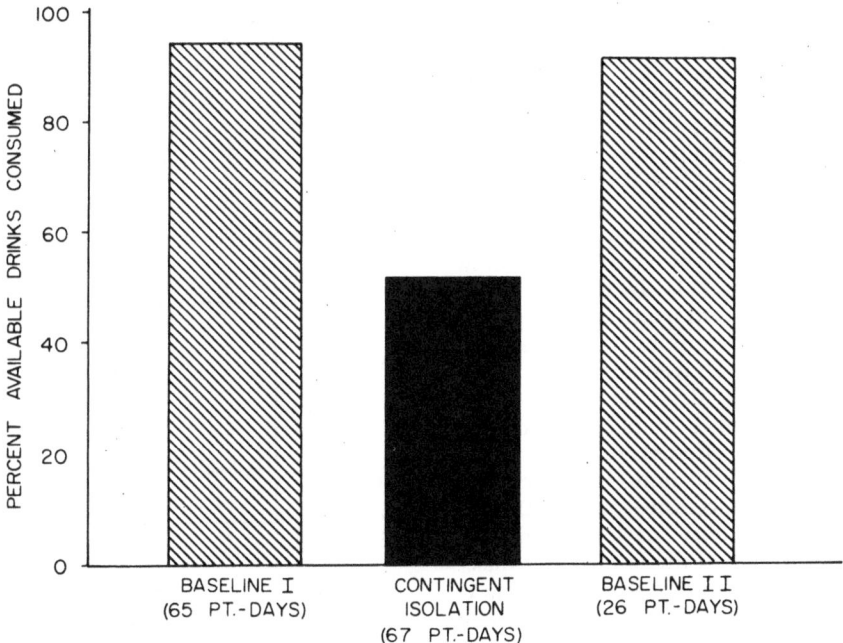

Figure 3. Summary of 10 subjects, showing the effect on ethanol self-administration of a period during which physical and social isolation for 10 or 15 min was imposed immediately contingent on receipt of each alcoholic drink.

ENHANCING BASELINE SENSITIVITY

All behavior analysis research must face the problem of establishing behavioral baselines sufficiently sensitive to reveal the effects of manipulated variables. When a manipulated variable shows no effect on a baseline performance it may mean that the variable is, in fact, irrelevant, or it may mean that the baseline is insensitive. The problem of evaluating controlling variables of ethanol self-administration is primarily a problem of developing sensitive baselines.

Procedures for establishing sensitive baselines are especially important in research dealing with behaviors that may be multiply con-

trolled. Development of sensitive baseline procedures permits the recognition of variables which may be weak individually but may exert powerful effects in combination with other variables. Two cases will be described in which alterations in baseline conditions have enhanced sensitivity to manipulated variables.

The first case, represented in Fig. 5, is from Bigelow, Liebson and Griffiths (4). This study, involving brief contingent isolation consequent upon receiving each alcoholic drink, was described above. Contingent isolation alone had a negligible effect on drinking in this subject. The requirement was then introduced that receipt of successive drinks be spaced at least 1 hour apart. Under this modified baseline

condition contingent isolation revealed a substantial suppressive effect of its own. It is important to
note that this is more than simply an
additive effect. This same interaction relationship was demonstrated
with a second subject in the same
study.

The second example is drawn
from a study by Griffiths, Bigelow
and Liebson (11). In this study the
effects of contingent loss of purely
social opportunities were assessed.
Subjects were given access to a maximum of 17 alcoholic drinks daily,
each containing 30 ml (1 fluid ounce)
of 95-proof ethanol. It was required
that at least 40 min elapse between
receipt of successive drinks. During
the contingency period receipt of
each drink had the immediate consequence of removal of all social interaction opportunities for 40 min. The
subject was not allowed to talk, or
engage in any form of social activity.
The effect of this contingency was
investigated under several baseline
conditions. Results for one subject
are shown in Fig. 6. Contingent
time-out from social opportunities
had no effect on drinking when the
subject retained access to all privileges on the research ward; however, when ward privileges were
restricted, the contingent social
time-out suppressed drinking. In
this case the baseline condition that
is manipulated—level of privileges
—shows no direct effect itself on
drinking. However, it does serve to
regulate the sensitivity of drinking
behavior to the social time-out contingency.

METHODOLOGICAL PRINCIPLES

The studies reviewed above comprise the major body of experimental

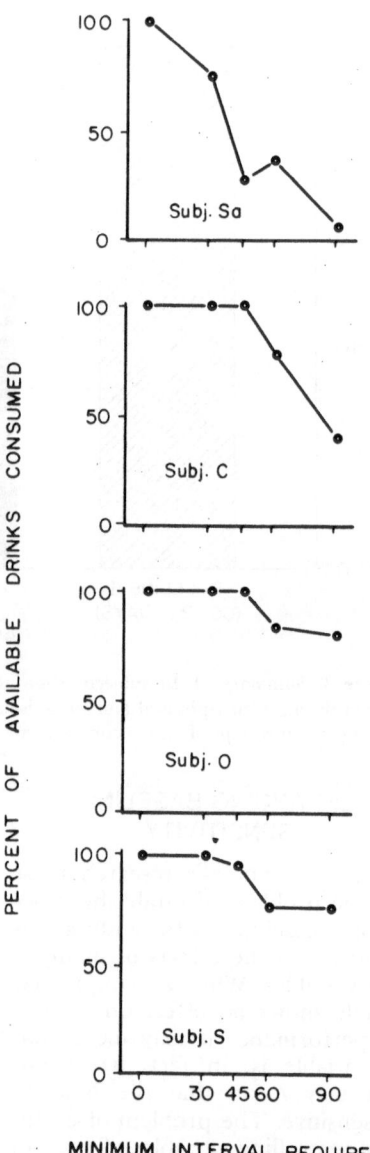

Figure 4. The effect on ethanol self-administration of varying the minimum interval required between receipt of successive alcoholic drinks.

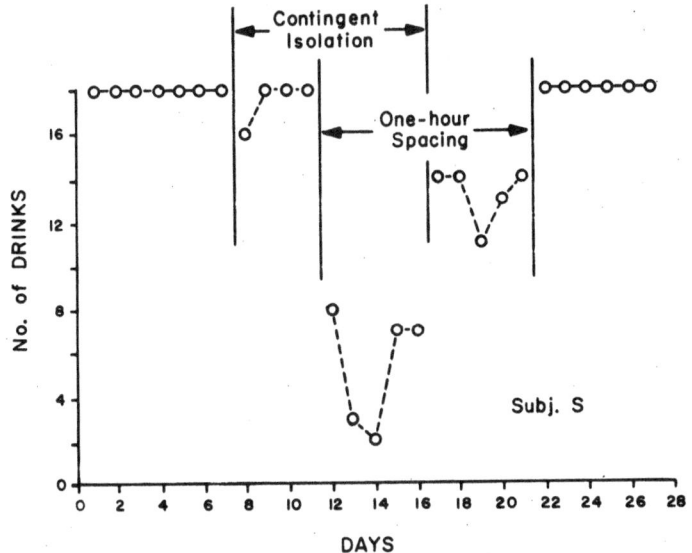

Figure 5. Ethanol self-administration is shown over successive days. Requirements of brief contingent physical and social isolation and of 1 hr minimum spacing of successive alcoholic drinks were imposed singly and in combination. The suppressive action of contingent isolation was evident only when combined with the spacing requirement.

work concerning the determinants of ethanol self-administration by alcoholics given continuing access to ethanol within residential laboratory settings. It seems appropriate at this time to attempt to glean from these studies some methodological principles that might prove useful in guiding future research concerning the determinants of human drug self-administration.

Optional drug self-administration

The foremost principle is that the behavior of drug self-administration should be optional, and free to vary. Variables thought to influence drug use should be directly assessed with respect to their effect on self-administration. It seems clear that the most powerful way to demonstrate that a variable influences a behavior is to manipulate the variable and record a change in the behavior.

Restricted drug access

Experimental paradigms permitting only restricted access to ethanol have been most successful in demonstrating controlling relationships. Baselines permitting relatively unrestricted access are quite unstable, varying between periods of high ethanol intake and periods of spontaneous abstinence. Such variability has made it difficult to evaluate experimentally manipulated variables. Whether this same relationship will hold with self-administration of other drugs is an empirical question. However, with ethanol it appears that conditions of restricted access yield baselines of greater stability and sensitivity.

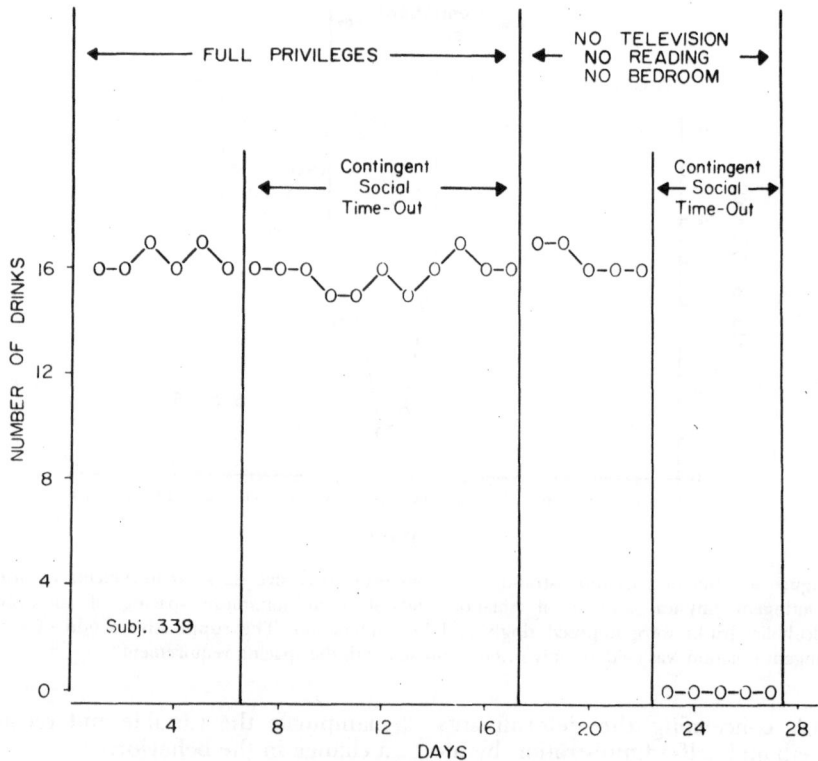

Figure 6. The effect on ethanol self-administration of contingent social time-out was examined under two baseline conditions. The appearance of a suppressive effect was dependent on baseline conditions of ward privilege availability.

Setting conditions versus behavioral consequences

In the studies reviewed, variables have been manipulated in at least two distinguishable modes—as setting conditions or as behavioral consequences. Setting conditions refer to those circumstances antecedent to the behavior, whereas behavioral consequences refer to the events which follow the behavior. These two focuses generally reflect different conceptions of where the determinants of behavior are to be found. It seems important to emphasize the distinction because of indi-

cations that the effects of a particular variable may be dependent on the mode in which it is manipulated. In the research reviewed here, manipulation of a variable (e.g., social access) as a behavioral consequence has been much more likely to modify ethanol self-administration than has manipulation as a setting condition. Unfortunately, these two modes of manipulating variables have not yet been systematically compared within a single experiment. However, the suggestion remains that at least in certain cases manipulation of behavioral consequences can offer the more powerful technique for modifying drug self-administration.

Successive exposure of subjects

Some studies have consisted of a single group of three or four subjects who have lived together and participated in all phases of the experiment simultaneously. Such simultaneous group exposure must substantially diminish the independence of subjects' performance. From the point of view of statistical evaluation and replicability, such simultaneous group experiments might more appropriately be considered as a sample of one rather than a sample of three or four.

Practice within our laboratory has evolved to the procedure of exposing subjects to a particular experimental procedure successively rather than simultaneously. Thus, several different experiments may be in progress on the research ward simultaneously. This procedure increases the independence of subjects' performance and therefore, because phenomena are replicated across several subjects under somewhat different ward contexts, our confidence in the generality of observed results is enhanced.

Independence of experimental conditions

In order properly to evaluate the effects of any within-experiment manipulation it is necessary that performances under the various experimental conditions be independent of one another. Some studies have utilized token economy procedures which permit subjects to earn points during one experimental condition and to spend them in a later one. Such a procedure dramatically reduces the independence

between experimental conditions and complicates the interpretation of results. With such a procedure, a subject's behavior cannot be brought strongly under the control of current conditions, because earnings carried over from past performance introduce an extraneous source of control. Consequently, for studies concerned with discovering the determinants of drug self-administration via experimental manipulation, appropriate constraints should be placed on token economy procedures so as to maintain the independence of experimental conditions.

Establishing adequate baselines

It should be emphasized that the task of assessing the effects of variables is largely a task of establishing adequate baselines. The research reviewed here has demonstrated the importance of baseline conditions. Baseline conditions have determined whether particular manipulations demonstrated an effect, and it has been shown that baseline conditions may be adjusted so as to enhance the responsivity of drinking to experimental manipulations. We feel that experimentally sensitive baselines of drug self-administration can be established by controlling and placing limitations on drug availability and on behavioral opportunities—i.e., by controlling possible sources of extraneous or error variation. Such limitations may result in the experimental model of drug self-administration becoming less naturalistic. However, the strength of basic research lies in the discovery of systematic relationships by eliminating the confusing variability observed in "natural" systems. Clearly, in conducting experiments the establishing of adequate baselines should not be sacrificed for naturalism.

CONCLUSION

This paper has reviewed the procedural evolution that has occurred in the experimental study of human ethanol self-administration over the past 10 years. Human experimental models of alcoholism have been established, and methods have been developed to modify the ethanol self-administration of alcoholic subjects. It seems likely that similar experimental developments will follow for other varieties of drug abuse.

The extent to which procedures used within the experimental laboratory can truly serve as models for procedures to be applied within the natural environment to modify human drug self-administration remains an open question. Certain controlling relationships demonstrated and shown effective within the laboratory may seem impractical or irrelevant when considered with respect to behavioral determinants in the natural ecology. However, such skepticism may reflect the extent to which our conceptualizations of the determinants of alcoholism and drug abuse are divorced from the empirical determinants of such drug self-administration behavior.

REFERENCES

1. ALLMAN, L. Group drinking during stress: Effects on alcohol intake and group process. *Int. J. Addict.* 8: 475, 1973.
2. ALLMAN, L. R., H. A. TAYLOR AND P. E. NATHAN. Group drinking during stress: Effects on drinking behavior, affect, and psychopathology. *Am. J. Psychiatry* 129: 669, 1972.
3. BIGELOW, G., AND I. LIEBSON. Cost factors controlling alcoholic drinking. *Psychol. Rec.* 22: 305, 1972.
4. BIGELOW, G., I. LIEBSON AND R. GRIFFITHS. Alcoholic drinking: Suppression by a brief time-out procedure. *Behav. Res. Ther.* 12: 107, 1974.
5. COHEN, M., I. A. LIEBSON AND L. A. FAILLACE. The role of reinforcement contingencies in chronic alcoholism: An experimental analysis of one case. *Behav. Res. Ther.* 12: 107, 1974.
6. COHEN, M., I. A. LIEBSON AND L. A. FAILLACE. A technique for establishing controlled drinking in chronic alcoholics. *Dis. Nerv. Syst.* 33: 46, 1972.
7. COHEN, M., I. LIEBSON AND L. FAILLACE. Controlled drinking by chronic alcoholics over extended periods of free access. *Psychol. Rep.* 32: 1107, 1973.
8. COHEN, M., I. A. LIEBSON, L. A. FAILLAICE AND R. P. ALLEN. Moderate drinking by chronic alcoholics: A schedule-dependent phenomenon. *J. Nerv. Ment. Dis.* 153: 434, 1971.
9. COHEN, M., I. A. LIEBSON, L. A. FAILLAICE AND W. SPEERS. Alcoholism: Controlled drinking and incentives for abstinence. *Psychol. Rep.* 28: 575, 1971.
10. GOLDMAN, M., H. TAYLOR, M. CARRUTH AND P. NATHAN. Effects of group decision-making on group drinking by alcoholics. *Q. J. Stud. Alcohol.* 34: 807, 1973.
11. GRIFFITHS, R., G. BIGELOW AND I. LIEBSON. Suppression of ethanol self-administration in alcoholics by contingent time-out from social interactions. *Behav. Res. Ther.* 12: 327, 1974.
12. LIEBSON, I. A., M. COHEN, L. A. FAILLACE AND R. F. WARD. The token economy as a research method in alcoholism. *Psychiatr. Q.* 45: 574, 1971.
13. McNAMEE, H. B., N. K. MELLO AND J. H. MENDELSON. Experimental analysis of drinking patterns of alcoholics: Concurrent psychiatric observations. *Am. J. Psychiatry* 124: 1063, 1968.
14. MELLO, N. Behavioral studies of alcoholism. In: *The Biology of Alcoholism, Vol. 2*, edited by B. Kissin and H. Begleiter. New York: Plenum, 1972, p. 219.
15. MELLO, N. K., H. B. McNAMEE AND J. H. MENDELSON. Drinking patterns of chronic alcoholics: Gambling and motivation for alcohol. In: *Clinical Research in Alcoholism*, edited by J. O. Cole. Washington, D.C.: Psychiatric Research Report No. 24, Am. Psychiatr. Assoc. 1968, p. 83.
16. MELLO, N. K., AND J. H. MENDELSON. Operant analysis of drinking patterns of chronic alcoholics. *Nature* 206: 43, 1965.
17. MELLO, N. K., AND J. H. MENDELSON. Experimentally induced intoxication

in alcoholics: A comparison between programmed and spontaneous drinking. *J. Pharmacol. Exp. Ther.* 173: 101, 1970.

18. MENDELSON, J. H. (editor). Experimentally induced chronic intoxication and withdrawal in alcoholics. *Q. J. Stud. Alcohol, Suppl. 2* 1964.

19. MENDELSON, J., J. LADOU AND P. SOLOMON. Experimentally induced chronic intoxication and withdrawal in alcoholics: III. Psychiatric findings. *Q. J. Stud. Alcohol, Suppl. 2* 1964, p. 40.

20. MENDELSON, J. H., AND N. K. MELLO. Experimental analysis of drinking behavior of chronic alcoholics. *Annu. N.Y. Acad. Sci.* 133: 828, 1966.

21. MENDELSON, J. H., N. K. MELLO AND P. SOLOMON. Small group drinking behavior: An experimental study of chronic alcoholics. *The Addictive States, Res. Publ. Assoc. Res. Nerv. Ment. Dis.* 46: 399, 1968.

22. NATHAN, P., AND J. O'BRIEN. An experimental analysis of the behavior of alcoholics and nonalcoholics during prolonged experimental drinking: A

necessary precursor of behavior therapy? *Behav. Ther.* 2: 455, 1971.

23. NATHAN, P., N. TITLER, L. LOWENSTEIN, P. SOLOMON AND A. ROSSI. Behavioral analysis of chronic alcoholism. *Arch. Gen. Psychiatry* 22: 419, 1970.

24. NATHAN, P., N. ZARE, E. FERNEAU AND L. LOWENSTEIN. Effects of congener differences in alcoholic beverages on the behavior of alcoholics. *Q. J. Stud. Alcohol, Suppl. 5* 1970, p. 87.

25. SCHUSTER, C., AND T. THOMPSON. Self administration of and behavioral dependence on drugs. *Annu. Rev. Pharmacol.* 9: 483, 1969.

26. SIDMAN, M. *Tactics of Scientific Research.* New York: Basic Books, Inc. 1960.

27. STEINGLASS, P., S. WEINER AND J. MENDELSON. Interactional issues as determinants of alcoholism. *Am. J. Psychiatry* 128: 275, 1971.

28. WEINER, S., J. TAMERIN, P. STEINGLASS AND J. MENDELSON. Familial patterns in chronic alcoholism: A study of father and son during experimental intoxication. *Am. J. Psychiatry* 127: 1646, 1971.

On neurochemistry and behavior

PETER L. CARLTON

Department of Psychiatry, CMDNJ –Rutgers Medical School
Piscataway, New Jersey 08854

I have been asked to provide an introduction to the papers to be presented in this section—and to that end, have elected to outline briefly some of the historical antecedents of the research to be described in the hope that such an outline will provide some perspective on that research.

There can be no doubt that among the most important events that began the business to be discussed is the fact that, in the mid-fifties, the previously ever-increasing number of patients maintained in psychiatric hospitals began to decline and has continued to do so. This decline was in part due to a greater sophistication in, and public acceptance of, mental health care. But it was in greater part due to the introduction of chlorpromazine and reserpine into psychiatric practice—the first-time reversal in the trend of one of our major public health problems came about because of the first-time availability of drugs truly useful in the management of psychotic behavior.

It is difficult to overestimate the impact of this turnaround on society in general and on science in particular. Rather than attempt to outline the full range of this impact, I have chosen to describe only those outcomes that are most pertinent to the discussion that is to follow. There are three of these.

The first has to do with psychiatry. Psychiatrists had for years described themselves as physicians who were afraid of the sight of blood. But with the mid-fifties, psychiatrists had a "pill" and came, it must be supposed, to think more of themselves as "real doctors."

More important than this much needed inflation of self-esteem, is the kind of thinking this event naturally engendered. Some psychiatrists had long been "biological" in outlook, although it must also be conceded that the biology was often sunk in a sea of ids and egos, theoretical concepts that became reified in a way that Freud probably never intended them to be. But the advent of a "pill" that could control otherwise refractory behavior led directly to thoughts about faulty chemistry, from these directly to thoughts about faulty enzymes and finally, and with equal directness, to thoughts about faulty genes. Thus, the prospect of a truly Biological Psychiatry became a viable one.

And it is very much a viable discipline today. Research—biological research by and for psychiatrists—abounds; some of it will be discussed today. Theories of psychosis—biochemical theories—are very much with us. And, I think *not* coincidentally, the supposition of an important genetic component in psychosis is a widely accepted one, perhaps as much because of an ambience of plausibility as because of the data themselves.

There are those who criticize this newly favored Biological Psychiatry on the grounds that it is *too* biological and insufficiently social. I want to comment briefly about that criticism by way of two digressions.

First, it can safely be supposed that schizophrenia, for example, is not merely a social disease, certain vocal protestants notwithstanding—and it can be supposed, with equal safety, that total preoccupation with a simplistic biology will never do justice to the etiology of that disease. It is worth noting that that other social disease is not only a product of a very "biological" organism that can be controlled by a drug, but the disease also has behavioral manifestations and is just as due to the organism as it is due to behavioral interactions of a certain kind in a certain milieu. Here the analog of the much-maligned schizophrenigenic mother is the syphilitic prostitute or lover, male or female—an analogy that, after all, has a certain Freudian flavor to it. Unfortunately, the complex of the interaction of the biological with the social in schizophrenia is not understood, whereas the complement in the analog is. That is my first digression.

The second is only a brief observation. Critics of Biological Psychiatry seem to overlook the obvious fact that, without successful pharmacological management, the *social* aspects of the disorder would only rarely be amenable to medical intervention.

That is, the Community Mental Health movement, for all its budgetary vicissitudes, would simply not be conceivable without the kind of management that does permit patients to go home and into society; without that management, a kind of treatment that *does* consider the whole patient, not only his biology but his family and social surround as well, could not have arisen. Care would continue to be custodial rather than social in any meaningful sense of the latter term. And that ends my second digression.

The first outcome of the introduction of drugs efficacious in controlling behavioral disorders is what has come to be called Biological Psychiatry. The second has come to be called Behavioral Pharmacology.

The behavior that brought Behavioral Pharmacology to the forefront was that seen in so-called conditioned avoidance situations. In these experiments, a rat was given a warning signal that foretold impending shock—if the animal emitted some prescribed response during the warning (avoided), shock was not delivered; if it failed in that, a painful electric shock was delivered and the rat could escape by emitting that same prescribed response.

The thing about chlorpromazine was that it—uniquely, it was believed—reduced avoidance behavior while leaving escape behavior intact. This suggested that chlorpromazine specifically attenuated something called "fear," a suggestion that subsequent data have rendered very unlikely. But the fact remains that the relatively selective effect on avoidance did, and still does, provide a useful technique for detecting phenothiazine activity in the laboratory, whether or not this be more predictive of anti-schizophrenic or parkinsonian side-effects or, for that matter, whether these two can ever be separated. Thus, behavior came to the pharmaceutical industry, an industry frantically searching for yet another phenothiazine in what cynics, with only partial justification, came to call "molecular roulette."

In the 1950's, the techniques for assaying this and other behaviors rapidly became widely used, vastly expanded, and refined, largely because of the availability of an operant technology that had been lurking in the wings of Harvard and Walter Reed. The advocacy of these procedures—most notably by Dews and his co-workers and by Brady and his—brought a degree of precision and reliability that ultimately permitted a true Behavioral Pharmacology.

Not that these advances were without their precedents. For example, Dews himself had experimented with motor activity in 1953; Skinner and Heron had reported on the effects of caffeine in 1937 and Macht had reported a wholly adequate and quite modern behavioral study of the anticholinergics in the 1923 volume of the *Journal of Pharmacology and Experimental Therapeutics*. And the extraordinary power of behavioral control techniques had been known at least since Skinner had trained a pigeon to guide the 1940's version of a guided missile. But it took chlorpromazine to bring all that into the foreground, quite simply because what it acted on was behavior and the way to understand it was to study behavior.

It is worth examining the phrase "Behavioral Pharmacology." There are two meanings implied that operate at two different levels, one of which is explicit, the other implicit.

At the implicit level, Behavioral Pharmacology has come to imply a largely empirical stress on behavioral phenomena, a stress that combines the persuasions of those raised in the Skinnerian tradition with the traditional persuasion of Pharma-

cology itself. Thus, supposedly explanatory concepts like "fear" and "drive" and "anticipation" have been washed away in a largely healthy—if sometimes unduly constraining—wave of empirical description. The cleansing has been largely healthy, and certainly necessary, because of the propensity for quasimythical concepts about behavior to be given a dignified reification such that they *appear* to account for phenomena when, in fact, they are largely circular and therefore ultimately sterile.

The second level of meaning in "Behavioral Pharmacology" is the explicit one. As a discipline, it is just that—a Pharmacology in which the end point is behavior, in which the assay is the behavior of an intact animal, and in which that assay is demonstrably as reliable and can be useful as the twitch of an isolated muscle or the change in cardiac output of an anesthetized dog.

The operative term in Behavioral Pharmacology is Pharmacology—it sets as its goal the characterization of drug action in a behavioral language largely stripped of excess meaning even to the point, for example, of questioning the utility of such not-entirely-unloaded terms as "stimulant" and "depressant."

But just as "Pharmacology" is the operative term in "Behavioral Pharmacology," there is another discipline in which the operative term is Behavior—and this is the third major outcome from the turnabout effected by chlorpromazine.

For lack of a better phrase, I would call this discipline the Pharmacology of Behavior—a phrase that would literally mean "pharmacological knowledge of behavior." The phrase is not only cumbersome but it is somewhat off-target; it *should* mean "how we can understand behavior by understanding drugs." Thus, emphasis is on Behavior and implicit in the phrase is the idea—and the hope—that we can understand normal behavior by understanding the abnormalities that drugs induce. The logic is the same as that in Neurology, experimental and otherwise; if we can understand what went wrong in an abnormal, lesioned brain, we can understand what goes right in a normal one. This is the basic concept of the Pharmacology of Behavior.

By way of elaborating this concept, I have selected two examples from the history of this discipline because they each make the same important point in different ways.

First, some years ago I suggested that the study of the behavioral effects of the anticholinergics, or any other drug for that matter, could tell us something about how a normal rat brain regulates learned behavior. Now, there are two notable things about this suggestion. First, the paper in which the suggestion was made appeared over 11 years ago—therefore, by the simple processes of subtraction and of intensive denial, I am led to conclude that I wrote it at the noteworthy age of 13. Second, and somewhat more important, the suggestion itself has foundered; the suggestion itself has foundered but not the basic concept, as this symposium abundantly testifies.

My idea was that inferences about anticholinergics could reveal the normal activities of brain-ACh—the basic notion was that the action of ACh was largely an inhibitory one, a notion whose dubiety looms daily more noticeable. But, regardless of what the ultimate validity of that idea may prove to be, the fact remains that it embodies only a *guess* about a mechanism in the CNS that, in turn, is based on the peripheral nervous system. But in the absence of actual data about the CNS itself, the idea can be no more than speculative.

Unfortunately, our knowledge about the biochemical mechanisms of the anticholinergics in the CNS remains small. This is true because techniques for manipulating and measuring ACh activity in the brain remain meager. This, in turn, is true because elucidation of the relevant biochemistry has yet to blossom. But without that growth, guesses like mine cannot go very far—they remain only guesses without empirical base. And that brings me to my second example.

Contrast this meager state of biochemical knowledge with that germane to the catecholamines. Active interest in the *behavioral* role of the catecholamines began with attempts to unravel the mechanism of action of chlorpromazine's cohort, reserpine. Pursuit of that inquiry merged with an older, separate biochemical concern that not only eventuated in a Nobel Prize but provided us with a truly astonishing armamentarium of techniques—today we know vast amounts about the mechanisms of catecholamine activity in brain; we also know about their anatomical distribution and have at our disposal an enormous array of pertinent drugs whose mechanism is known. Thus, we can manipulate amine activity and behavior in a knowledgeable and meaningful way. With respect to the catecholamines—and to a lesser extent, serotonin—, a genuine Pharmacology of Behavior is truly at hand. Perhaps we should describe the discipline—still more cumbersomely—as the Biochemical Pharmacology of Behavior.

And that is what this symposium is all about; it is an updating on some of the most significant advances in this new field. And, lest my comments seem without purpose, let me now highlight some of the things you are to read: The potential relevance of all this to Biological Psychiatry is, I think, self-evident; the role of Behavioral Pharmacology is too—not only for its introduction of behavioral technique but also for its insistence on an empiricism that eschews metaphysical inference and therefore inevitably leads to the pursuit of mechanism, in this case biochemical mechanism; furthermore, Behavioral Pharmacology gave rise to the concept that the details of the behavior being measured can importantly determine the effect of a drug so that drug action not only controls behavior but behavior controls drug action—that notion is today extended to include the biochemical step that necessarily lies within this dynamic interplay; I briefly mentioned Neurology and the conceptual base for analysis that it provides—here, we will not only encounter some of what Anatomy can tell a Biochemical Pharmacology of Behavior but, probably more important, what Biochemistry can tell Anatomy.

A final note: Two large-scale studies of the development of crucial scientific advance have both recently indicated that the lag between the inception of a discipline and its breakthrough is some 20 years. This necessarily means that basic research is uncommonly difficult to sell to those who insist on "relevance"; the delay of reinforcement, for them, is enormous. It also means that, if chlorpromazine is the starting point, we are very near our own point of breakthrough—as this symposium makes clear, this is indeed a very likely prospect.

Behavioral correlates of serotonin depletion[1]

JOHN A. HARVEY, ARTHUR J. SCHLOSBERG
AND LIBBY M. YUNGER

Department of Psychology, University of Iowa, Iowa City, Iowa 52242

ABSTRACT

Depletion of telencephalic serotonin (5-HT) content by medial forebrain bundle lesions, which interrupt the ascending serotonergic pathways or by DL-p-chlorophenylalanine produces an increased sensitivity to pain as measured by the flinch–jump, stabilimetric, or hot-plate methods. Examination of the effects of a number of other lesions and drugs indicated that dopamine, norepinephrine and acetylcholine are not involved in pain sensitivity. Dosages of 75 mg/kg DL-5-hydroxytryptophan(5-HTP), 37.5 mg/kg L-5-HTP or 50 mg/kg Ro 4-4602 (N^1-(DL-seryl)-N^2-(2,3,4-trihydroxybenzyl)hydrazine) plus 37.5 mg/kg L-5-HTP administered to medial forebrain bundle lesioned rats returned both the telencephalic content of 5-HT and the pain threshold to normal values. Injection of 37.5 mg/kg of D-5-HTP or an equimolar dose of L-dopa had no effect on pain threshold. Normal animals display increased sensitivity to pain and decreased 5-HT contents in frontal pole, hippocampus, and amygdala during dark as compared to light hours. All three of these telencephalic areas are innervated by the ascending serotonergic pathways, and cells in these areas show inhibition of firing following the iontophoretic application of 5-HT. Taken together, these data suggest that the serotonergic system normally acts to inhibit the effects of painful stimuli. A review of a variety of behavioral effects of 5-HT depletion including an enhanced response to lysergic acid diethylamide and amphetamine suggests that the ascending serotonergic system may have a general role in the inhibition of arousal, rather than a specific role with respect to various categories of behavior.—HARVEY, J. A., A. J. SCHLOSBERG AND L. M. YUNGER. Behavioral correlates of serotonin depletion. *Federation Proc.* 34: 1796–1801, 1975.

[1] Supported by Public Health Service Grant MH-16841 and National Institute of Mental Health Research Scientist Award MY-21849 to J.A.H. A.J.S. was supported by predoctoral fellowship MH08333 and L.M.Y. by predoctoral fellowship MH10641.

Abbreviations: 5-HT, serotonin; 5-HTP, 5-hydroxytryptophan; MFB, medial forebrain bundle; p-CP, DL-p-chlorophenylalanine; Ro 4-4602, N^1-(DL-seryl)-N^2-(2,3,4-trihydroxybenzyl)hydrazine; AMPT, α-methyl-p-tyrosine; ACh, acetylcholine.

The first demonstration that sero-tonin (5-HT) was associated with a distinct fiber system in the brain came from lesion studies in the rat. These studies demonstrated that destruction of the medial forebrain bundle (MFB) or of areas associated with this fiber system (dorsomedial midbrain tegmentum, ventral midbrain tegmentum, and septal area) significantly decreased whole brain content of 5-HT (22). Unilateral lesions in the MFB, a fiber system having a primarily unilateral distribution, decreased 5-HT content of the brain ipsilateral to the lesion only; also, maximal depletion did not occur until approximately 10 days after lesion placement. These results suggested that 5-HT was localized in small diameter fibers within the MFB (16).

The development of the Falck and Hillarp (10) histochemical fluorescence method verified the findings of the lesion studies and provided a more precise anatomical description of the serotonergic pathways in the brain (8,44). It is now clear that serotonergic cell bodies are located in three regions of the midbrain: the B-7 cell group in the dorsal raphe nucleus; the B-8 cell group in the medial raphe nucleus; and the B-9 cell group in the ventromedial midbrain tegmentum. Axons from these cell groups innervate the cerebellum, diencephalon, and, via the MFB, project diffusely to the telencephalon.

Our major interest has been to determine the role of this serotonergic system in behavior. In 1965, Harvey and Lints (17) reported that MFB lesions in the rat produced both a decrease in 5-HT content of the brain and an increased sensitivity to pain, as measured by the flinch–jump technique of Evans (9). In this procedure, an animal is placed in a chamber with a grid floor through which electric shock is delivered to the paws in an alternating series of ascending and descending shock intensities. After delivery of each foot-shock (0.1 sec duration) the animal is given one of three scores: *1*) no response; *2*) flinch response, any reflexive movement of the body except that both rear paws do not leave the grids; and *3*) jump response, a reflexive movement in which both rear paws leave the grids. Each shock intensity is delivered 10 times. Both flinch and jump thresholds are then calculated for each animal as the milliamperage at which an animal exhibited that response 5 out of 10 trails. Figure 1 (left side) presents the mean number of jump responses for control rats and rats with MFB lesions as a function of shock intensity. It can be seen that the curve for the MFB lesioned animals is shifted to the left. The horizontal line represents the jump threshold and it can be seen that this is an adequate measure of the separation of the two curves (see Fig. 1 legend for more detail). These changes in jump threshold were not associated with any change in the flinch threshold, suggesting that the lesion had not affected the animal's detection threshold (17, 29, 50). In addition, the lesion was not found to affect noise-elicited startle (50). Although our flinch–jump procedure is conducted so that the rater is unaware of the identity of the animal being tested, it is well known that subtle cues can influence the judgment of an observer. To test this we employed a stabilimetric method for assessing the effects of MFB lesions on an animal's response to foot-shock (50). In this method the animal's response to foot-shock was recorded on a polygraph using an astatic cartridge attached to the testing chamber as the transducer. The results obtained with this method are presented in Fig. 1 (right side) as the magnitude of the animal's

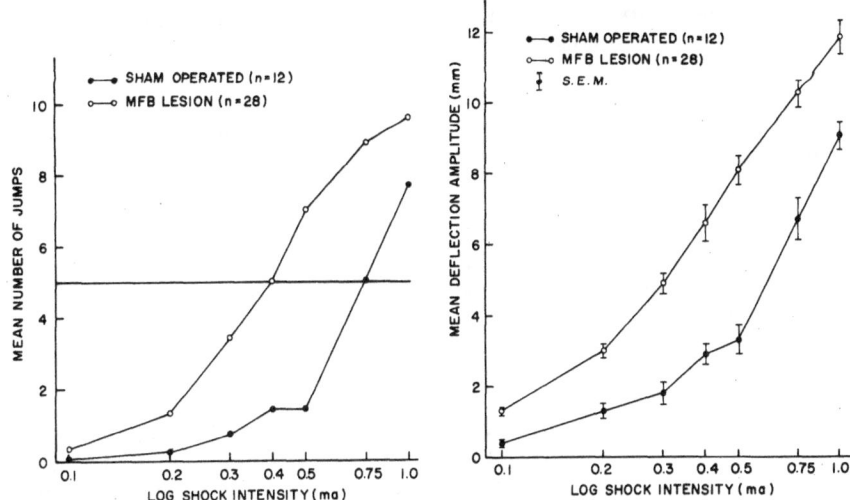

Figure 1. The response of a group of animals to foot shock was simultaneously recorded by the stabilimetric and flinch–jump methods (see text). The left side presents the mean number of jump responses as a function of shock intensity. The horizontal line represents the measure of jump threshold (5 out of 10 responses). The jump thresholds for lesioned and control animals taken from this figure would be 0.40 and 0.75 mA respectively. In practice, a separate jump threshold is calculated for each rat. The mean jump thresholds calculated in this manner for lesioned and control rats were (mean ± SEM): 0.38 ± 0.02 and 0.68 ± 0.03 mA respectively. The right figure presents the mean response magnitude as a function of shock intensity. Response magnitude is expressed as mm of pen deflection. For more detail see Yunger and Harvey (50).

response as a function of shock intensity. It can be seen that the results paralleled those of the flinch–jump method in that the curve for the MFB lesioned animals is shifted to the left. Finally, we also examined the effects of MFB lesions on pain sensitivity by means of the hot-plate method (50). In this procedure animals were placed on a copper surface (52 or 55.5 C) and their latency to lick a paw was recorded. Again, rats with MFB lesions demonstrated significantly shorter paw-lick latencies (19, 20, 50).

The behavioral analysis described above indicates that MFB lesions produce an increased sensitivity to painful stimuli as measured by the flinch–jump, stabilimetric, and hot-plate methods. In addition, we found a significant correlation between the

percentage decrease in jump threshold and brain content of 5-HT produced by MFB lesions. (29). We concluded that increased pain sensitivity following MFB lesions was due to the interruption of the ascending serotonergic pathways, and that in the normal animal 5-HT might function to inhibit the effects of painful stimuli. However, lesions in the MFB also decrease brain content of norepinephrine (21) and dopamine (2). Ascending axons from noradrenergic cell bodies in the brain stem innervate cerebellum, diencephalon, and, via the MFB, the entire telencephalon. Dopaminergic cell bodies located in and about the pars compacta of the substantia nigra also project rostrally through the MFB and adjacent internal capsule to innervate basal telencephalic areas and the caudate nucleus (2, 8, 21, 44). It is clear therefore that, depending on placement, brain lesions can produce differential decreases in brain content of all three monoamines (40). Table 1 summarizes the results of several studies in which we placed lesions at different levels of the ascending serotonergic system and in regions that would affect the ascending noradrenergic neurons. Rats were then examined for their changes in jump threshold and in whole brain content of 5-HT and norepinephrine. Every lesion placed in the serotonergic pathway, from the cell bodies in the raphe nuclei to the septal area, produced a significant reduction in both brain content of 5-HT and jump threshold. There was no consistent relationship between the effects of a lesion on jump threshold and brain content of norepinephrine. Table 1 also shows that depletion of 5-HT content by DL-p-chlorophenylalanine (p-CP) produced a decreased jump threshold

as originally reported by Tenen (43). In contrast, intraventricular injection of 6-hydroxydopamine had no effect on jump threshold, though norepinephrine content was reduced by 78%, and (not shown in the table) dopamine content was reduced by 58% (48).

Although a significant decrease in jump threshold was only seen if a lesion also produced a significant decrease in brain content of 5-HT, there was not a good correlation between these two measures. Thus, p-CP produced a 90% decrease in brain content of 5-HT, but only a 38% decrease in jump threshold, while MFB lesions produced only a 34% decrease in 5-HT with a 46% decrease in jump threshold (Table 1). In part, this lack of correlation is due to the fact that the lesions produce a decrease primarily in telencephalic 5-HT concent while p-CP produces its decreases throughout the brain Table 2). Table 2 also demonstrates that there are no additive effects of

TABLE 1. Effect of CNS lesions and drugs on jump thresholds and whole brain content of 5-HT and norepinephrine

| | Percentage change from control values | | |
| | Jump threshold | Whole brain content | |
Drug or lesion group		5-HT	norepinephrine
Control + p-CP, 300 mg/kg	−38*	−90*	−18
Dorsomedial tegmentum (raphe)	−24*	−49*	−20*
Medial forebrain bundle	−46*	−34*	−36*
Nucleus accumbens[a]	−22*	−17*	−22
Septal area	−33*	−13*	− 8
Caudate nucleus	− 5	0	—
Midbrain reticular formation	− 5	−10	−30*
Ventrolateral tegmentum	+ 2	− 8	−40*
Medial hypothalamus	− 6	− 3	+ 7
Control + 6-hydroxydopamine, 200 μg[a]	+ 6	− 2	−78*

Each value is the mean of 4–33 animals. The data are taken from the following refs: (14, 17, 20, 22, 25, 29, 31, 48). DL-p-CP (100 mg/kg) was injected on 3 consecutive days and rats tested 24 hr after the third injection. 6-Hydroxydopamine was injected intraventricularly and animals tested 7 weeks later.   * Asterisk indicates a value significantly different from controls ($P < 0.05$).   [a] 5-HT and norepinephrine decreases are for telencephalon.

TABLE 2. Effect of p-CP (300 mg/kg) and MFB lesions on jump thresholds and brain content of 5-HT

Experimental group	Jump threshold, mA	Serotonin content, nmole/g	
		Telencephalon	Brain stem
Control	0.66 ± 0.04	3.75 ± 0.11	5.68 ± 0.37
Control + p-CP	0.41 ± 0.06*	0.40 ± 0.04*	0.80 ± 0.06*
MFB lesion	0.39 ± 0.02*	0.91 ± 0.09*	5.34 ± 0.04
MFB lesion + p-CP	0.32 ± 0.04*	0.34 ± 0.06*	0.68 ± 0.06*

All values are given as the mean ± SEM. Values are based on 4–16 animals. p-CP was injected as described in Table 1.   * Asterisk indicates a mean value significantly different from controls ($P < 0.01$).

p-CP and MFB lesions on the jump threshold, suggesting that both procedures are acting via a common mechanism, depletion of 5-HT in the telencephalon.

To further examine these relationships, we compared the effects of drugs and lesions on: pain sensitivity, as measured by the hot-plate method; telencephalic content of 5-HT, norepinephrine, and acetylcholine (ACh); and caudate content of dopamine. Again, the only consistent relationship was between decreases in 5-HT content of telencephalon and decreases in paw-lick latency (Table 3). In agreement with previous studies (27, 34), septal lesions produced a significant decrease in telencephalic content of ACh (Table 3). Lesions in the MFB and nigrostriatal bundle had no effect on telencephalic content of ACh, though both lesions did decrease paw-lick latencies (Table 3). Previous studies have also reported that septal lesions decrease whole brain content of ACh (35, 42) while MFB lesions do not (41). These results suggest that the cholinergic systems of the brain (28) are not involved in determining an animal's sensitivity to painful stimuli. Effects of lesions or of α-methyl-p-tyrosine (AMPT) on telencephalic content of

TABLE 3. Effect of lesions, p-CP and AMPT on pain sensitivity and on brain content of putative synaptic transmitters

Drug or lesion group	Percent change from control				
	Paw-lick latency	Telencephalon			Caudate
		5-HT	Norepinephrine	ACh[a]	Dopamine
Control + p-CP, 300 mg/kg	−48*	−89*			
Medial forebrain bundle	−62*	−77*	−28*	− 4	−32*
Nigrostriatal bundle	−44*	−45*	−46*	0	−55*
Septal area	−68*	−22*	−12	−21*	+20*
Control + AMPT, 250 mg/kg	− 7		−72*		

Each value is the mean of 4–8 animals. Paw-lick latencies were determined on a surface at 52 C, except for animals given AMPT which were tested at 55.5 C. DL-p-CP was injected in one dosage (300 mg/kg) intraperitoneally and animals tested 3 days later. DL-α-methyl-p-tyrosine, methyl ester HCl (AMPT) was injected in two 125 mg/kg dosages spaced 4 hr apart and animals tested 4 hr after the second injection. Data are taken from the following sources: (13, 19, 23, 41, 51).   * Asterisk indicates a significant difference from controls ($P < 0.05$).   [a] Data taken from Oltmans, G. A., J. P. Sorensen and J. A. Harvey, unpublished data.

norepinephrine, or caudate levels of dopamine, also bore no relationship to the occurrence of a decrease in paw-lick latency. However, there was again no consistent relationship between the magnitude of the 5-HT decrease in the telencephalon and the decrease in paw-lick latency. The most rostral lesion in the ascending serotonergic pathway, the septal lesion, produced as large an increase in the sensitivity to pain as the other lesions, but only a 22% decrease in 5-HT content. This suggested that the decrease in 5-HT content responsible for the increased pain sensitivity might be occurring in some restricted regions within the telencephalon rostral to the septal area. Evidence for this view came from work with normal animals.

Rats demonstrate a diurnal rhythm in 5-HT content of the brain with the highest levels occurring during the light hours (the period of quiescence for the rat) and the lowest levels occurring during dark hours (the period of maximal activity) (36). This suggested to us that the rats might be more sensitive to pain at night. These variations in 5-HT content only occur in specific brain regions. In the telencephalon, significant diurnal rhythms have been reported to occur

in the frontal cortex, hippocampus and amygdala, areas receiving strong input from the septum (36, 37, 39). In addition, septal lesions produce significant decreases in 5-HT content of neocortex and hippocampus though not in amygdala (23).

Table 4 demonstrates that rats exhibit a 55% decrease in paw-lick latency during dark hours as compared with light hours. While there is no detectable change in telencephalic content of 5-HT between light and dark hours, frontal pole, hippocampus and amygdala demonstrate significant decreases in 5-HT content during the night. This provides additional evidence that low content of 5-HT is related to an increased sensitivity to pain. These data also suggest that the absence of a strong correlation between telencephalic content of 5-HT and pain sensitivity following lesions and drugs (as noted in Table 3) might be due to the fact that the critical changes in 5-HT are occurring in specific regions within the brain, rostral to the septal area. The diurnal changes in 5-HT and paw-lick latency noted in Table 4 also suggest that lesioned animals demonstrating a permanent reduction in 5-HT content should not demonstrate a diurnal change in paw-lick latency to the same

TABLE 4. Pain sensitivity and 5-HT content of brain during light and dark hours

Measurement	Light hours	Dark hours	Percent change
	Paw-lick latency, sec		
Hot-plate test	24.38 ± 1.94	10.97 ± 0.88	−55*
	5-HT content, nmole/g		
Telencephalon	3.07 ± 0.34	3.29 ± 0.17	+ 7
Frontal poles	4.43 ± 0.23	3.75 ± 0.45	−15*
Hippocampus	3.35 ± 0.06	2.61 ± 0.06	−22*
Amygdala	4.83 ± 0.28	3.52 ± 0.17	−27*

Values represent the mean ± SEM and are based on 4–15 animals. Light hours refers to 4 hr after light onset and dark hours 2 hr before light onset. Data are taken from ref. (19).   * Asterisk indicates a significant ($P < 0.05$) difference between light and dark hours.

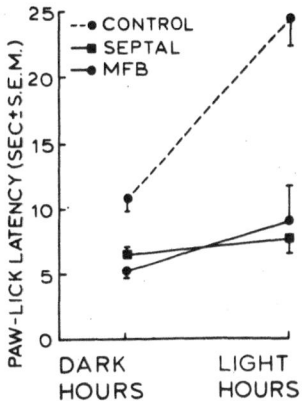

Figure 2. Paw-lick latencies (seconds) in control and lesioned rats tested during dark hours (2 hours prior to light onset) and light hours (4 hours after light onset). Each point represents the mean of 5–15 animals. Temperature of the hot-plate was set at 52 C (19). There was no significant difference between the paw-lick latencies of lesioned rats obtained during the light and dark hours.

extent as controls. Rats with septal or MFB lesions and controls were tested on the hot-plate during light and dark hours (Fig. 2). It can be seen that the lesioned rats do demonstrate a flat curve compared to controls, presumably because they cannot increase their 5-HT content during the light hours.

If the increased pain sensitivity of lesioned animals is indeed related to lowered 5-HT content in the brain, then it might be possible to inject 5-HTP (the immediate precursor of 5-HT) and return both the 5-HT content of the brain and pain sensitivity to normal values. Table 5 presents the results of several experiments that investigated this possibility in MFB lesioned rats (30, 48). Lesioned rats injected with saline demonstrated a 41% decrease in the jump threshold, and this was not affected by injections of D-5-HTP, L-dopa, or Ro 4-4602.

The dosage of Ro 4-4602 employed (50 mg/kg) has been shown to produce inhibition of peripheral but not central 1-aromatic amino acid decarboxylase (3). Injection of 75 mg/kg DL-5-HTP or of 37.5 mg/kg L-5-HTP returned the jump thresholds to within −5 and +18%, respectively, of control jump thresholds. These effects were not due to the peripheral formation of 5-HT since injection of L-5-HTP in combination with Ro 4-4602 also produced this reversal. In the latter case, telencephalic content of 5-HT which had been reduced 68% by the MFB lesion was increased to the level of controls (−7% of control value) (48). Similarly, Tenen (43) has demonstrated that the effects of p-CP on jump threshold can be reversed by administration of DL-5-HTP.

Figure 3 presents the dose-effect relationship between jump threshold and telencephalic content of 5-HT following administration of DL-5-HTP to MFB lesioned rats. The data for the jump thresholds and

TABLE 5. Reversal by L-5-HTP of increased pain sensitivity in MFB lesioned rats

Drug	No.	Jump threshold, percent change from control
Saline	6	−41*
D-5-HTP, 37.5 mg/kg	6	−39*
L-Dopa, 35.0 mg/kg	6	−35*
Ro 4-4602, 50 mg/kg	4	−43*
DL-5-HTP, 75 mg/kg	8	− 5
L-5-HTP, 37.5 mg/kg	6	+18
Ro 4-4602, 50 mg/kg + L-5-HTP, 37.5 mg/kg	8	+ 1

DL-5-HTP was suspended in 0.3% (wt/vol) gum tragacanth and injected in a volume of 1 ml/kg. All other drugs were dissolved in 0.9% (wt/vol) NaCl. Drugs were injected intraperitoneally 30 min before testing, except for Ro 4-4602 which was injected 60 min prior to testing. Data are taken from refs (30, 48). * Asterisk indicates a jump threshold significantly lower than that of controls ($P < 0.05$).

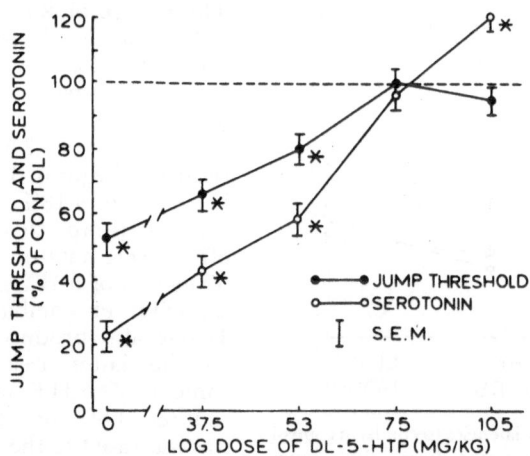

Figure 3. Effect of DL-5-HTP on jump threshold and telencephalic content of serotonin in rats with medial forebrain bundle lesions. DL-5-HTP was injected intraperitoneally 30 min prior to testing of animals. At the termination of testing (80 min after 5-HTP injection) animals were decapitated and brains analyzed for serotonin content. For details see Harvey and Lints (18). Each point represents the average of 4–5 animals. Asterisks indicate a significant difference from control values ($P < 0.05$).

5-HT content were obtained in the same animals (18). There is an orderly relationship between the increase in telencephalic content of 5-HT and the return of the jump threshold to normal values. Increasing 5-HT above normal values, as occurred after a dosage of 105 mg/kg DL-5-HTP, did not further increase the jump threshold above control values. Injection of 37.5 to 105 mg/kg of DL-5-HTP also had no analgesic effect in control rats (18).

The ability of 5-HTP to reverse both the behavioral and chemical effects of MFB lesions provides further evidence that 5-HT normally functions to inhibit an animal's response to painful stimuli. It is not clear, however, how either the biochemical or behavioral reversal occurs. Since the majority of serotonergic neurons have been destroyed and have degenerated, it is unlikely that the 5-HT being formed from 5-HTP in the brains of lesioned animals could simply be derived from decarboxylation within the few remaining serotonergic neurons or that this would produce a complete reversal of the behavioral effects. The experiments in which the peripheral decarboxylase activity is inhibited by Ro 4-4602 also suggest that the increase in 5-HT responsible for the reversal is not occurring within the periphery or within the cells (pericytes) lining the capillaries of the brain. One possibility derives from the suggestion by Heller and Moore (23) that the decrease in 5-HT content of telencephalon is due to a trans-synaptic effect on serotonergic neurons within the telencephalon. However, recent evidence does not support such a view. For example, the percentage decrease in 5-HT content of various telencephalic structures produced by lesions is proportional to the loss of tryptophan-5-hydroxylase activity and to the decrease in the uptake of serotonin into brain slices or crude synaptosomal fractions (15, 20, 26, 49). The changes in uptake are due to a decrease in V_{max}; K_m is not affected (48). The septal area provides one exception in that neither p-CP nor raphe lesions appear to affect its tryptophan-5-hydroxylase activity, although 5-HT is depleted (15).

Another source of decarboxylase activity occurs within the noradrenergic and dopaminergic neurons in the brain. We have already speculated on the possibility that 5-HT might be formed within these neurons and may be released as a "false" transmitter at the same sites normally innervated by serotonergic neurons (20). Evidence for this has been suggested on the basis of histochemical fluorescence methods (7). Additional evidence for such a conclusion has been provided by Yunger (48) who found that MFB lesioned rats pretreated with 6-hydroxydopamine do not demonstrate a reversal of the increased pain sensitivity when injected with L-5-HTP (37.5 mg/kg).

The results of these investigations suggest that the behavioral effects of a lesion can be related to its chemical consequences, in this case a decrease in telencephalic content of serotonin. Furthermore, the behavioral effects of a lesion can be reversed by appropriate treatment with precursors that reverse the neurochemical changes. This reversal does not appear to be a reinstatement of previous conditions since it appears that it may be due to the production of 5-HT in catecholaminergic neurons. At least two areas of the telencephalon (frontal poles and hippocampus) may be intimately involved in the behavioral changes seen after lesions or drugs that deplete 5-HT. Normal animals demonstrate a diurnal variation in pain sensi-

tivity and 5-HT content of frontal poles and hippocampus that is in the same direction as that predicted by the lesion studies (low 5-HT being associated with increased pain sensitivity). Lesions in the raphe nuclei, MFB or septal area and injection of p-CP produce decreases in 5-HT content of the frontal poles and hippocampus (15, 26, 49). Furthermore, microiontophoretic application of 5-HT to neurons in these regions produces inhibition of firing (4). Taken together these data suggest that decreases in 5-HT content of telencephalon result in the removal of an inhibitory control over an animal's response to painful stimuli.

Increased pain sensitivity is not the only behavioral consequence of 5-HT depletion. This raises the question as to whether 5-HT plays a specific role in pain sensitivity or whether the results reported above reflect some more general role for this compound in a variety of behavioral processes. With respect to pain, all that we can conclude is that depletion of 5-HT enhances the animal's reflexive response to painful stimuli, but this could be due to other factors than simply a hyperalgesia. Various investigators have reported that depletion of 5-HT results in increased locomotor activity in the rat (32, 33), increased wakefulness in the cat (24), enhanced response to both pleasant and unpleasant tastes in the rat (47), and a potentiation of the effects of d-lysergic acid diethylamide (LSD) in the rat (11, 12). LSD is of special interest in this regard since it has been shown to produce both an increased responsiveness to stimulus input (11, 12) and an inhibition of neuronal activity in the ascending serotonergic system (1). Another common effect of p-CP or of lesions that decrease 5-HT content of the brain is to produce an enhanced re-

TABLE 6. Effect of MFB and raphe lesion or of p-CP on drug action

Drugs	Ratio to control effect or control ED-50		
	MFB-lesion	Raphe-lesion	DL-p-CP
D-Amphetamine sulfate, 1 mg/kg	3.06*	4.08*[a]	2.83*
Morphine sulfate, 10 mg/kg	1.20	0.90	0.88
Na thiopental, 20 mg/kg	1.43	3.29*	
Chlorpromazine, 2 mg/kg	1.04	0.81	
Reserpine, ED-50	2.78*	1.21	

Values given in the table are the ratio of the drug effect in lesioned or p-CP treated rats to the drug effect of controls at the dosages indicated. For reserpine the ratios given are based on the control and lesioned ED-50's. Data were taken from the following sources: (13, 14, 19, 32, 33, 38, 40). * Asterisk indicates a drug effect significantly different from control ($P < 0.05$). [a] Data obtained with 1 mg/kg DL-amphetamine sulfate.

sponse to amphetamine (Table 6) as reflected either in increased locomotor activity (32, 33) or increased responding for water reward (13). The enhanced response to amphetamine produced by p-CP can be reversed by injection of 5-HTP (32). In contrast, there is no consistent relationship between 5-HT depletion and an animal's response to the depressant effects of thiopental, chlorpromazine, and reserpine or to the analgesic effects of morphine (Table 6).

Each of the reported changes in behavior or drug sensitivity produced by depletion of 5-HT appears to reflect a common process which can be best described as an increased arousal. Serotonin may not, therefore, play any specific role with respect to various categories of behavior such as sleep, pain, and the like, but rather may act as an inhibitory transmitter that reduces the effects of a variety of arousing stimuli including drug induced arousal produced by amphetamine. This conclusion is identical to the original proposal by Brodie and Shore (6) and Brodie and Costa (5), who suggested that 5-HT acted as a synaptic transmitter in an inhibitory system.

REFERENCES

1. AGHAJANIAN, G. K., AND H. J. HAIGLER. In: *Serotonin and Behavior*, edited by J. Barchus and E. Usdin. New York: Academic, 1973, p. 263.
2. ANDÉN, N. E., A. CARLSSON, A. DAHLSTRÖM, K. FUXE, N. Å. HILLARP AND K. LARSSON. *Life Sci.* 3: 523, 1964.
3. BARTHOLINI, G., W. P. BURKARD, A. PLETSCHER AND H. M. BATES. *Nature* 215: 852, 1967.
4. BLOOM, F. E., B. J. HOFFER, C. N. NELSON, Y. SHEU AND G. R. SIGGINS. In: *Serotonin and Behavior*, edited by J. Barchus and E. Usdin. New York: Academic, 1973, p. 249.
5. BRODIE, B. B., AND E. COSTA. *Psychopharmacol. Serv. Cent. Bull.* 2: 1, 1962.
6. BRODIE, B. B., AND P. SHORE. *Ann. N. Y. Acad. Sci.* 66: 631, 1957.
7. BUTCHER, L. L., J. ENGEL AND K. FUXE. *Brain Res.* 41: 387, 1972.
8. DAHLSTRÖM, A., AND K. FUXE. *Acta Physiol. Scand.* 62, Suppl. 232: 1, 1964.
9. EVANS, W. O. *Psychopharmacologia* 2: 318, 1961.
10. FALCK, B., N. Å. HILLARP, G. THIEME AND A. TORP. *J. Histochem. Cytochem.* 10: 348, 1962.
11. FREEDMAN, D. X. *J. Pharmacol. Exp. Ther.* 134: 160, 1961.
12. FREEDMAN, D. X., AND N. J. GIARMAN. In: *EEG and Behavior*, edited by G. H. Glaser. New York: Basic, 1963, p. 198.
13. GREEN, T. K., AND J. A. HARVEY. *J. Pharmacol. Exp. Ther.* 190: 109, 1974.
14. HARVEY, J. A. *J. Pharmacol. Exp. Ther.* 147: 244, 1965.

15. HARVEY, J. A., AND E. M. GÁL. *Science* 183: 869, 1974.
16. HARVEY, J. A., A. HELLER AND R. Y. MOORE. *J. Pharmacol. Exp. Ther.* 140: 103, 1963.
17. HARVEY, J. A., AND C. E. LINTS. *Science* 148: 250, 1965.
18. HARVEY, J. A., AND C. E. LINTS. *J. Comp. Physiol. Psychol.* 74: 28, 1971.
19. HARVEY, J. A., A. J. SCHLOSBERG AND L. M. YUNGER. In: *Advances in Biochemical Psychopharmacology*, edited by E. Costa, G. L. Gessa and M. Sandler. New York: Raven, 1974, vol. 10, p. 233.
20. HARVEY, J. A., AND L. M. YUNGER. In: *Serotonin and Behavior*, edited by J. Barchus and E. Usdin. New York: Academic, 1973, p. 179.
21. HELLER, A., AND J. A. HARVEY. *Pharmacologist* 5: 264, 1963.
22. HELLER, A., J. A. HARVEY AND R. Y. MOORE. *Biochem. Pharmacol.* 11: 859, 1962.
23. HELLER, A., AND R. Y. MOORE. In: *Advances in Pharmacology*, edited by S. Garattini and P. A. Shore. New York: Academic, 1968, vol. VI, part A, p. 191.
24. JOUVET, M. In: *Serotonin and Behavior*, edited by J. Barchus and E. Usdin. New York: Academic, 1973, p. 385.
25. KOE, B. K., AND A. WEISSMAN. *J. Pharmacol. Exp. Ther.* 154: 499, 1966.
26. KUHAR, M. J., G. K. AGHAJANIAN AND R. H. ROTH. *Brain Res.* 44: 165, 1972.
27. KUHAR, M. J., V. H. SETHY, R. H. ROTH AND G. K. AGHAJANIAN. *J. Neurochem.* 20: 281, 1973.
28. LEWIS, P. R., AND C. C. D. SHUTE. *Brain* 90: 521, 1967.
29. LINTS, C. E., AND J. A. HARVEY. *J. Comp. Physiol. Psychol.* 67: 23, 1969.
30. LINTS, C. E., AND J. A. HARVEY. *Physiol. Behav.* 4: 29, 1969.
31. LORENS, S. A., J. P. SORENSEN AND J. A. HARVEY. *J. Comp. Physiol. Psychol.* 73: 284, 1970.
32. MABRY, P. D., AND B. A. CAMPBELL. *Brain Res.* 49: 381, 1973.
33. NEILL, D. B., L. D. GRANT AND S. P. GROSSMAN. *Physiol. Behav.* 9: 655, 1972.
34. PEPEU, G., A. MULAS AND M. L. MULAS. *Brain Res.* 57: 153, 1973.
35. PEPEU, G., A. MULAS, A. RUFFI AND P. SOTGIU. *Life Sci.* 10: 181, 1971.
36. QUAY, W. B. *Life Sci.* 4: 379, 1965.
37. QUAY, W. B. *Am. J. Physiol.* 215: 1448, 1968.
38. ROTH, B. F., AND J. A. HARVEY. *J. Pharmacol. Exp. Ther.* 161: 155, 1968.
39. SCAPAGNINI, U., G. P. MOBERG, G. R. VAN LOON, J. DE GROOT AND F. GANONG. *Neuroendocrinology* 7: 90, 1971.
40. SIMPSON, J. R., F. GRABARITS AND J. A. HARVEY. *Pharmacologist* 9: 213, 1967.
41. SORENSEN, J. P., JR., AND J. A. HARVEY. *Federation Proc.* 30: 1279, 1971.
42. SORENSEN, J. P., JR., AND J. A. HARVEY. *Physiol. Behav.* 6: 723, 1971.
43. TENEN, S. S. *Psychopharmacologia* 10: 294, 1967.
44. UNGERSTEDT, U. *Acta Physiol. Scand. Suppl.* 367: 1, 1971.
45. UNGERSTEDT, U. *Acta Physiol. Scand. Suppl.* 367: 69, 1971.
46. VON VOIGTALANDER, P. F., AND K. E. MOORE. *J. Pharmacol. Exp. Therapeut.* 184: 542, 1973.
47. WEISSMAN, A. In: *Serotonin and Behavior*, edited by J. Barchus and E. Usdin. New York: Academic, 1973, p. 235.
48. YUNGER, L. M. (Ph.D. Thesis.) Iowa City: Univ. of Iowa, 1974.
49. YUNGER, L. M., AND J. A. HARVEY. *Federation Proc.* 32: 3165, 1973.
50. YUNGER, L. M., AND J. A. HARVEY. *J. Comp. Physiol. Psychol.* 83: 173, 1973.
51. YUNGER, L. M., J. A. HARVEY AND S. A. LORENS. *Physiol. Behav.* 10: 909, 1973.

Neurochemical changes associated with schedule-controlled behavior[1]

SHELDON B. SPARBER

Department of Pharmacology, University of Minnesota Minneapolis, Minnesota 55455

ABSTRACT

Studies of monoamine metabolites within the cerebrospinal fluid compartment have indicated that this approach may be useful in examining central metabolic changes in vivo. By combining the technologies of radioisotope chemistry, operant behavior control and modification, and brain perfusion with push-pull cannulas, we have been able to examine minute to minute changes in the disposition of radiolabeled monoamine transmitter candidates and their metabolites. These substances appear to co-vary with changes in complex behavior maintained by operant schedules of reinforcement and affected by changes in schedules or administration of psychotropic drugs. In agreement with other perfusion studies, we have observed changes in fractional distribution of radiolabeled urea, a so-called extracellular marker, along with shifts in monoamines; but the former appear more transient. These observations nevertheless support the concept of dynamic changes within the extracellular environment of the CNS that may be part of a hormone-like communicating system with functional significance. Furthermore, the presence of peaks and/or troughs, in perfusates, of [^{14}C]urea or similar substances should not be taken as a priori evidence for nonspecificity of the technic, since selective release or inhibition of release of monoamines can be shown with appropriate drugs that are thought to act through these aminergic systems. Destruction of catecholamine nerve terminals with 6-hydroxydopamine likewise attenuates the signal-locked release of radiolabeled norepinephrine by a conditioned stimulus after conditioning occurs. No such release is seen on presentation of the to-be-conditioned neutral stimulus in control or 6-hydroxydopamine treated rats. These initial studies indicate the availability of a powerful tool for the study of drug-neurochemical-behavioral interactions using subjects as their own controls for extended periods of time so that phenomena of plasticity, tolerance and dependence may likewise be examined.—SPARBER, S. B. Neurochemical changes associated with schedule-controlled behavior. *Federation Proc.* 34: 1802–1812, 1975.

[1] Partially supported by Public Health Service grants MH-08565 and DA-00532.

The concept of individual differences underlies, to a great extent, the philosophy of an operant approach to the analysis of behavior. Accordingly, when studying the behavioral effects of manipulating environmental variables or administering drugs which may or may not alter behavior by interacting with those variables identified as eliciting or maintaining behavior (i.e., deprivation, stimulus control, reinforcing character and density, and so on) it would likewise appear advantageous to adhere to the aforementioned philosophy (30).

Until recently, the closest approximation to an examination of the biochemical changes that occur within the central nervous system (CNS) consequent to manipulating external and/or internal environmental variables has centered around a technology that depends on the instantaneous or at least the rapid sacrifice of experimental subjects during or after behavioral sessions. Although much information on the mechanism of action of drugs, at the biochemical level, has been gleaned by this approach, much information about the close temporal relationship between biochemistry and behavior is lost. In addition, the destruction of an animal that required much time and effort to train is a costly way of obtaining the desired information. I should like here to describe some work in which we have been engaged that attempts to circumvent many of the obvious disadvantages just stated. We have attempted to superimpose on the technology that has evolved around the study of operant behavior, the additional technologies associated with radioactive isotope labeling of neurotransmitter candidates and their precursors, as well as that of perfusion of individual organs or parts thereof so that in vivo con-

ditions might be approached. I am convinced that, together with other, more traditional approaches to psychopharmacological problems, the utilization of the technics described herein will add much to our understanding of the biochemical substrate of CNS function and ultimately behavior and drug action on such function and behavior.

BIOGENIC AMINES AND CONDITIONED BEHAVIOR

The evidence for involvement of the biogenic amines (dopamine, norepinephrine, and serotonin (5-HT)) in the maintenance of conditioned behavior stems from variations of original experiments first performed by Carlsson and co-workers (11, 50, 51) and Brodie and co-workers (10) who found that behavioral depression coincided with the administration of drugs that produced depletion of these amines. Specific inhibitors of the synthesis of the catecholamines norepinephrine and dopamine, which likewise depress operant behavior, as well as studies in which depressed behavior has been at least partially reversed by the administration of the precursor of dopamine and norepinephrine, l-dopa, have led many to conclude that it is mainly the catecholamines that are involved in the behavioral depression seen after the administration of drugs that lower steady-state levels of both phenethylamines and indolealkylamines (12, 21, 47). Findings contrary to the prediction that destruction of catecholamine containing neurons with 6-hydroxydopamine would lead to a deficit in performance of operant behavior (42, 48) have increased the level of ambiguity regarding the specific involvement of the amine transmitter candidates in the mediation of operant behavior.

Because of an important pharmacological maxim—that drugs can do nothing that the organism (or parts thereof) is incapable of in the absence of drugs—we set about to determine if catecholamines could be shown systematically to co-vary with behavior controlled by a specific reinforcement schedule. This allowed us to manipulate behavior by changing the consequences of that behavior (18) without having to introduce the *confounding* drug variable. To determine if the radioactive norepinephrine within the push-pull perfusate we were collecting during performance of an operant was released selectively, control experiments were performed with the so-called extracellular marker, [14C]urea; in addition we compared the effects of drugs with apparently different mechanisms of action, in dosages that disrupted the operant behavior to an equal extent (58). One of these drugs, *d*-amphetamine, appears to act via facilitation of release of catecholamines and, additionally, through prevention of reuptake by nerve terminals, while no good evidence exists to suggest that mescaline, the second drug, acts in a similar fashion. We subsequently have examined the relationship between equieffective doses of *d*-amphetamine, mescaline and *d*-lysergic acid diethylamide (*d*-LSD-25) as they relate to changes in the diffusible extracellular content of norepinephrine, 5-HT and some of their metabolites during operant sessions. The relationship between tolerance to the operant behavioral action of *d*-amphetamine and its ability to alter the disposition of radiolabeled norepinephrine and 5-HT introduced into the brains of rats prior to the operant sessions is likewise discussed.

I will also touch briefly on some newer perfusion experiments that allow a finer dissection of the relationship between isomers of drugs like the amphetamines and their effects on conditioned behavior.

CHOICE OF METHODS

Having witnessed the use of Gaddum's push-pull cannula (23) in the laboratory of Dr. F. Sulser (56), it became immediately apparent that a potentially powerful tool was available for psychopharmacological research. The main obstacles to be overcome included the design of a system that would allow perfusion, with chronic indwelling cannulas, of brains of animals who were engaged in complex behavior tasks that would be sensitive to specific and nonspecific stimuli. The sensitivity to specific stimuli would enable the experimenter to manipulate, at will, those variables thought to maintain the behavior. Sensitivity to nonspecific stimuli would allow for a built-in control against changes in intracranial pressure, tissue integrity, and the like, which have been suggested as possible contaminants of experiments in which brain tissue is perfused (13, 29, 43).

Most of our work to date has been centered around perfusion of the lateral ventricular space in the rat. The decision to stay within the ventricular space was prompted in part by the aforementioned objections to the nonphysiological conditions produced by perfusion of solid tissues, as well as the theoretical questions raised about the research and clinical use of cerebrospinal fluid constituents as a measure of periventricular brain metabolism (37).

In addition, the relatively small size of the rat ventricular system allows for more rapid diffusion and mixing of substances released from periventricular structures with the perfusion medium delivered by the tip of a relatively large device (cannula) inserted into the space. Finally, this species has been used in our laboratory, and many others, for studies of effects of psychotropic drugs on operant behavior and it would be foolhardy to attempt to manipulate too many variables without a strong base of information on which one can build. As you will see, this systematic approach has allowed us to compare and contrast data generated in our own laboratories, as well as others, on the behavioral effects of drugs of separate pharmacologic classes and drugs within a class.

I will not spend any time on the specific perfusion methodology, other than to refer the interested individual to the excellent chapter of Myers' (39) and the original publications from our own laboratory cited at appropriate places within the text of this communication. We have subsequently modified our cannulas, collection procedures, and separation technics as described in a forthcoming communication.

Manipulating schedule parameters and perfusion of lateral ventricle

[³H]norepinephrine injected into the lateral ventricle of rats is taken up into catecholamine-containing neurons and is distributed in a parallel fashion to endogenous norepinephrine (1, 26, 27). Several laboratories have taken advantage of these observations to combine push-pull perfusion technics with isotope pulse labeling procedures in order to follow the release of transmitter candidates after electrical stimulation of brain (28, 55), handling and injection (thought to act as stressor stimuli) (60), drug administration (56) or stimulation of olfactory neurons with odorous substances (13). We reasoned that if substances injected into the

cerebrospinal fluid (CSF) gained access to brain structures in a selective way, and substances within CSF reflected metabolism of brain structures in a fairly selective manner (38), sampling of the ventricular space would allow us to infer metabolic or other dynamic changes occurring concurrently with operant behavior output or drug action. Our initial studies were designed to determine whether changes in operant behavior that would be compared to an *emotional* reaction would show changes in the disposition of [³H]norepinephrine that had been previously injected and allowed to equilibrate with endogenous stores. Withholding of positive reinforcement after a history of low fixed ratio schedule control of behavior was chosen as the manipulation because of its behavioral consequences, which have been interpreted within an emotional and autonomic framework. Possible CNS catecholamine involvement in emotional stress in many species has been reported (7, 8, 46).

After slowly injecting [³H]norepinephrine or [¹⁴C]urea into the lateral cerebral ventricles of rats trained to bar-press for 45 mg Noyes food pellets on a relatively low fixed ratio schedule, the site of injection within the lateral ventricle was perfused at a rate of approximately 20 μl/min (57). Total radioactivity in aliquots of 2 min collection samples was then determined. Unlike some reports in the literature that claim a lack of *release* of extracellular marker in push-pull or ventriculocisternal perfusion (12, 28, 56) we observed slight increases in [¹⁴C]urea in the perfusate as a consequence of handling for injection (Fig. 1) or changing bar-pressing rates by imposing an extinction period during fixed ratio performance (Fig. 2) (54). Our initial observations indicated a definite peak in [³H]norepi-

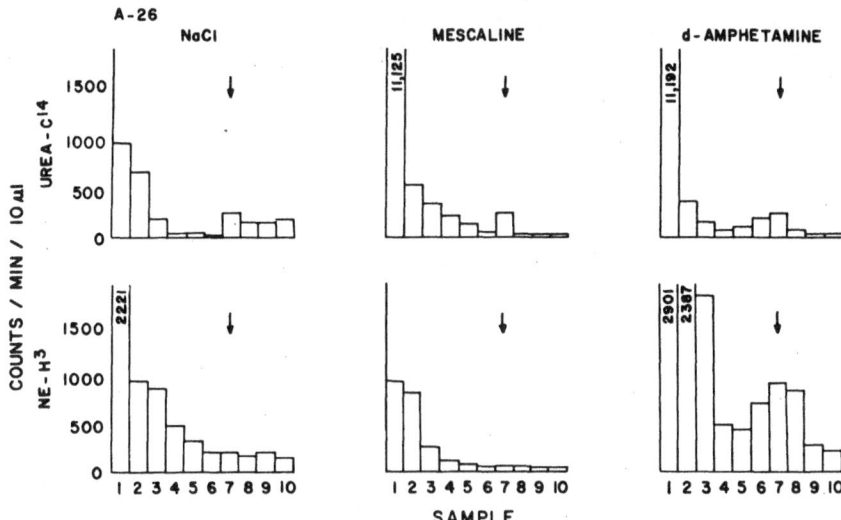

Figure 1. Radioactivity histograms for one rat showing slight increases in [^{14}C]urea efflux in push-pull perfusate resulting from injection of saline (NaCl), mescaline HCl (15 mg/kg) or (+) amphetamine SO$_4$ (2.5 mg/kg). Only (+) amphetamine produced an increase in [^{3}H](from [^{3}H]norepinephrine (^{3}H-NE)) in the ventricular perfusate. Arrows indicate approximate time of injection of drug. Samples were collected every 2 min during the operant session which started 1 hr after injecting [^{3}H]norepinephrine into the lateral ventricle. (Reprinted with permission (54).)

nephrine (and metabolites) only in cases in which extinction resulted in large increases in responding (post-extinction bursts). When the behavioral consequences (bursts) of extinction were minimal, only slight changes in the profile of sample to sample contents of ^{3}H could be seen (Fig. 3). However, these data also confirmed the suggestion by Chase and Kopin (13) that perfusion experiments may be confounded to an extreme extent by nonspecificity of *release* and/or that "local changes within the extracellular compartment may attend neural stimulation." At about the same time we were doing these experiments Winson and Gerlach (60) were observing similar results with extracellular marker substances but their data, like those of Stein and Wise (55) and Chase and Kopin (13),

were derived from experiments in which cannula size and/or flow rates through solid tissue were so large as to question the specificity of release based on these criteria alone (39). However, Winson and Gerlach did make an interesting observation which reinforced our own thoughts regarding the necessity and sufficiency of showing changes in putative transmitter profiles in the absence of extracellular marker profiles. Although they could not cause *release* into push-pull perfusates of the hypothalamus by *stressing* their subjects, which did cause changes in radioactivity from [^{3}H]norepinephrine, ^{3}H$_2$O or [^{14}C]urea origin in the amygdala perfusates, they were able to partially replicate the work of Sulser et al. (56) by injecting desmethylimipramine and reserpine, which did increase [³H]norepinephrine content in hypothalamic push-pull perfusate. Unlike Sulser et al. (56) however, Winson and Gerlach did report increases of [¹⁴C]urea. It occurred to us that in the few cases in which extracellular marker substances were not *released*, the perfused subjects were either unconscious or paralyzed. In those cases in which immobilized subjects did show so-called nonspecific release, the severe perfusion parameters could just as likely have caused movement of substances through extracellular

channels by bulk flow and damage alone (9). What had to be attempted was an experimental protocol that utilized drugs as tools to determine if specificity of release could be demonstrated. From our own studies (52), and those of others (6, 20), with fixed ratio schedule control and effects of psychotomimetic and psychomotor stimulants, it was becoming increasingly clear that similarities in the various drugs' disrupting effects on operant behavior, maintained under some schedules of reinforcement, could not be explained through

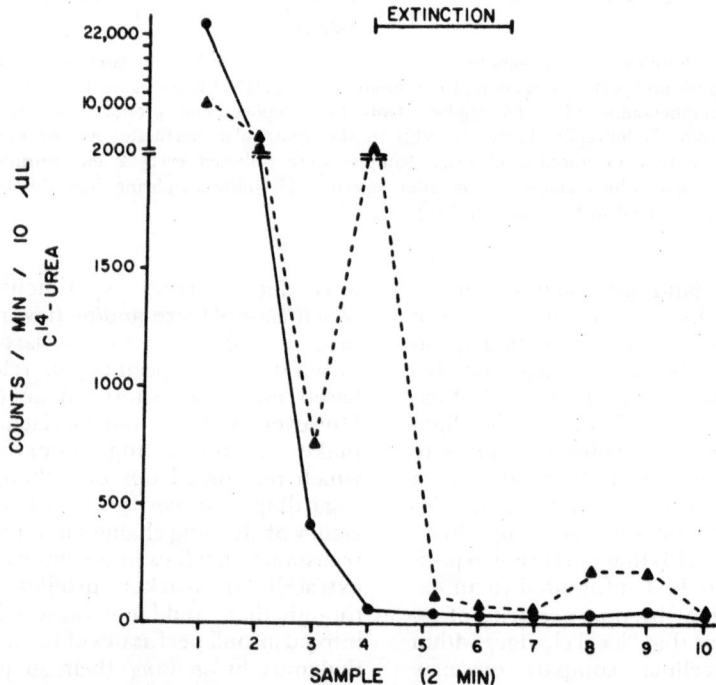

Figure 2. Increases in radioactivity from [¹⁴C]urea resulting from extinction. Each point represents the average of the three rats during control sessions (circles) and 5 min of extinction (triangles) while bar-pressing for food on a fixed ratio (FR) 5 schedule of reinforcement. In all three cases there were postextinction increases in bar-pressing (see Fig. 3 for

identical records) followed by pauses in this behavior prior to resumption of bar-pressing on reinstallation of reinforcement. The appearance of the second peak is suggested as resulting from increased motor activity and lowering of the rats' heads to the food delivery trough. (Reprinted with permission (54).)

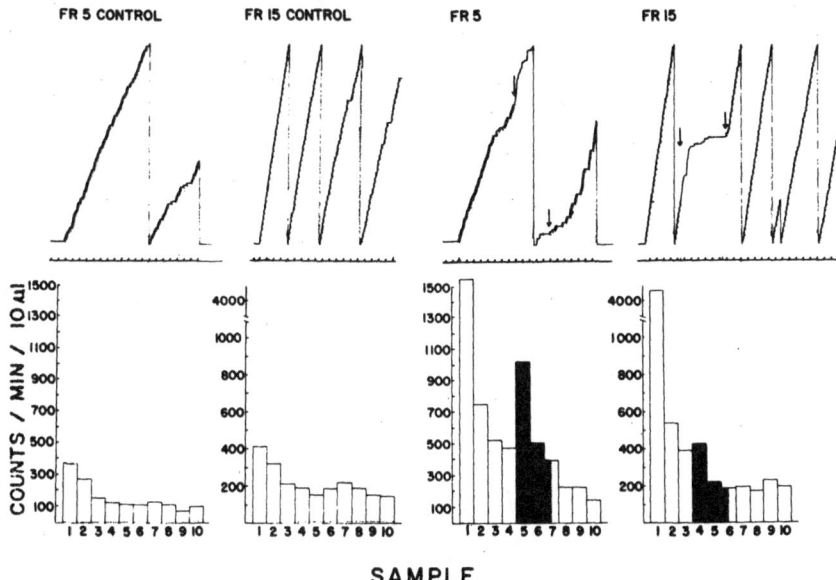

SAMPLE

Figure 3. Cumulative records and radioactivity histograms for control and extinction (arrows) sessions for rat A28. Tracings of event pens beneath the cumulative records indicate a 1 min time base and each bar of the histograms represents the radioactivity, in 10 μl of perfusate, from [7-³H]norepinephrine injected 1 hr prior to the sessions. Perfusion rates were 20 μl/min, using artificial CSF as the medium. Shaded areas indicate the extinction-perfusion samples. (Reprinted with permission (54).)

similarities in biochemical mechanisms (22, 24, 32, 35, 41). Therefore, a series of experiments was designed in which the fixed ratio was kept constant and the effect of d-amphetamine, mescaline and d-LSD-25 upon the disposition of [³H]norepinephrine and 5-[¹⁴C]HT in push-pull perfusate could be examined over short periods of time, along with such variables as latency to the onset and duration of the behavioral effects of these drugs (58). Additionally, we attempted an analysis of aliquots of perfusate by thin layer chromatography to determine if appropriate metabolites could be observed (25).

Figure 4 shows the type of behavioral response to drugs we were dealing with. By defining disruption as the point in time, after drug admin-

istration (52), at which 2 min have elapsed without the organism receiving a reinforcer, we were able to choose equieffective dosages for further study. Table 1 shows the relationships between the various doses of d-amphetamine and mescaline, as well as a dose of d-LSD of approximate equieffectiveness on fixed ratio behavior, and latency to disruption. Figures 5 and 6 show the effects of the two phenethylamines on ³H from [³H]norepinephrine and ¹⁴C from 5-[¹⁴C]HT injected into the lateral ventricles prior to perfusion and drug administration. Interestingly, this dose (0.4 mg/kg) of d-LSD caused an increase in neither ³H nor ¹⁴C in the perfusates during its course of action, as was seen with d-amphetamine or mescaline, respectively. On the other

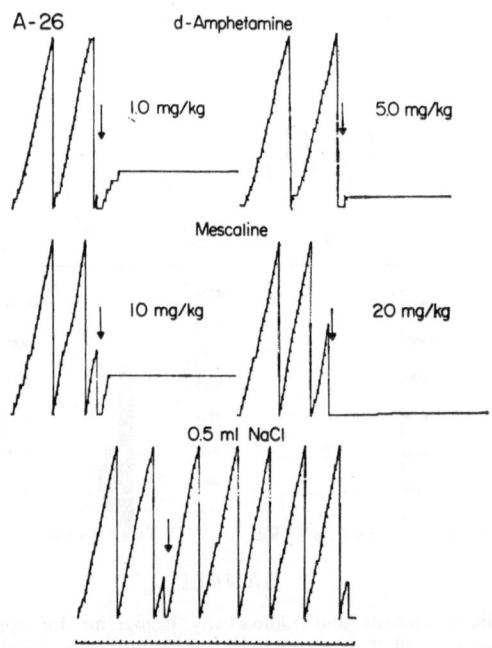

Figure 4. Sample cumulative records from one rat showing dose-related disruption of fixed ratio (FR) 30 responding. Mescaline, (+) amphetamine or NaCl were injected i.p. 10–15 min into the session (↓). Each excursion of the event pen at the bottom of the figure indicates a 1 min time base and the recorder pen automatically reset after 550 responses. The rate of responding before drug injection is similar to that obtained before and after NaCl, indicating that the perfusion procedure and injection per se did not affect food-reinforced responding. (Reprinted with permission (58).)

hand, it produced a significant decrease in the ¹⁴C (from 5-[¹⁴C]HT) in perfusates subsequent to injection and during its behavioral action (Table 2). The data in Tables 3 and 4 indicate that changes in the disposition of total radioactivity in push-pull perfusates could also be attended by changes in apparent metabolism of these putative transmitters. When *d*-amphetamine was injected, there followed an almost immediate significant increase in the percentage of unchanged [³H]norepinephrine in perfusate, as well as a significant increase in the percentage of radioactivity in the spot where authentic normetanephrine co-chromatographed. This was accompanied by a corresponding decrease in other spots on the thin layer chromatograms that are essentially *O*-methylated-deaminated metabolites and assorted unidentified compounds. At no time did *d*-amphetamine, in the doses used, cause significant changes in 5-[¹⁴C]HT or metabolites in acute experiments of this type. However 5-[¹⁴C]HT and its metabolite(s) were significantly shifted by the two higher doses of mescaline (15 and 20 mg/kg). Unlike the effect on [³H]norepinephrine produced by *d*-amphetamine, mescaline produced a significant de-

TABLE 1. Latency to onset of behavioral disruption after drug administration

Treatment	Latency, min ± SEM
Mescaline	
10 mg/kg (7)	4.5 ± 0.3
15 mg/kg (7)	3.6 ± 0.2
20 mg/kg (13)	2.9 ± 0.2
d-Amphetamine	
1.0 mg/kg (7)	6.5 ± 0.8
2.5 mg/kg (7)	3.9 ± 0.2
5.0 mg/kg (13)	2.7 ± 0.2
d-LSD	
0.4 mg/kg (6)	2.8 ± 0.2

Rats were injected intraperitoneally (i.p.) with drugs 10 to 15 min into the perfusion session. Latency to drug action is defined as the period between injection of the drug and the point in time when no reinforcers were delivered for 2 min. Each point is the mean ± SEM. Numbers in parentheses represent the number of experiments performed at that dose.

crease in the percentage of unchanged 5-[^{14}C]HT and a corresponding increase in its acid (5-[^{14}C]HIAA) and other metabolites. Even at the time when the peak in radioactivity was no longer evident (sample 15), there were continued significant differences in the ratio of the distribution of radioactivity on the chromatograms, indicating that changes in transmitter chemistry continued to accompany the protracted behavioral disruption. Interestingly, while d-LSD decreased the overall output of ^{14}C from 5-[^{14}C]HT, the distribution of 5-HT metabolites on the chromatogram remained unchanged, indicating that, at least within the period sampled, 0.4 mg d-LSD/kg seemed to decrease the rate of release (turnover?) of 5-HT without altering significantly its compartmentalization or metabolism. Our interpretation of these last data suggests a presynaptic releasing effect of mescaline, similar perhaps to

d-amphetamine and tyramine effects on catecholaminergic neurons, while d-LSD may be affecting mainly serotonergic receptors (postsynaptic?) directly and thereby decreasing serotonergic nerve function through some feedback inhibitory or direct presynaptic inhibitory action. Biochemical and electrophysiological evidence for this interpretation has likewise been offered (2–4, 19, 33, 45, 59). In a follow-up experiment we examined the effects of continued administration of d-amphetamine on [^{3}H]norepinephrine, 5-[^{14}C]HT and their metabolites (53). After about a week and a half of daily injections of 2.5 mg d-amphetamine sulfate/kg, the drug no longer produced its disruptive effect on the operant. On the other hand, mescaline (15 mg/kg) did produce the characteristic abrupt cessation of fixed ratio responding, indicating no or incomplete cross-tolerance between the two phenethylamines (Fig. 7). Table 5 shows the continued release of [^{3}H]norepinephrine (and metabolites) into the perfusate to about the same extent both during disruption of the operant and after tolerance to the disruptive action of d-amphetamine. However, unlike the absence of an effect by d-amphetamine on 5-[^{14}C]HT prior to tolerance, there was a significant increase in 5-[^{14}C]HT (and metabolites) in the perfusates subsequent to d-amphetamine administration (Fig. 8). Additionally, although 15 mg of mescaline HCl/kg was able to increase 5-[^{14}C]HT in perfusates to 65% from an average of 45% in samples subsequent to NaCl injection prior to tolerance to d-amphetamine, the same dose of mescaline (15 mg/kg) increased it further, to an average of 85% of the sample prior to injection, indicating perhaps greater reactivity of serotonergic neurons to pharmacological manipulation. The interpretation of these data can include the possibility that refractoriness of catecholaminergic receptors has developed and/or indicate a functional antagonism between catecholaminergic and serotonergic central pathways. Alternately, repeated doses of d-amphetamine could mimic large acute doses in their ability to alter brain serotonin and could be linked to some aspects of toxic psychosis produced by high doses of amphetamine (5, 16, 17). If mobilization of serotonin is part of a homeostatic mechanism, blockade of synthesis of 5-HT or its receptors should inhibit tolerance to at least some of amphetamine's actions.

With the availability of chemicals that can specifically destroy presynaptic catecholamine nerve terminals, experiments could be attempted to support or refute the concept of specificity of release from appropriate nerve terminals when [³H]norepinephrine (and metabolites) appears after manipulation of behavior by antecedent environ-

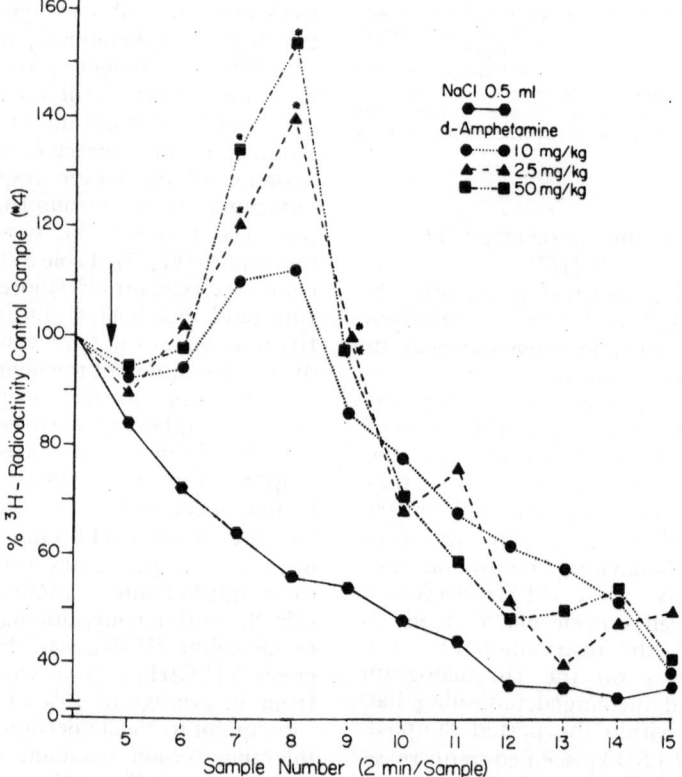

Figure 5. Release of radioactivity from [³H]norepinephrine into the perfusate from the ventricles after the i.p. injection of various doses of (+) amphetamine or NaCl. Results are expressed as a percentage of radioactivity compared to sample no. 4 (100%). Each point is the mean of three experiments for the two lower doses and six for the 5.0 mg/kg dose and the NaCl experiments. The * indicates a significant difference from NaCl levels using a matched pair t-test, $P < 0.05$. The ↓ indicates the approximate point of injection. (Reprinted with permission (58).)

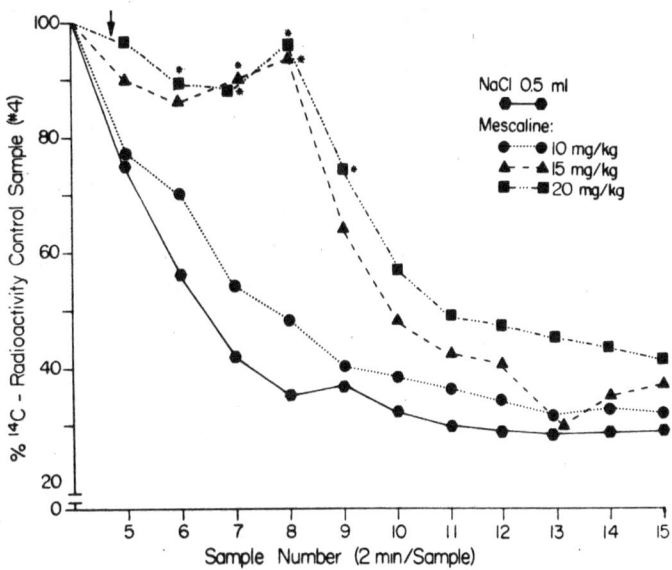

Figure 6. Release of radioactivity from 5-[^{14}C]HT into the perfusate from the ventricles after the i.p. administration of various doses of mescaline or NaCl. Results are mean percentage of radioactivity compared with sample no. 4 taken as 100%. Four animals were used for the experiments involving lower doses of mescaline and seven for the 20 mg of mescaline per kg and NaCl experiments. The * indicates a significant difference from NaCl levels using a matched pair t-test, $P < 0.05$. The ↓ indicates the approximate point of injection. (Reprinted with permission (58).)

TABLE 2. Effect of intraperitoneal drug administration on appearance of radioactivity in the ventricular perfusate

Treatment	Mean % radioactivity of sample no. 4	
	^{14}C	^{3}H
NaCl	38.0 ± 2.8 (7)	51.1 ± 5.3 (6)
d-Amphetamine		
1.0 mg/kg	42.6 ± 1.5 (4)	72.6 ± 0.7^a (3)
2.5 mg/kg	47.1 ± 5.4 (4)	78.5 ± 3.2^a (3)
5.0 mg/kg	41.9 ± 1.8 (7)	89.3 ± 8.6^a (6)
Mescaline		
10 mg/kg	46.2 ± 3.4 (4)	56.0 ± 4.5 (3)
15 mg/kg	65.7 ± 6.5^a (4)	51.4 ± 3.9 (3)
20 mg/kg	68.9 ± 5.7^a (7)	56.8 ± 4.7 (6)
d-LSD		
0.4 mg/kg	28.7 ± 3.6^a (3)	42.6 ± 6.3 (3)

[^{3}H]-norepinephrine or 5-[^{14}C]HT were infused intraventricularly 1 hr or 30 min before perfusion sessions, respectively, and drugs, or NaCl were injected i.p. 12 to 15 min into the session. Results are expressed as mean ± SEM percentage of radioactivity in samples subsequent to the injection relative to sample no. 4. Number of subjects is in parentheses. ^{a} Significantly different from NaCl value, matched pair t test, $P < 0.05$.

mental variables or drug administration. To this end, animals were trained to avoid a shock by jumping onto a platform during presentation of a previously neutral visual stimulus which was subsequently paired with a 2 mA shock delivered to their feet through a stainless steel grid floor. Prior to training and implantation of push-pull perfusion cannulas, half the animals were injected intraventricularly with about 250 μg of 6-hydroxydopamine in an ascorbic acid solution while the other half were injected with only the ascorbic acid solution (0.1%). Two weeks later they were implanted with push-pull cannulas according to methods previously described (54). When perfused, samples were collected every 4 min instead of every 2 min. Aliquots of samples 4 (before experimental manipulation), 7 and 10 (12 and 20 min after stimuli presentation, respectively) were spotted on cellulose

thin layer chromatography (TLC) plates and co-chromatographed with authentic norepinephrine and normetanephrine.

In the first experiment (I), the animals' brains were pulse labeled with d,l-[^{14}C]norepinephrine 1 hr prior to perfusion and their lateral ventricles were perfused prior to conditioning of the visual stimulus-unconditioned shock (CS-US) pairings. The purpose of this portion of the study was to determine if a) injection with 6-hydroxydopamine altered the disposition of total ^{14}C (from [^{14}C]norepinephrine) or the ratio of unchanged norepinephrine to normetanephrine in perfusates generally, and b) presentation of a neutral visual stimulus would cause alterations in the parameters described in a). The second experiment (II) examined the effects of the US (shock) on those parameters examined in the first experiment. Be-

TABLE 3. Effect of intraperitoneal administration of drugs on percentage of [^{3}H] norepinephrine and [^{3}H] metabolites of total radioactivity in the perfusate

Treatment	Sample no.	Mean percent-total radioactivity			R = DOM/ (NE + NM)
		% NE	% NM	% DOM	
NaCl, 0.5 ml	4	28.2 ± 2.4	9.7 ± 2.8	63.9 ± 1.6	1.78 ± 0.18
	8	32.8 ± 5.2	8.8 ± 0.6	58.4 ± 5.8	1.52 ± 0.40
	15	29.4 ± 2.2	8.8 ± 1.0	62.0 ± 2.0	1.65 ± 0.15
d-Amphetamine, 5.0 mg/kg	4	26.9 ± 0.7	8.2 ± 2.0	67.4 ± 2.9	2.07 ± 0.08
	8	55.4 ± 3.9[a]	21.9 ± 2.4[a]	22.8 ± 2.7[a]	0.30 ± 0.05[a]
	15	33.5 ± 5.1	38.3 ± 10.9[a]	28.2 ± 9.4[a]	0.45 ± 0.21[a]
Mescaline, 20 mg/kg	4	28.8 ± 0.9	5.5 ± 1.5	65.8 ± 1.0	1.93 ± 0.09
	8	29.6 ± 2.3	5.8 ± 1.2	64.6 ± 2.6	1.86 ± 0.22
	15	30.1 ± 0.8	5.6 ± 0.7	64.3 ± 1.8	1.81 ± 0.10
d-LSD, 0.4 mg/kg	4	33.6 ± 1.8	9.0 ± 0.6	60.1 ± 2.6	1.53 ± 0.17
	8	32.5 ± 1.1	4.0 ± 1.2	63.6 ± 2.3	1.76 ± 0.16
	15	30.1 ± 0.5	5.7 ± 1.5	63.4 ± 1.3	1.74 ± 0.09

Samples of perfusate before and after injection of psychoactive substances were analyzed by TLC for percentage of [^{3}H] norepinephrine (NE) and [^{3}H] normetanephrine (NM) of total counts per minute on the TLC plate. All other counts on the plate are [^{3}H]DOM. DOM = deaminated-O-methylated metabolites. Results are mean ± SEM of percentage of total radioactivity and each value is the mean of the same three subjects. [a] Significantly different from predrug value of same treatment, matched pair t test, $P < 0.05$.

TABLE 4. Effect of drugs administered intraperitoneally on percent 5-[¹⁴C]HT and 5-[¹⁴C]HIAA of total radioactivity in the perfusate

Treatment	Sample no.	Mean percent-total radioactivity		
		% 5-HT	% 5-HIAA	% other
NaCl, 0.5 ml	4	62.3 ± 3.3	3.9 ± 0.4	33.8 ± 3.6
	8	59.9 ± 6.7	1.6 ± 0.1	48.7 ± 6.5
	15	61.5 ± 1.3	1.2 ± 0.1	37.3 ± 1.2
d-Amphetamine, 5.0 mg/kg	4	57.2 ± 3.2	3.9 ± 0.1	38.9 ± 3.2
	8	63.0 ± 3.6	2.3 ± 0.5	34.7 ± 6.3
	15	60.9 ± 5.3	2.8 ± 0.7	36.3 ± 4.6
Mescaline, 20 mg/kg	4	62.6 ± 1.5	5.4 ± 1.8	31.9 ± 0.3
	8	35.5 ± 0.4ᵃ	8.5 ± 0.9ᵃ	55.4 ± 3.5ᵃ
	15	30.5 ± 3.7ᵃ	16.4 ± 2.6ᵃ	53.0 ± 5.7ᵃ
d-LSD, 0.4 mg/kg	4	54.8 ± 7.2	3.1 ± 0.5	43.9 ± 6.9
	8	55.9 ± 3.0	2.8 ± 0.3	41.4 ± 2.7
	15	55.0 ± 8.9	2.9 ± 0.8	42.3 ± 7.8

Samples of perfusate before and after i.p. injection of psychoactive substances were analyzed by TLC for percentage of 5-[¹⁴C]HT or 5-[¹⁴C]HIAA. Results are mean ± SEM of percentage of total radioactivity and each value is the mean of the same three subjects.   ᵃ Significantly different from preinjection level of same treatment, matched pair t test, $P < 0.05$.

tween the second and third experiments the rats were given 10 escape-avoidance trials a day for 5 days with the platform available. By the third day all rats in both groups were making an avoidance response during the 40 seconds of CS presentation.

The third and fourth experiments (III, IV) were essentially identical to the first experiment except that only the CS was presented (without shock or the opportunity to make the avoidance response by jumping onto the platform). In addition, the fourth experiment was carried out after the avoidance response to the CS was extinguished by forcing the rats to experience the presentation of the CS without its being paired with the US. Ten daily sessions were utilized for the extinction procedure. Perfusions were not carried out during training or extinction sessions.

This dose of 6-hydroxydopamine did not affect the rate at which conditioned avoidance behavior was acquired. By the end of the second

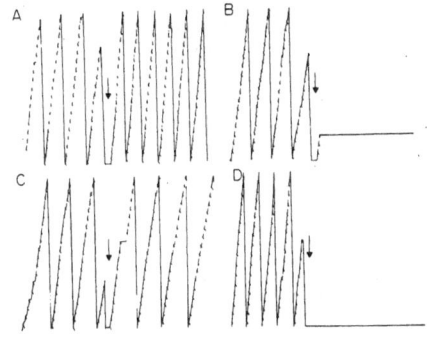

Figure 7. Sample cumulative records showing tolerance to the behavioral effects of 2.5 mg (+) amphetamine SO₄/kg (i.p.) and lack of cross tolerance to mescaline HCl (15 mg/kg). Intraperitoneal isotonic NaCl did not affect fixed ratio (FR) 30 responding (A), while (+) amphetamine initially produced abrupt cessation of responding (B), followed 8–12 days later by tolerance to the daily injections of (+) amphetamine (C). Mescaline, the following day, produced behavioral disruption (D). Arrows (and recorder reset) indicate the point of injection and the event marks are a 1 min time base. (Reprinted with permission (53.)

TABLE 5. Effect of tolerance to the behavioral suppressant action of d-amphetamine and attempted cross-tolerance with mescaline on the appearance of radioactivity in the perfusate

Treatment	% Radioactivity relative to sample #4[a]	
	[³H] norepinephrine (and metabolites)	5-[¹⁴C]HT (and metabolites)
NaCl	42.2 ± 2.6	45.6 ± 9.7
d-Amphetamine (2.5 mg/kg)		
Pre-tolerance	69.5 ± 2.1[b]	49.6 ± 8.4
Tolerance	64.2 ± 4.3[b]	70.1 ± 10.4[b]
Mescaline (15 mg/kg)		
Prior to d-amphetamine tolerance	51.4 ± 3.9	65.7 ± 6.5[b]
After tolerance to d-amphetamine	54.0 ± 3.1	83.1 ± 3.8[c]

[a] Results are expressed as mean ± SEM of three subjects in subsequent samples related to sample #4, taken as 100%.   [b] P < 0.05, matched pair t-test.   [c] P < 0.01, matched pair t-test.

daily session of 10 trials, both groups of rats were avoiding on the average of 90%, remaining above 90% avoidance responding on the subsequent three sessions. Likewise, the profiles of total ¹⁴C-radioactivity in 4 min perfusion samples were very similar for both groups during experiments *I* and *II*. For both groups, radioactivity at the sites on the TLC plates corresponding to authentic norepinephrine and normetanephrine showed a distribution of 24% and 8% of the total counts on the

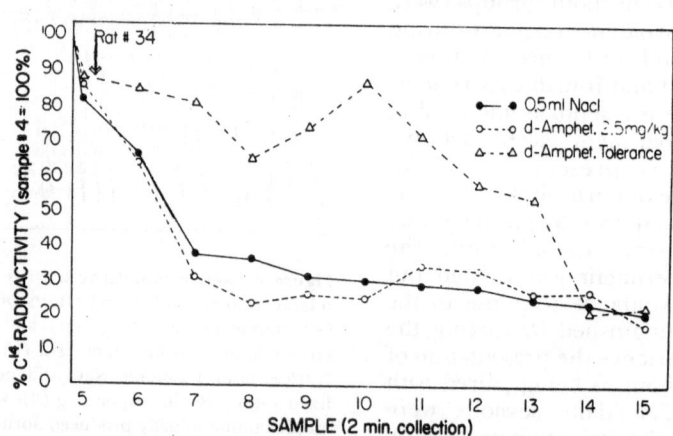

Figure 8. Releasability of 5-[¹⁴C]HT (and metabolites) by (+) amphetamine after tolerance. Before tolerance to (+) amphetamine the profile of 5-[¹⁴C]HT is identical to that after an injeciton of NaCl and decreases in a multiexponential fashion. The arrow denotes the point of injections and radioactivity, expressed as percent of radioactivity in sample #4, derived by counting 10 μl aliquots of 2 min samples (20 μl/min perfusion rate). (Reprinted with permission (53).)

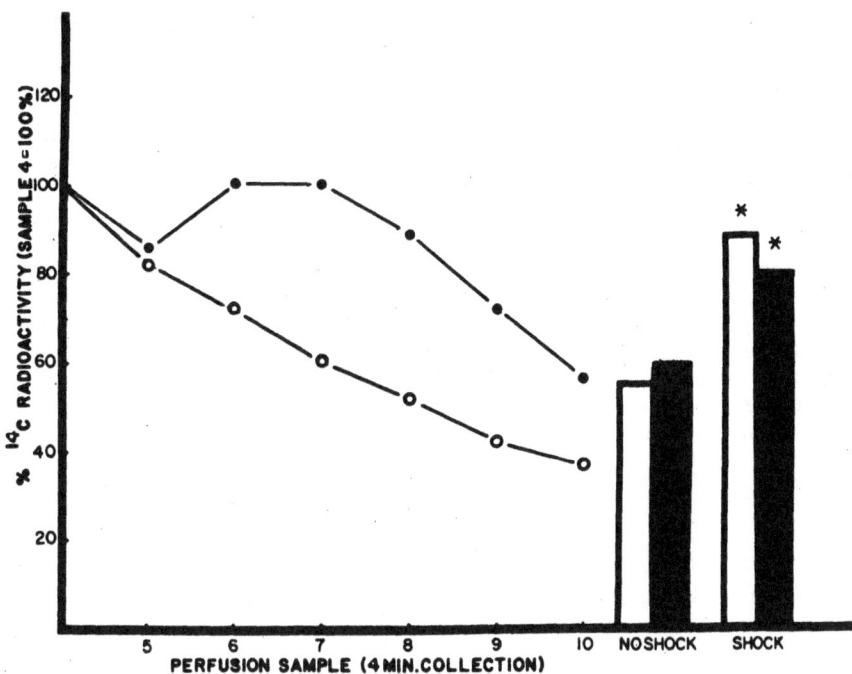

Figure 9. Distribution of [^{14}C]norepinephrine (and metabolites) in 10 μl aliquots of perfusion samples after presentation of a neutral visual stimulus for 60 sec (o —— o) in experiment I or a neutral visual stimulus for 60 sec, the last 20 sec of which were paired with a 2 mA shock to rats' feet (● —— ●) in experiment II. The stimuli presentation was between samples 4 and 5. Both ascorbic acid vehicle and 6-hydroxydopamine groups are combined (N = N = 4). The open histograms represent the control group's average and the cross hatched histograms represent the 6-hydroxydopamine group's average radioactivity in samples after presentation of stimuli, relative to sample 4. Shock presentation resulted in a statistically reliable increase in the average amount of [^{14}C] in samples 5–10 relative to sample 4 (P <0.05, correlated 2-tailed t-test). Analyzed separately, both groups likewise showed significantly elevated [^{14}C] activity in perfusates after shock presentation, *P < 0.05, correlated 2-tailed t-test. (Tilson, Rech and Sparber, unpublished observations.)

entire plate for sample number 4. The 6-hydroxydopamine-treated group tended to decrease in percent of norepinephrine and normetanephrine a little more rapidly by sample 10 than the vehicle controls (Fig. 10). Presentation of the unconditioned shock stimulus (US) produced a significant increase in ^{14}C radioactivity (Fig. 9), which was also seen on the TLC plates (Fig. 10).

However, although both groups showed a significant increase in ^{14}C at the norepinephrine spot, the increase in the percentage of radioactivity at this spot was significantly greater for controls than the 6-hydroxydopamine-treated group. To further support this suggestion are the data from experiment *III*, in which just the CS (visual stimulus) was presented under conditions in which avoidance was not possible. Figure 11 shows the significant increase in ^{14}C in samples 5–10, relative to sample 4, only in the control group, the experimental group's profile being very similar to that of experiment *I*. In addition, analysis of samples 4, 7 and 10 by TLC separation indicate significantly more norepinephrine and normetanephrine in the control group perfusate as a consequence of exposure to the CS compared with the 6-hydroxydopamine group (Fig. 10). Extinction sessions produced a washout curve of total radioactivity that paralleled that of experiment *I* for both groups.

These data indicate that periventricular limbic structures may be responsible for release of catecholaminergic transmitters in the absence of overt motor movement. The inability to show differences in avoidance behavior (after the 6-hydroxydopamine treatment regimen used) with this paradigm is not surprising, since many reports have appeared that indicate the relative resistance of conditioned and unconditioned behavior to severe alterations in catecholamine levels with the use of

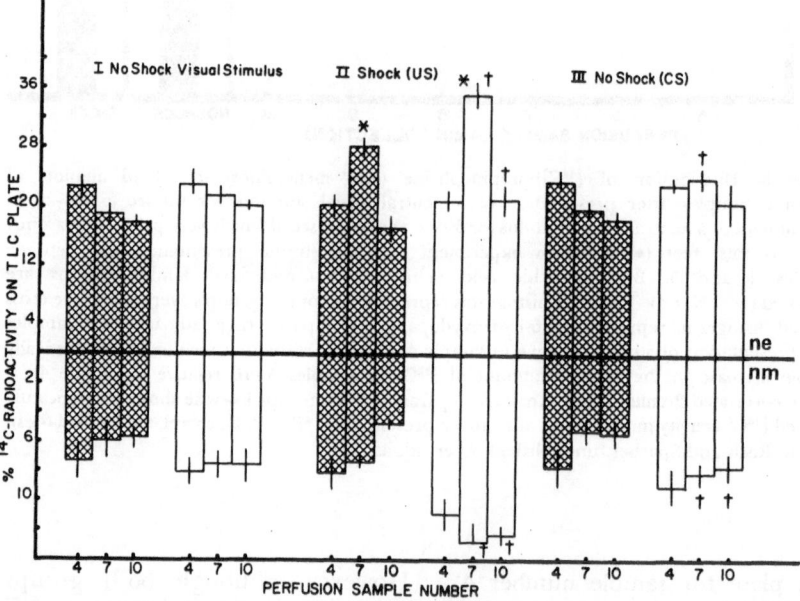

Figure 10. Radioactivity histograms for norepinephrine (ne) and normetanephrine (nm) in samples 4 (prior to presentation of neutral visual stimulus in experiment I, both neutral stimulus and shock, US, in experiment II, and the conditioned visual stimulus alone, CS, in experiment III), 7 (12 min after stimuli presentation) and 10 (20 min after stimuli presentation). Data are expressed as a % of [^{14}C] at spots where authentic ne and nm cochromatographed on the cellulose plates relative to the total amount of [^{14}C] on the plate. The * indicates a significant difference between that sample and its corresponding control (sample 4), $P < 0.05$ correlated 2-tailed t-test. The † indicates a significant difference between that sample and identical sample in the experimental group, $P < 0.05$, 2-tailed t-test. Vertical lines represent ±1 SEM for groups of 4 rats each. (Tilson, Rech and Sparber, unpublished observations.)

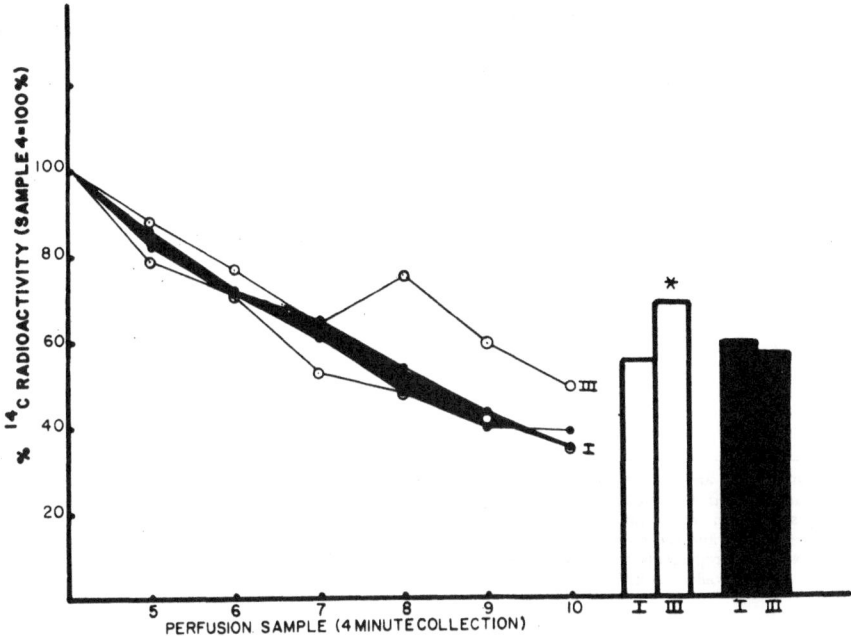

Figure 11. Effect of presentation of the CS (visual stimulus only), after avoidance learning had occurred, on the total radioactivity in each sample relative to sample 4 for the control (open area) and 6-hydroxydopamine group (cross-hatched area). In each case, the control group showed greater amounts of [^{14}C]norepinephrine (and metabolites) subsequent to presentation of the CS compared to experiment I, when the same visual stimulus was neutral and had no consequences. The average amount of radioactivity in samples collected after presentation of the CS is significantly greater only for the control group when compared to radioactivity collected in identical samples during experiment I. * $P < 0.05$ correlated t-test, N = N = 4. (Tilson, Rech and Sparber, unpublished observations.)

this compound (31, 49). It appears, however, that in these cases a more plausible explanation is that the behavioral measures were too insensitive to pick up the chronic effects of 6-hydroxydopamine (14, 15) as is also likely with the experimental paradigm employed in our study. Interestingly, Peterson and Sparber (unpublished observations) have found a significant increase in fixed ratio responding subsequent to 6-hydroxydopamine in which brain norepinephrine was significantly depleted while dopamine levels were hardly affected. Additionally, daily extinction sessions, in which positive reinforcement was completely withheld, resulted in a significant reduction in nonreinforced responding on the part of 6-hydroxydopamine treated (norepinephrine depleted) rats (Table 6). If one interprets these data as indicating enhanced learning of the consequences of barpressing in the absence of reinforcement, they are in agreement with those of Cooper et al. (14) who also showed enhancement of shuttle-box avoidance acquisition after preferential norepinephrine depletion with 6-hydroxydopamine. Regardless of

TABLE 6. Effects of intraventricular 6-hydroxydopamine (6-OHDA) on
responding during extinction

Time after treatment, (days)		Extinction behavior[a]		
		Complete suppression		Nonreinforced responding (5 min/day)
		(minutes)	(responses)	
12	Vehicle	19 ± 2^{b}	511 ± 125	
	6-OHDA	20 ± 2	561 ± 82	
13–17	Vehicle			607 ± 130
	6-OHDA			415 ± 57
107	Vehicle	28 ± 5	671 ± 141	
	6-OHDA	30 ± 7	637 ± 126	
108–112	Vehicle			620 ± 58
	6-OHDA			427 ± 58^{c}

[a] On days 12 and 107 animals were put in extinction after receiving 10 reinforcers (300 responses on fixed ratio-30) and the time and response output were measured to the point where responding was completely suppressed for 4 min. On 5 subsequent days (13–17 and 108–112) animals were placed into the operant chamber and the number of nonreinforced responses were measured during a 5 min period. The value presented represents the sum of each block of 5 sessions. [b] Mean ± SE for 9 animals in each group. [c] Two tailed t-test, $P < 0.05$.

the interpretation of the altered extinction performance by the 6-hydroxydopamine animals performing the operant, the data nevertheless support our contention that stimuli that have come to maintain conditioned behavior are in one way or another interacting with central catecholaminergic function (and/or other transmitters) that, in turn, may influence the degree to which those stimuli initiate or maintain that same behavior.

In ending, I might say that the question of whether or not useful information regarding the release or metabolism of transmitter substances with the use of technologies similar to that described here was really answered about 10 years ago when McLennan (36) showed release of endogenous acetylcholine and dopamine with perfusion technics and, more recently, by the demonstration by McKenzie and Szerb (34) that endogenous dopamine could be found in push-pull perfusate from striatal tissue.

Furthermore, variations of these technics in which homeostatic CNS mechanisms are used as a bioassay for active substances in perfusate (40) can be extended to the degree of elegance recently demonstrated in a report by Yaksh and Myers (61). Perfusates of monkeys' hypothalamuses taken during periods of food deprivation (hunger) were able to initiate fixed ratio responding after satiation, in the same monkeys during reperfusion, indicating a biologically active substance that is catecholamine-like, in that similar behavior could be reproduced by injection of norepinephrine into those same sites. It remains to be determined if the biologically active substance is indeed norepinephrine, as suggested by their bioassay, using blood pressure responses in a pithed rat. In any event, their data are extremely attractive regarding the probability that norepinephrine is the active substance.

Some of the many problems we have encountered while doing the perfusion work include the inability, so far, of maintaining a sustained level of newly synthesized catecholamines in perfusate after pulse labeling with tyrosine or l-dopa. This has been due in part to the physical-chemical and ^{3}H-exchange problems we have encountered when using high specific activity compounds. We have been able however, to separate appropriate metabolites when dopamine has been injected intraventricularly. There are virtually no norepinephrine metabolites to speak of in the perfusates. With time, unchanged dopamine diminishes and homovanillic and dihydroxyphenylacetic acid as well as 3-methoxytyramine tend to increase (Peterson and Sparber, unpublished observations). With better separation technics we have been able to pick up relatively increasing amounts of dihydroxyphenylglycol after pulse labeling with norepinephrine. It appears, therefore, that much more information might be forthcoming from ventricular perfusions in the rat, regarding drug-behavioral-neurochemical interactions, with this approach. Like Myers and co-workers, we have had success in perfusing solid tissue during performance of an operant and are currently examining the relationship between perfusate constituents from caudate and bilateral motor activity after unilateral nigroneostriatal lesioning. The utilization of a right-left lever operant discrimination schedule should enable us to more precisely study presumptive excitatory and inhibitory transmitter disposition and their relationship to normal and abnormal behavior.

REFERENCES

1. AGHAJANIAN, G. K., AND F. E. BLOOM. Electronmicroscopic localization of tritiated norepinephrine in rat brain: Effect of drugs. *J. Pharmacol. Exp. Ther.* 156: 407–416, 1967.

2. AGHAJANIAN, G. K., W. E. FOOTE AND M. H. SHEARD. Action of psychotogenic drugs on single midbrain raphe neurons. *J. Pharmacol. Exp. Ther.* 171: 178–187, 1970.

3. AGHAJANIAN, G. K., W. E. FOOTE AND M. H. SHEARD. Lysergic acid diethylamide: Sensitive neuronal units in the midbrain raphe. *Science* 161: 706–708, 1968.

4. ANDÉN, N.-E., H. CORRODI, K. FUXE AND T. HOKFELT. Evidence for a central 5-hydroxytryptamine receptor stimulation by lysergic acid diethylamide. *Br. J. Pharmacol.* 34: 1–7, 1968.

5. ANGRIST, B. M., B. SHOPSIN AND S. GERSHON. Comparative psychotomimetic effects of stereoisomers of amphetamine. *Nature (London)* 234: 152–153, 1971.

6. APPEL, J. B., AND D. X. FREEDMAN. Tolerance and cross-tolerance among psychotomimetic drugs. *Psychopharmacologia* 13: 267–274, 1968.

7. BLISS, E. L., J. AILION AND J. ZWANZIGER. Metabolism of norepinephrine, serotonin and dopamine in rat brain with stress. *J. Pharmacol. Exp. Ther.* 164: 122–134, 1968.

8. BLISS, E. L., AND J. ZWANZIGER. Brain amines and emotional stress. *J. Psychiatr. Res.* 4: 189–198, 1966.

9. BONDAREFF, W., A. ROUTTENBERG, R. NAROTSKY AND D. G. MCLONE. Intrastriatal spreading of biogenic amines. *Exp. Neurol.* 28: 213–229, 1970.

10. BRODIE, B. B., AND E. COSTA. Some current views on brain monoamines. *Psychopharmacol. Serv. Cent. Bull.* 2: 1–25, 1962.

11. CARLSSON, A., E. ROSENGREN, A. BERTLER AND J. NILSSON. Effect of reserpine on the metabolism of catecholamines. In: *Psychotropic Drugs*, edited by S. Garattini and V. Ghetti. Amsterdam: Elsevier, 1957, p. 363–372.

12. CARR, L., AND K. MOORE. Effects of amphetamine on the contents of norepinephrine and its metabolites in the effluent of perfused cerebral ventricles of the cat. *Biochem. Pharmacol.* 19: 2361–2374, 1970.

13. CHASE, T. N., AND I. J. KOPIN. Stimulus-induced release of substances from olfactory bulb using the push-pull cannula. *Nature* 217: 466–467, 1968.

14. COOPER, B. R., G. R. BREESE, J. L. HOWARD AND L. D. GRANT. Effect of central catecholamine alterations by 6-hydroxydopamine on shuttle box avoidance acquisition. *Physiol. Behav.* 9: 727–731, 1972.

15. COOPER, B. R., G. R. BREESE, J. L. HOWARD AND L. D. GRANT. Enhanced behavioral depressant effects of reserpine and α-methyltyrosine after 6-hydroxydopamine treatment. *Psychopharmacologia* 27: 99–110, 1972.

16. ELLINWOOD, E. H., JR. Amphetamine psychosis: I. Description of the individuals and process. *J. Nerv. Ment. Dis.* 144: 273–283, 1967.

17. ELLINWOOD, E. H., JR. Amphetamine psychosis: II. Theoretical implications. *Int. J. Neuropsychiatry* 4: 45–54, 1968.

18. FERSTER, C. B., AND B. F. SKINNER. In: *Schedules of Reinforcement.* New York: Appleton-Century-Crofts, 1957.

19. FREEDMAN, D. X. Effects of LSD-25 on brain serotonin. *J. Pharmacol. Exp. Ther.* 134: 160–166, 1961.

20. FREEDMAN, D. X., R. GOTTLIEB AND R. LOVELL. Psychotomimetic drugs and brain 5-hydroxytryptamine metabolism. *Biochem. Pharmacol.* 19: 1181–1188, 1970.

21. FUXE, K., AND L. C. F. HANSON. Central catecholamine neurons and conditioned avoidance behavior. *Psychopharmacologia* 11: 439–447, 1967.

22. FUXE, K., AND U. UNGERSTEDT. Histochemical studies on the effect of (+)-amphetamine, drugs of the imipramine group and tryptamine on central catecholamine and 5-hydroxytryptamine neurons after intraventricular injection of catecholamines and 5-hydroxytryptamine. *Eur. J. Pharmacol.* 4: 135–144, 1968.

23. GADDUM, J. Push-pull cannulae. *J. Physiol., London* 155: 1P–2P, 1961.

24. GIARMAN, N. J., AND D. X. FREEDMAN. Biochemical aspects of the actions of psychotomimetic drugs. *Pharmacol. Rev.* 17: 1–25, 1965.

25. GIESE, J., E. RUTHER AND N. MATUSSEK. Quantitative estimation of H³-norepinephrine and its metabolites by thin layer chromatography. *Life Sci.* 6: 1975–1982, 1967.

26. GLOWINSKI, J., AND L. IVERSON. Regional studies of catecholamines in the rat brain: III. Subcellular distribution of endogenous and exogenous catecholamines in various brain regions. *Biochem. Pharmacol.* 15: 977–987, 1966.

27. GLOWINSKI, J., I. KOPIN AND J. AXELROD. Metabolism of H³-norepinephrine in the rat brain. *J. Neurochem.* 12: 25–30, 1965.

28. HOLLOWAY, J. A. Release of norepinephrine and serotonin from the amygdala during rewarding median forebrain bundle stimulation. Paper presented at the Second Annual Meeting of the Society for Neuroscience, October 8–11, 1972, Houston, Texas.

29. IZQUIERDO, I., AND J. A. IZQUIERDO. Effects of drugs on deep brain centers. *Annu. Rev. Pharmacol.* 11: 189–209, 1971.

30. KELLEHER, R. T., AND W. H. MORSE. Determinants of the specificity of behavioral effects of drugs. *Ergeb. Physiol. Biol. Chem. Exp. Pharmakol.* 60: 1–56, 1968.

31. LAVERTY, R., AND K. M. TAYLOR. Effects of intraventricular 2,4,5-trihydroxyphenylethylamine (6-hydroxydopamine) on rat behavior and brain catecholamine metabolism. *Br. J. Pharmacol.* 40: 836–846, 1970.

32. LEONARD, B. E., AND S. R. TONGE. The effects of some hallucinogenic drugs upon the metabolism of noradrenaline. *Life Sci.* 8: 815–826, 1969.

33. LIN, R. C., S. H. NGAI AND E. COSTA. Lysergic acid diethylamide: Role in conversion of plasma tryptophan to brain serotonin (5-hydroxytryptamine). *Science* 166: 237–239, 1969.

34. MCKENZIE, G. M., AND J. C. SZERB. The effect of dihydroxyphenylalanine, pheniprazine and dextroamphetamine on the in vivo release of dopamine from the caudate nucleus. *J Pharmacol. Exp. Ther.* 162: 302–308, 1968.

35. MCLEAN, J. R., AND M. MCCARTNEY. Effect of d-amphetamine on rat brain noradrenaline and serotonin. *Proc. Soc. Exp. Biol. Med.* 107: 77–79, 1961.

36. MCLENNAN, H., The release of acetylcholine and of 3-hydroxytyramine from the caudate nucleus. *J. Physiol., London* 174: 152–161, 1964.

37. MOIR, A. T. B., G. W. ASHCROFT, T. B. B. CRAWFORD, D. ECCLESTON AND H. C. GULDBERG. Cerebral metabolites in cerebrospinal fluid as a biochemical approach to the brain. *Brain* 93: 357–368, 1970.

38. MYERS, R. D. Chemical mechanisms in the hypothalamus mediating eating and drinking in the monkey. *Ann. N. Y. Acad. Sci.* 157: 918–933, 1969.

39. MYERS, R. D., editor. *Methods in Psychobiology.* London and New York: Academic 1972, vol. 2.

40. MYERS, R. D. Transfusion of cerebrospinal fluid and tissue bound chemical factors between the brains of conscious monkeys: A new neurobiological assay. *Physiol. Behav.* 2: 373–377, 1967.

41. PAASONEN, M. K., AND M. VOGT. The effects of drugs on the amounts of substance P and 5-hydroxytryptamine

in mammalian brain. *J. Physiol., London* 131: 617–626, 1956.

42. PETERSON, D. W., AND S. B. SPARBER. Increased fixed ratio rates following norepinephrine depletion by 6-hydroxydopamine (6HDA). *Federation Proc.* 32: 753, 1973.

43. PORTIG, P. G., AND M. VOGT. Release into the cerebral ventricles of substances with possible transmitter function in the caudate nucleus. *J. Physiol., London* 204: 687–715, 1969.

44. RECH, R. H., L. A. CARR AND K. E. MOORE. Behavioral effects of α-methyltyrosine after prior depletion of brain catecholamines. *J. Pharmacol. Exp. Ther.* 160: 326–335, 1968.

45. ROSECRANS, J. A., R. A. LOVELL AND D. X. FREEDMAN. Effects of lysergic acid diethylamide on the metabolism of brain 5-hydroxytryptamine. *Biochem. Pharmacol.* 16: 2011–2021, 1967.

46. SCHILDKRAUT, J. J., AND S. S. KETY. Biogenic amines and emotions. *Science* 156: 21–30, 1967.

47. SCHOENFELD, R. I., AND L. S. SEIDEN. Effect of α-methyltyrosine on operant behavior and brain catecholamine levels. *J. Pharmacol. Exp. Ther.* 167: 319–327, 1969.

48. SCHOENFELD, R. I., AND N. J. URETSKY. Operant behavior and catecholamine-containing neurons: Prolonged increase in lever-pressing after 6-hydroxydopamine. *Eur. J. Pharmacol.* 20: 357–362, 1972.

49. SCHOENFELD, R. I., AND M. J. ZIGMOND. Effect of 6-hydroxydopamine (HDA) on fixed ratio (FR) performance. *Pharmacologist* 12: 227, 1970.

50. SEIDEN, L. S., AND A. CARLSSON. Brain and heart catecholamine levels after *l*-dopa administration in reserpine treated mice: Correlations with a conditioned avoidance response. *Psychopharmacologia* 5: 178–181, 1964.

51. SEIDEN, L. S., AND A. CARLSSON. Temporary and partial antagonism by *l*-dopa of reserpine-induced suppression of a conditioned avoidance response. *Psychopharmacologia* 4: 418–423, 1963.

52. SPARBER,.S. B., AND H. A. TILSON. Environmental influences upon drug-induced suppression of operant behavior. *J. Pharmacol. Exp. Ther.* 179: 1–9, 1971.

53. SPARBER, S. B., AND H. A. TILSON. The releasability of central norepinephrine and serotonin by peripherally administered *d*-amphetamine before and after tolerance. *Life Sci.* 11: 1059–1067, 1972.

54. SPARBER, S. B., AND H. A. TILSON. Schedule controlled and drug induced release of norepinephrine-7-³H into the lateral ventricle of rats. *Neuropharmacology* 11: 453–464, 1972.

55. STEIN, L., AND C. WISE. Release of norepinephrine from hypothalamus and amygdala during rewarding medial forebrain bundle stimulation and amphetamine. *J. Comp. Physiol. Psychol.* 67: 189–198, 1969.

56. SULSER, F., M. L. OWENS, S. J. STRADA AND J. V. DINGELL. Modification by desimipramine (DMI) of the availability of norepinephrine released by reserpine in the hypothalamus of the rat in vivo. *J. Pharmacol. Exp. Ther.* 168: 272–282, 1969.

57. TILSON, H. A., AND S. B. SPARBER. On the use of the push-pull cannula as a means of measuring biochemical changes during ongoing behavior. *Behav. Res. Methods Instrum.* 2: 131–134, 1970.

58. TILSON, H. A., AND S. B. SPARBER. Studies on the concurrent behavioral and neurochemical effects of psychoactive drugs using the push-pull cannula. *J. Pharmacol. Exp. Ther.* 18: 387–398, 1972.

59. TONGE, S. R., AND B. E. LEONARD. The effect of some hallucinogenic drugs on the aminoacid precursors of brain monoamines. *Life Sci.* 9: 1327–1335, 1970.

60. WINSON, J., AND J. L. GERLACH. Stressor-induced release of substances from the rat amygdala detected by the push-pull cannula. *Nature London New Biol.* 230: 251–253, 1971.

61. YAKSH, T. L., AND R. D. MYERS. Neurohumoral substances released from hypothalamus of the monkey during hunger and satiety. *Am. J. Physiol.* 222: 503–515, 1972.

Serotonergic and cholinergic mechanisms during disruption of approach and avoidance behavior[1]

M. H. APRISON, J. N. HINGTGEN, AND W. J. McBRIDE

Section of Neurobiology, The Institute of Psychiatric Research and Departments of Psychiatry and Biochemistry Indiana University Medical Center, Indianapolis, Indiana 46202

ABSTRACT

Injections of D,L-5-hydroxytryptophan (D,L-5-HTP) into pigeons and rats' working on approach schedules produce a period of behavioral depression that is temporally correlated to *increased* levels of total serotonin (5-HT) in the telencephalon and diencephalon. Administration of α-methyl-*meta*-tyrosine (α-MMT) also results in depressed responding; however, the temporal correlation is with *decreased* levels of total 5-HT in brain. Our hypothesis to explain these two apparent opposite biochemical states which result in similar behavioral disruptions is that in both cases more 5-HT is released within certain key serotonergic synapses mediating this behavior. Evidence from subcellular studies supports this concept. Not only are the levels of 5-HT significantly higher in preparations of nerve endings isolated from the telencephalon and diencephalon of pigeons given injections of D,L-5-HTP, but in vitro studies also show that low concentrations of L-5-HTP significantly *increased* the release of radioactive 5-HT from serotonergic nerve endings. On the other hand, L-5-HTP in much higher concentrations had no effect on the release of labeled dopamine or norepinephrine. A major metabolite of α-MMT, α-methyl-*meta*-tyramine, also caused a significant *increase* in the release of labeled 5-HT from similar preparations of nerve endings. Whereas serotonin appears to be

---

[1] Supported in part by Public Health Service Research Grant No. MH-03225 from the National Institute of Mental Health.

Abbreviations: try, tryptophan; 5-HTP, 5-hydroxytryptophan; 5-HT, 5-hydroxytryptamine (serotonin); 5-HIAA, 5-hydroxyindoleacetic acid; MAO, monoamine oxidase; dopa, 3,4-dihydroxyphenylalanine; DA, dopamine (3,4-dihydroxyphenylethylamine); NE, norepinephrine; ACh, acetylcholine; AChE, acetylcholinesterase; α-MMT, α-methyl-*meta*-tyrosine; α-MMTA, α-methyl-*meta*-tyramine; TBZ, tetrabenazine; Mult FR 50 FI 10, multiple fixed-ratio 50, fixed-interval 10 minutes.

involved in the disruption of approach behavior, another series of studies have indicated that acetylcholine may play a role in excitation during avoidance behavior. Behavioral excitation observed following administration of tetrabenazine 18 hr after iproniazid pretreatment to rats working on shock-avoidance schedules was temporally correlated with lowered levels of acetylcholine in the telencephalon. Pretreatment with 0.8 mg/kg of atropine blocked excitation whereas one-eighth of this dose increased the duration. Excitation in these rats was shortened by 50% following bilateral septal lesions, which lowered brain acetylcholine levels. Mechanisms to explain these neurochemical correlates of behavior are discussed.—APRISON, M. H., J. N. HINGTGEN, AND W. J. MCBRIDE. Serotonergic and cholinergic mechanisms during disruption of approach and avoidance behavior. *Federation Proc.* 34: 1813–1822, 1975.

If specific drugs or precursors are administered alone or in combination to animals trained to emit stable patterns of learned behavior, it is possible to induce depressed or excited states of responding. One approach to the study of neurochemical correlates of behavior involves the investigation of these abnormally produced states with concomitant measures of behavior and transmitter levels in specific areas of the brain. If higher than normal levels of certain putative central nervous system (CNS) transmitters such as acetylcholine (ACh) and serotonin (5-HT) were released from nerve endings at specific sites within the brain, it could possibly result in abnormal functioning of certain neural networks which in turn could have profound effects on behavior. Since it is possible to experimentally manipulate the animal's behavior with certain injected compounds and to determine the transmitter levels under various behavioral states, we can attempt to correlate changes in the cellular pools or compartments of these agents with changes in behavior. Such studies are important for our understanding of the interrelationships of chemistry and behavior and for a better understanding of how the nervous system functions in both normal and atypical states. To this end, a research program was developed in our labora-

tories in 1959 to investigate relationships between CNS transmitters and behavior (10, 11).

Neurotransmitters have major effects on the behavior of animals by their individual and grouped action in the CNS. This concept serves as a basis for our research strategy in developing new experiments in the field of neurochemical correlates of behavior (10). It is assumed that an important factor in the stable emission of a specific learned response is the change in concentrations of the neurotransmitters within the synaptic cleft which act at the postsynaptic membrane of a group of important synapses involved in key neuronal pathways utilized for the performance of that particular behavior. Our models of the biochemical mechanisms functioning at such specific nerve endings were proposed about 15 years ago. Using these models, it became apparent that one could easily study how abnormal levels of such transmitter candidates can have major effects on behavior by quantitatively measuring both the transmitter levels in different structures of the brain with concomitant changes in the behavior of the organism. In this review, we will present the evidence that we have accumulated indicating that serotonergic neurons may be involved in certain types of depression during food-reinforced approach behavior and cholinergic neurons may be involved in drug-induced excitation of shock avoidance behavior. We will present our models and show a) how each has helped in formulating our experiments and b) how the behavioral disruptions seen in these animals are explained.

TEMPORAL RELATIONSHIPS BETWEEN INCREASED LEVELS OF SEROTONIN AND DEPRESSION OF APPROACH BEHAVIOR

When 5-hydroxytryptophan (5-HTP), the immediate precursor of 5-HT, is administered to pigeons working on a multiple fixed-ratio 50, fixed-interval 10 min (Mult FR 50 FI 10) schedule of food reinforcement, behavioral depression occurs. Intramuscular doses of 25, 50 and 75 mg/kg of D,L-5-HTP have increasingly disruptive effects on response rates during this approach behavior, with 50 mg/kg lowering responding for an average duration of 180 min (3, 4). However, it is possible that the 5-HTP effect observed may be due to a peripheral rather than central action. One way to differentiate between central and peripheral effects of injected 5-HTP would be to utilize an inhibitor of the CNS catabolic enzyme of 5-HT. It was hypothesized that if the depression was due mainly to a central mechanism involving free or physiologically effective 5-HT (11), then iproniazid, which is a much better central than peripheral monoamine oxidase (MAO) inhibitor in vivo, should enhance the behavioral effect of 5-HTP.

A single injection of iproniazid (50 mg/kg) had no effect on the approach behavior of pigeons. Three doses of iproniazid spaced equally over a period of 40 hr followed by a 50 mg/kg dose of D,L-5-HTP resulted in a period of behavioral disruption of approximately 10 hr (compared to 3 hr for 5-HTP alone). Successive 5-HTP administration without further iproniazid treatment yielded a gradual return to the previous period of depression of 3 hr in approximately 30 days after the last injection of iproniazid. In another group of pigeons given the same series of iproniazid injections, brain and liver MAO activity was measured. Normal MAO activity in brain was restored in about 30 days whereas the MAO activity in liver returned to normal in about 12 days. The return of the behavioral effect from the longer periods of depression to the shorter periods seen after 5-HTP alone followed the same time course as the recovery of MAO activity in brain (Fig. 1). In other studies where the dose of 5-HTP was varied and the dose of iproniazid held constant, it was found that at any level of brain MAO activity (during the recovery period), the greatest behavioral effect was obtained at the highest 5-HTP dose injected. Of even more importance, these data clearly showed that at any given dose of 5-HTP, the greatest behavioral effect was obtained at the lowest brain MAO level (4, 11). The results of these experiments are best explained in terms of elevated cerebral 5-HT levels resulting from reduced MAO activity in brain (4, 5).

To verify further the importance of 5-HT in behavioral depression, it was necessary to study the kinetic relationship between the behavioral disruption following 5-HTP and the variation of 5-HT content in brains of pigeons not given an MAO inhibitor. Total 5-HT was measured in four specific brain areas (telencephalon, diencephalon plus optic lobes, cerebellum, and pons-medulla oblongata) as well as in liver, heart, lung, and blood of birds sacrificed at various time intervals during the period of atypical behavior following injection of 50 mg/kg D,L-5-HTP (13). The behavioral depression was temporally related to the three-to fourfold increases in 5-HT content in only the telencephalon and diencephalon (Fig. 2). In 1965, the enzyme that synthesizes 5-HT, 5-HTP-decarboxylase, was thought to be the same as 3,4-dihydroxyphenylalanine (dopa) decarboxylase, the enzyme that synthesizes dopamine, the precursor of norepinephrine. It therefore became imperative to measure dopamine and norepinephrine levels in a similar group of 5-HTP injected pigeons (6). No consistent significant changes were noted in dopamine and norephinephrine contents in

the telencephalon and diencephalon. The levels of norepinephrine and dopamine in the pons-medulla oblongata and cerebellum appeared to change slightly with respect to normal levels; these changes did not appear to be correlated with the observed behavioral depression.

Rats, working on a variable-ratio 40 schedule of food reinforcement, were also tested to determine if 5-HTP had any behavioral effects in animals other than pigeons. Similar temporal correlations were found between the increase in 5-HT in the telencephalon and the period of behavioral disruption. Furthermore, norepinephrine levels were not significantly altered under these conditions (8).

We have repeated the 5-HTP experiments using L-tryptophan. Al-

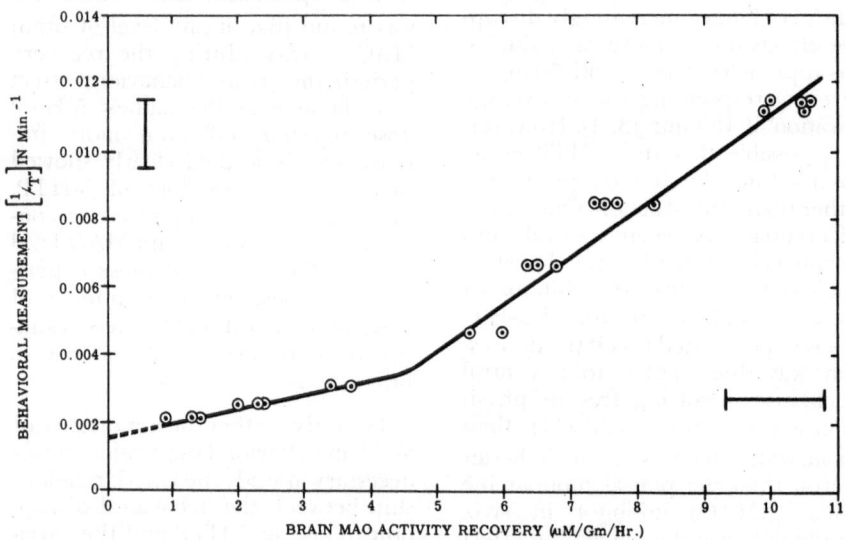

Figure 1. Correlation of the behavioral effect observed after an injection of 50 mg/kg of D,L-5-HTP into pigeons with brain MAO activity present at the time of the injection of 5-HTP. The behavioral measurement was expressed as 1/T, the reciprocal of the time during which the bird's performance was depressed. Brain MAO activity was expressed as μmoles NH_3/g per hr. The range of the behavioral measurement and brain MAO activity are given in the upper left and lower right hand part of the figure, respectively. (From Aprison and Ferster, (4), with permission. Copyright 1961 by Pergamon Press, New York).

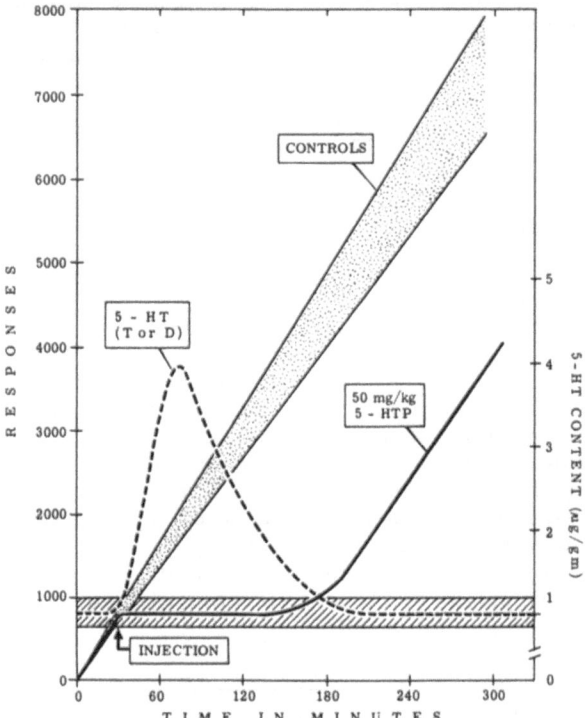

Figure 2. Diagrammatic presentation of the effect of an intramuscular (i.m.) injection of 50 mg/kg D,L-5-HTP on the approach response rate and 5-HT content in the telencephalon (T) or diencephalon plus optic lobes (D) of the pigeon. After the injection of 5-HTP, the 5-HT level increases, passes through a maximum and then decreases to normal; the behavior of the experimental animal was disrupted for a period of time (mean = 154 min) and then returned to normal at approximately the same time as the 5-HT levels return to normal. The speckled band indicates the extreme range of response rates during control sessions (with or without saline injecitons). The band at the bottom represents the range of the 5-HT concentration in the telencephalon from control pigeons. The broken line shows the change in 5-HT concentration in this brain part at different times after the 5-HTP injection. The data from the diencephalon were similar. (From Aprison and Hingtgen (6), with permission. Copyright 1965 by Pergamon Press, New York).

though injections of large doses of L-tryptophan were required, similar behavioral data were obtained. Furthermore, pretreatment with iproniazid produced enhanced behavioral disruptions after L-tryptophan administration (19a).

Additional evidence for a central 5-HT mechanism is given by the fact that anorexia has been ruled out as a cause for the lowered reponse rates following 5-HTP administration. Pigeons and rats were given free access to food during the period of 5-HTP induced depression. Both groups consumed the same amount of food during this period as they did during control sessions. Thus, a simple loss of appetite related to peripheral effects was not demonstrated in these animals (6, 8).

TEMPORAL RELATIONSHIPS BETWEEN DECREASED LEVELS OF SEROTONIN AND DEPRESSION OF APPROACH BEHAVIOR

Since *increases* in brain 5-HT are followed by periods of behavioral depression in our animals, then what behavioral effects would occur if brain 5-HT levels were *decreased*? Using α-methyl-*meta*-tyrosine (α-MMT), an amino acid which causes a temporal differential depletion of 5-HT, norepinephrine, and dopamine levels in brain, Aprison and Hingtgen (7) investigated the relationship between quantitatively measured behavioral changes and changes in the brain

levels of these biogenic amines. The behavioral response rates of pigeons working on a Mult FR 50 FI 10 schedule after an injection of a dose of 100 mg/kg α-MMT were lowered to about 20% of normal after 3 hr and gradually returned to preinjection levels after 9 hr (16). Another group of pigeons was decapitated at various times following α-MMT injection and whole brain tissue was assayed for 5-HT, norepinephrine, dopamine, and the decarboxylation products of α-MMT, namely, α-methyl-*meta*-tyramine (α-MMTA) and metaraminol (aramine). Amine levels varied as follows: *a*) 5-HT was decreased 30% in 3 hr and returned to normal levels at about 9 hr, *b*) norepinephrine was decreased 50% in 12 hr and then returned to normal by 5–7 days, *c*) dopamine decreased and remained below

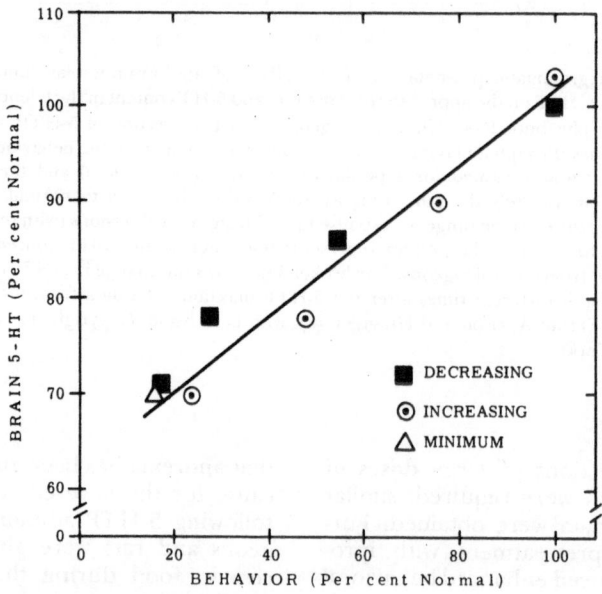

Figure 3. Correlation between changes in the 5-HT content of brain and changes in the response rate of pigeons working on a Mult FR 50 FI 10 approach schedule of reinforcement after an i.m. injection of 100 mg/kg α-MMT.

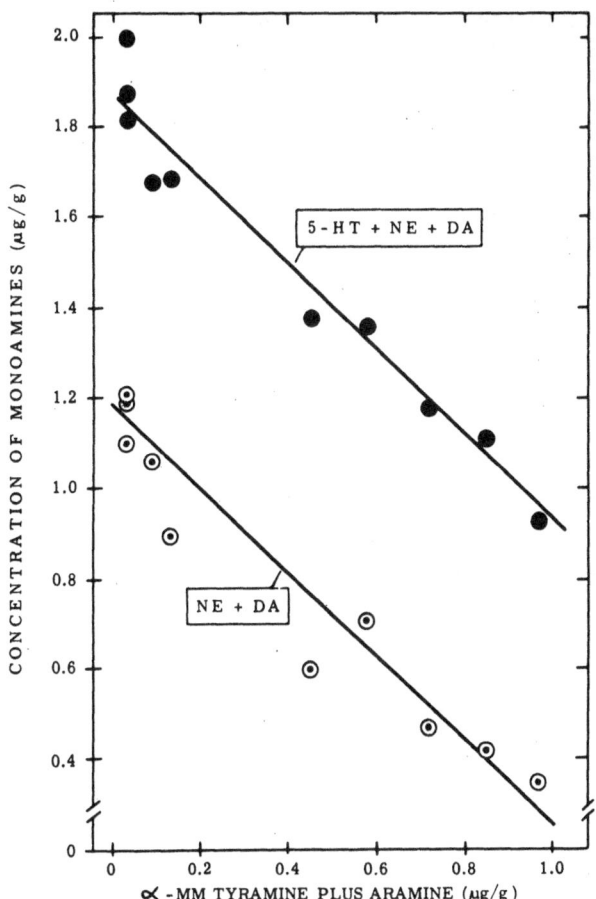

Figure 4. Correlation between the change in levels of α-methyl-*meta*-tyramine plus metaraminol (aramine) and the change in levels of 5-HT plus norepinephrine plus dopamine in whole brain of pigeons following i.m. injections of 100 mg/kg α-MMT.

50% of normal for 12 hr, returning to normal by 48 hr, and *d*) α-methyl-*meta*-tyramines increased and returned to normal by 2 to 4 days. The data indicated that the time course of the behavioral depression paralleled the time course of the changes in 5-HT content but not that of the other amines measured (7). The correlation of the changes in 5-HT content and changes in behavior are seen in Fig. 3. Further-more, it was noted that the increase in α-methyl-*meta*-tyramine and metaraminol was best correlated with the fall in 5-HT plus norepinephrine plus dopamine (Fig. 4). Finally, as in the case of 5-HTP, an anorexic side effect for α-MMT was also ruled out, since pigeons working on the approach schedule as well as naive birds consumed equivalent amounts of food after injections of saline or α-MMT (17).

PROPOSED MECHANISM FOR THE INVOLVEMENT OF SEROTONIN IN THE DEPRESSION OF APPROACH BEHAVIOR AND SOME NEUROCHEMICAL DATA TO SUPPORT THIS HYPOTHESIS

After injection of D,L-5-HTP, the brain tissue levels of 5-HT *increased* whereas after injection of α-MMT, these levels *decreased.* Our hypothesis to explain these two opposite biochemical states which result in the same type of behavioral change in the same animal species is that in both cases there is an increased release of 5-HT from nerve endings, and it is this 5-HT acting on certain "key" synapses which causes the observed depression in the animal's behavior. Recent studies using isolated nerve ending preparations provided evidence consistent with the theory that in vivo both L-5-HTP and α-MMT could cause an increased release of 5-HT.

Preparations of nerve endings (P₂) isolated from the telencephalon and from the diencephalon plus optic lobes of pigeons and from the telencephalon of rats were used to study the effects of 5-HTP on the content of 5-HT in nerve endings as well as the release of radio-

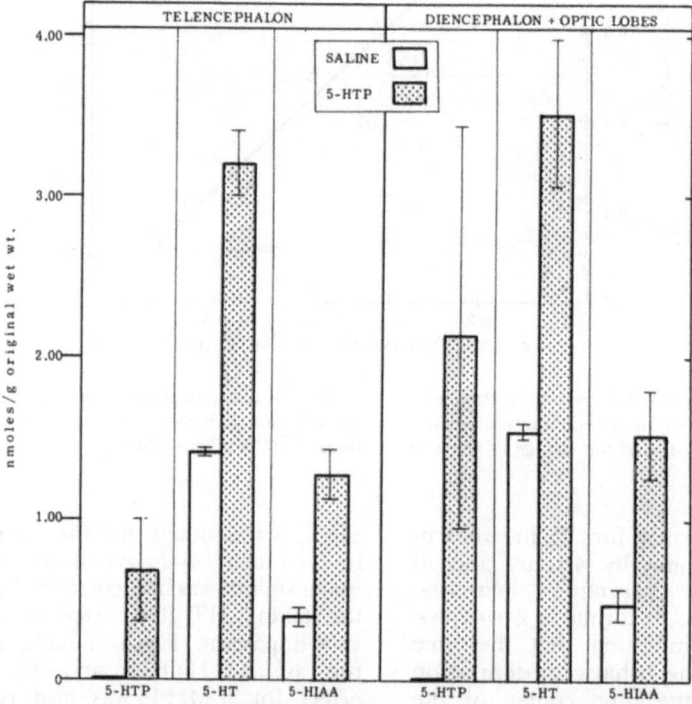

Figure 5. The levels of 5-HTP, 5-HT, and 5-HIAA in a nerve ending fraction (P₂) isolated from the telencephalon and diencephalon plus optic lobes of pigeons killed 60 min after i.m. injections of saline or 50 mg/kg D,L-5-HTP in saline. The data represent the means ± SEM of 4 animals in each group. All values for 5-HTP treated pigeons were significantly higher ($P < 0.05$) than control values. Adapted from McBride et al. (23).

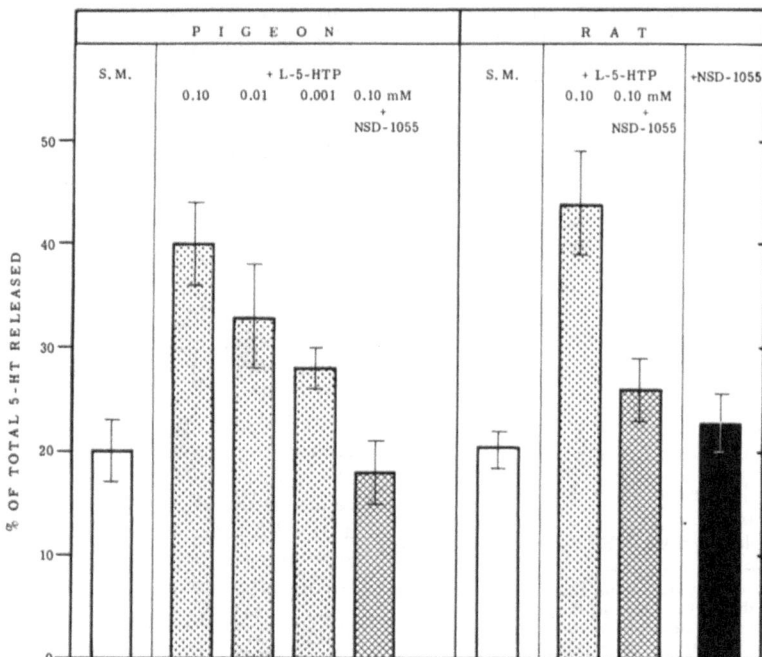

Figure 6. Effects of L-5-HTP on the release of 5-[³H]hydroxytryptamine from preparations of nerve endings (P₂) isolated from the telencephalon of the pigeon and rat. Efflux studies were carried out for 10 min at 37 C in the presence or absence of the above compounds. NSD-1055 (*m*-hydroxy-*p*-bromobenzyl oxyamine) is a decarboxylase inhibitor and was used at a concentration of 0.20 mM. Data represent the means ± SD of at least 4 determinations each. In all cases where L-5-HTP was present without NSD-1055 the amount of 5-[³H]HT released was significantly higher ($P < 0.05$) in comparison to control values. Adapted from McBride et al. (23).

actively labeled 5-HT from nerve endings (23). The levels of 5-HTP, 5-HT, and 5-hydroxyindoleacetic acid (5-HIAA) were significantly higher in the nerve ending fraction isolated from the telencephalon and from the diencephalon plus optic lobes of pigeons given intramuscular injections of 50 mg/kg D,L-5-HTP in comparison to control values (Fig. 5). These data provided direct biochemical evidence to indicate that the injected 5-HTP could accumulate within neurons and significantly raise the levels of 5-HT within the nerve endings.

Using amine concentrations (25 nM) which were supposed to label specifically serotonergic or catecholaminergic terminals, McBride et al. (23) found that L-5-HTP increased the release of 5-[³H]HT from serotonergic nerve terminals isolated from the telencephalon of the pigeon and rat (Fig. 6). Since this effect by L-5-HTP was blocked by NSD-1055, a decarboxylase inhibitor, it would appear that it was not the L-5-HTP itself that caused the release of ·5-[³H]HT but instead it was the newly formed 5-HT (from the added L-5-HTP) that resulted in the release of the labelled amine. Furthermore, L-5-HTP had no apparent effect on catecholaminergic terminals since L-5-HTP, at a concentration of 1.5 mM,

TABLE 1. Effects of L-5-hydroxytryptophan
on the release of [³H]norepinephrine
and [³H]dopamine

Incubation conditions	% of total amine released
Pigeon	
[³H]norepinephrine	
control	23.6 ± 8.1 (4)
plus 1.5 mM L-5-HTP	22.9 ± 8.1 (3)
[³H]dopamine	
control	4.5 ± 3.0 (4)
plus 1.5 mM L-5-HTP	3.7 ± 1.8 (4)
Rat	
[³H]norepinephrine	
control	25.5 ± 9.2 (4)
plus 1.5 mM L-5-HTP	17.7 ± 5.9 (3)
[³H]dopamine	
control	11.4 ± 4.0 (4)
plus 1.5 mM L-5-HTP	14.1 ± 0.7 (4)

The crude synaptosomal fraction (P₂) was isolated from the telencephalon of the rat and pigeon and incubated as described by McBride. Aprison and Hingtgen (23). In these experiments [³H]norepinephrine or [³H]dopamine was used in place of 5-[³H]hydroxytryptamine. Data represent the means ± SD of the number of determinations given in parentheses. (Adapted from McBride et al. (23).)

did not increase the release of [³H]-norepinephrine or [³H]-dopamine from nerve endings prepared from the telencephalon of the rat or the pigeon (Table 1).

Studies similar to those carried out with L-5-HTP were also undertaken to determine the effects of α-MMT on the release of labeled 5-HT from isolated preparations of nerve endings (21, 22). In the initial studies it was determined that α-MMT itself had no apparent effect either on the uptake (22) or efflux (Table 2) of 5-HT. However, since the content of α-methyl-*meta*-tyramines had also increased in pigeon brain after injection of α-MMT (7), resulting in the relationship seen in Fig. 4, and since α-MMTA and metaraminol had been reported to cause a release of norepinephrine in the heart and brain (15, 26, 28, 32), these two amines were tested to determine if they had any effects on the release of 5-[³H]HT from isolated nerve endings. In preparations from either the

TABLE 2. Effects of α-methyl-*meta*-tyrosine, α-methyl-*meta*-tyramine and metaraminol on the release of serotonin from nerve endings isolated from the telencephalon

Incubation conditions	% of total released in 5 min (mean ± SD; no. = 4)	
	5-HT	5-HIAA
Pigeon		
control	1.94 ± 1.84	22.0 ± 0.44
+2.55 mM α-MMT	0	21.0 ± 1.00
control	2.98 ± 0.25	14.3 ± 1.23
+0.012 mM α-MMTA	10.7 ± 1.50[a]	13.5 ± 1.55
control	3.05 ± 1.49	16.0 ± 2.70
+0.012 mM metaraminol	3.78 ± 1.49	21.2 ± 1.28[a]
Rat		
control	3.05 ± 2.15	11.9 ± 0.78
+0.012 mM α-MMTA	8.71 ± 0.63[a]	12.8 ± 0.70
control	4.96 ± 0.29	10.9 ± 0.64
+0.012 mM metaraminol	5.36 ± 1.71	13.0 ± 0.52[a]

[a] Statistical significance of difference between control and drug-treated preparations are indicated as follows: $P < 0.05$. Adapted from McBride et al. (22) and McBride and Aprison (21).

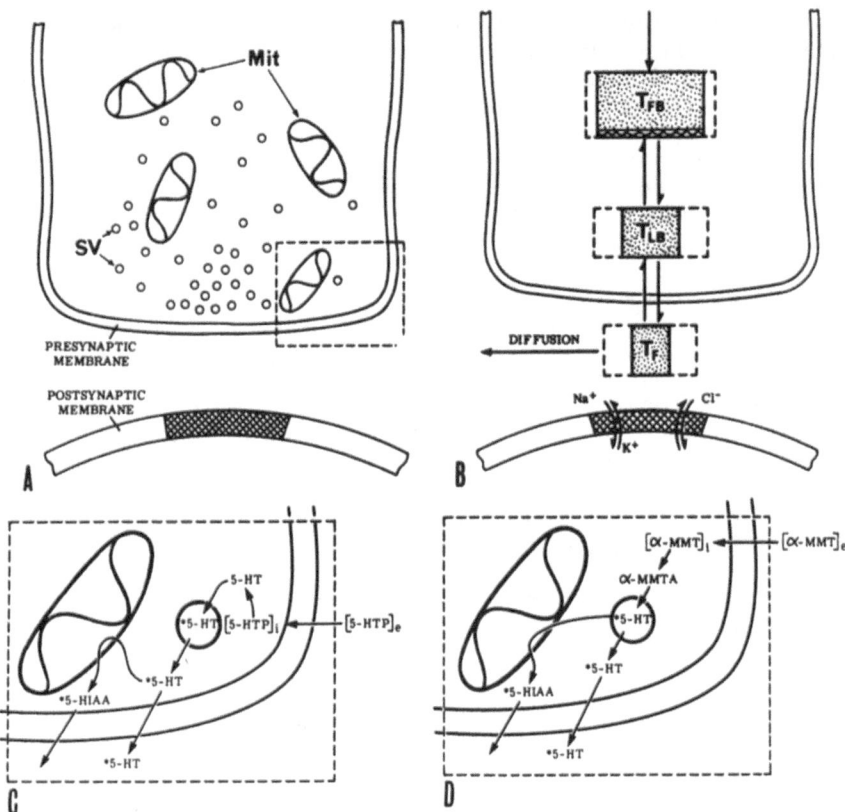

Figure 7. Part *A* shows a simplified drawing of a nerve ending. Part *B* shows our hypothetical model of a typical synapse. A more complete description of our model serotonergic synapse has been published (11). Part *C* shows a simplified diagram of the proposed mechanism by which 5-HTP can cause an increased accumulation and release of 5-HT from serotonergic nerve terminals. Part *D* shows a simplified diagram of the proposed mechanism by which α-MMT can cause an increased release of 5-HT from such nerve terminals.

pigeon or rat, α-MMTA, at concentraions as low as 0.012 mM, caused a significant release of 5-[³H]HT from serotonergic nerve endings (Table 2). On the other hand, metaraminol at low concentrations had no apparent effect on the release of 5-[³H]HT. Therefore, on the basis of the combined *a*) behavioral data, *b*) correlative in vivo neurochemical data, and *c*) data obtained with isolated nerve endings, we drew the following conclusions concerning the mechanism of

the involvement of 5-HT in producing depression in pigeons or rats working with specific learned behavioral tasks. In the case of studies where total 5-HT levels *increased*, we believe that significant amounts of the injected 5-HTP were taken up by serotonergic nerve endings and there decarboxylated to 5-HT; the levels of 5-HT then increase in the nerve endings (Fig. 7). Some of the 5-HT is converted to 5-HIAA by mitochondrial MAO and is then released.

However, significant amounts of 5-HT are stored (presumably in synaptic vesicles) and then released or significant amounts are accumulated outside the synaptic vesicles and then released. It is possible that both mechanisms function in vivo causing an increased release of 5-HT. In the case where total 5-HT levels decreased following injections of α-MMT, this amino acid is taken up and then decarboxylated; one of the decarboxylation products, α-MMTA, also causes an increased release of 5-HT (Fig. 7). The overall effect of this increased release of 5-HT is a net decrease in the tissue levels of 5-HT.

The combination of data provides strong evidence to support the hypothesis of Hingtgen and Aprison (16) and Aprison and Hingtgen (7, 10, 11) to explain why they observed the same behavioral effect in pigeons working on a Mult FR 50 FI 10 schedule of reinforcement when the levels of total 5-HT changed in opposite directions. They postulated that in both cases the physiologically effective pool of 5-HT (T_{LB} plus T_F, i.e., that available for release into the synaptic cleft plus that in the cleft; see B of Fig. 7 and ref 11) increased and it was these molecules of 5-HT acting on certain "key" synapses that caused the observed depression in the animal's behavior. Thus, we envisage the following biochemical relationships under the usual physiological conditions:

a) plasma try (bound) $\rightleftharpoons$ plasma try (free) $\rightleftharpoons$ brain try (free) $\rightarrow$ synaptosomal try

$\hookrightarrow$ brain tryptamine

$\hookrightarrow$ protein

b) synaptosomal try $\rightarrow$ 5-HTP $\rightarrow$ 5-HT$_{LB}$ $\rightleftharpoons$ 5-HT$_{FB}$

$\hookrightarrow$ R

5-HT (free in nerve ending) $\rightarrow$ 5-HIAA (nerve ending)

5-HT (free in cleft or T_F in Fig. 7)

c) 5-HIAA (nerve ending) $\rightleftharpoons$ 5-HIAA (extraneuronal) $\rightarrow$ 5-HIAA (blood)

Reactions shown in b) occur in the region of the nerve endings. In this relationship, b), R refers to other products such as 5-methoxytryptamine, N-acetyl-5-HT and 5-HT-O- SO$_3$, present in very small amounts at any one time. It is indeed possible that in some forms of functional illness, the relationships in b) are stressed to the point that 5-HT (free in the cleft) is too high at certain synapses or too low at other synapses. Depending on whether we are dealing with excitation or inhibition (this is determined by the type of ions moving across the receptor site in the postsynaptic membrane since in many cases the same transmitter can be either excitatory or inhibitory) at an important set of synapses, the behavioral effect will be determined. In the cases just discussed, we note behavioral depression; however, we cannot tell whether the predominant effect was more neurophysiological inhibition or more neurophysiological excitation. Neural networks can be constructed for each case to explain the behavior. Thus additional studies are necessary to identify the specific neural pathways being utilized in the brains of these animals. Only then, after additional work, will we gain still greater insight to this problem.

TEMPORAL RELATIONSHIPS BETWEEN DECREASED LEVELS OF ACETYLCHOLINE AND EXCITATION

It is well known that in vivo levels of ACh in brain can vary inversely with the state of an organism's functional activity (30). Therefore, it is logical to suspect that ACh may play a role in the underlying mechanism involved in certain types of drug-induced behavioral excitation. In addition, it is possible that neural systems interact such that the predominant neurotransmitter or modulator of one affects the postsynaptic interactions of another (2). Thus, for these two reasons, we decided to measure ACh in the brain of rats injected with drugs known to effect 5-HT and norepinephrine levels and behavior. Toru et al. (31) measured ACh in brain parts of rats that emitted

elevated response rates following an injection of tetrabenazine (TBZ) (2 mg/kg) after iproniazid (50 mg/kg) pretreatment (16–18 hr before TBZ). Although these rats were killed during the early stage of excitation, it was found that ACh contents in all three brain parts were lower than control values. In subsequent studies, these early experiments were extended to include the complete period of excitation and the measurement of ACh, 5-HT, and norepinephrine levels in three major brain areas (telencephalon, diencephalon plus midbrain, and pons-medulla oblongata) as well as changes in response rates in the same rats. After stable behavior was obtained on a Sidman shock-avoidance schedule (RS 40; SS 20), rats were preinjected with iproniazid and given TBZ 30-60 min after the start of the session. Behavioral excitation (response rates at least 1.5 times greater than control rates) typically occurred for a period of time following iproniazid-TBZ administration. No behavioral excitation or depression was noted after giving a comparable single dose of iproniazid alone nor two such injections 12 hr apart. Rats were killed by the near-freezing method of Takahashi and Aprison (30); the brains were removed and dissected into the three parts, which were then assayed for total content of ACh, 5-HT, and norepinephrine (12).

Within 30 min following the injection of TBZ into the iproniazid pretreated rats, the average rate of avoidance responding doubled from a normal level of 5 responses/min (see Fig. 8). After passing through a period of maximum responding, the rate decreased to normal levels within 90 min. A comparison of the behavioral measure of excitation to the content of ACh in the telencephalon, diencephalon plus midbrain, and pons-medulla oblongata showed that although the transmitter content decreased in all three areas, only the changes in ACh content in the telencephalon appeared to follow the same time course as the behavioral measure. Acetylcholine levels in both the diencephalon plus midbrain and

pons-medulla oblongata returned to normal long before the behavior returned to normal.

During the first part of the excitation period, an inverse relationship was found between the ACh content in the telencephalon and the behavior (until the latter doubled in value). However, as the behavior

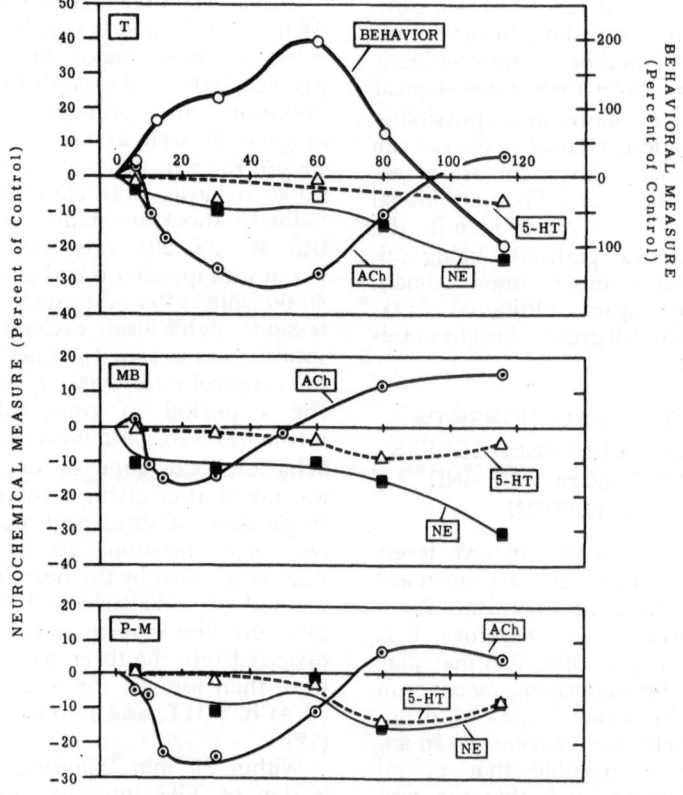

Figure 8. Temporal variations in ACh, 5-HT, and norepinephrine content in the telencephalon (T), diencephalon plus mesencephalon (MB), and pons-medulla oblongata (P-M) and avoidance response rates in rats preinjected with 50 mg/kg iproniazid 16 hr before being given 2 mg/kg TBZ. Each point represents the biochemical or behavioral measure obtained from the same group of rats killed at a specific time after injection. The abscissa axis refers to time (in min) after the TBZ injection. Acetylcholine levels in the telencephalon are most closely related to the behavioral excitation following this drug administration. After the period of excitation, response rates fall to normal levels and then the rats go into a period of behavioral depression (see top portion of figure) which is again followed by a return to normal response rates (not shown on figure). (From Aprison and Hingtgen (10), with permission. Copyright 1970 by Academic Press, New York).

TABLE 3. Content of acetylcholine in telencephalon (T), diencephalon plus mesencephalon (D-MB), and pons-medulla oblongata (P-M)

Condition	Acetylcholine, nmoles/g (mean ± SD)		
	T	D-MB	P-M
Excited	11.4 ± 1.7^a (16)	17.6 ± 1.9^a (16)	12.2 ± 1.5^a (16)
Nonexcited	14.8 ± 2.6 (5)	19.9 ± 3.8 (5)	14.2 ± 2.4 (5)
Naive control	15.6 ± 1.8 (15)	20.3 ± 1.9 (16)	15.6 ± 2.1 (16)
Iproniazid control	15.2 ± 1.2 (4)	19.5 ± 0.6 (4)	15.0 ± 1.4 (4)

Data were obtained from naive and iproniazid control rats and rats exhibiting behavioral excitation or no behavioral excitation following iproniazid (50 mg/kg) and tetrabenazine (2 mg/kg) administration. Values in parentheses refer to number of rats in each group. Adapted from Aprison and Hingtgen (9). $^a P < 0.01$ compared to iproniazid and naive control groups. No significant differences were found between ACh levels in the three brain parts of control rats compared to the brain parts of the nonexcited rats or the iproniazid control rats.

was increasing from two to three times greater than control values, there was no further significant decrease in ACh levels (Fig. 8). In rats treated with iproniazid alone, large increases in both 5-HT and norepinephrine were observed in the three brain parts. During the period of excitation following iproniazid-TBZ administration, the mean norepinephrine content in the telencephalon and pons-medulla oblongata as well as the mean 5-HT content in the telencephalon and diencephalon plus midbrain did not differ from the corresponding values for iproniazid control rats. At 80 min of excitation (a point near the end of the excitation period), the content of norepinephrine in the diencephalon plus midbrain as well as the content of 5-HT in the pons-medulla area was lower.

It was found that a small number of rats did not exhibit excitation following the iproniazid-TBZ drug treatment. This group of rats was studied more closely (9). As was predicted, when these nonexcited rats were killed at times comparable to the times when excited rats were killed, the levels of ACh in the three brain areas did not vary from control values (Table 3).

In the cholinergic system, the fall in content of transmitter can be explained as being due to the destruction of free or neuronally released ACh by membrane bound acetylcholinesterase (AChE). Therefore, four pieces of evidence strongly implicate ACh in this type of drug-induced excitation: a) the fall in ACh at the beginning of the period of behavioral excitation; b) the inverse relationship that exists between ACh content and the behavioral measure until the latter reaches twice normal levels; c) the rise toward normal levels of ACh in the telencephalon near the end of the excitation period; and d) the lack of change in ACh levels in brain from rats receiving the drugs but exhibiting no excitation. However, the possible interactions of 5-HT and norepinephrine in this mechanism cannot yet be ruled out (Fig. 8).

To further clarify the role of ACh in drug-induced behavioral excitation, we initiated a series of experiments to study the effect of atropine on the iproniazid-TBZ treated rats (18). Atropine, a well known anticholinergic, can act in the CNS, whereas its methyl derivative is believed to act only peripherally. Hence, a behavioral effect of atropine that is not duplicated by methyl atropine can most reasonably be attributed to the action of atropine at central cholinergic synapses. Since ACh in the CNS appears to be involved in the behavioral excitation observed in our rats, we predicted that atropine treatment of the rats during the regular iproniazid-TBZ injection sequence should block excitation, whereas treatment with methyl atropine should have little or no effect.

Rats were preinjected with 50 mg/kg iproniazid about 16 hr before a shock avoidance session. Thirty minutes following the start of the session, they were injected with either saline or varying doses of atropine alkaloid 0 to 120 min before an injection of 2 mg/kg TBZ. The iproniazid-TBZ injection interval was fixed at 18 hr. The results indicated that 0.1 mg/kg atropine given 60 to 120 min before TBZ administration and following iproniazid pretreatment increased the duration of behavioral excitation by more than 50%. During the excitation period itself, the response rates increased from one and a half to four times greater than control rates. This intensity of excitation was not changed by atropine treatment; only the duration of excitation was affected. Since atropine alone, atropine with TBZ alone, or atropine with iproniazid alone produced no behavioral effect, it must be assumed that it was the combination of atropine with iproniazid and TBZ that resulted in the extended periods of excitation.

When the injection time is held constant (60 min before TBZ) and the dose of atropine is increased, higher doses (0.8 mg/kg) result in a complete blocking of the excitation effect (Table 4). The fact that equal molar doses of methyl atropine nitrate do not have any significant effect on the duration of excitation suggests that

the action of atropine is occurring centrally. Since the 0.8 mg/kg dose of atropine completely blocks excitation, this finding supports our suggestion that ACh is implicated in the production of drug-induced excitation. Apparently this dose is high enough to block ACh action at the specific cholinergic receptors normally utilized in the generation of this type of avoidance behavior. If this explanation of our data is correct, then why didn't the lower doses of atropine also block excitation? In fact not only was excitation not blocked, the duration of excitation was enhanced by about 50%.

In light of our previous results, these data suggest that this effect is due to one or more of four possible atropine actions: a) it affects the metabolism of TBZ; b) it inhibits AChE activity; c) it causes an increase in the release of ACh; or d) it blocks the uptake of ACh. The first possibility does not appear likely, since atropine did not interfere with TBZ-induced depression in rats not pretreated with iproniazid. The second suggestion has never been demonstrated in vitro to our knowledge. Further, since it is usually assumed that released ACh is destroyed by AChE localized in the pre- and/or postsynaptic membrane, and since early data (with eserine) showed little ACh uptake (14), the concept that some is taken up into the pre- or postsynaptic neuron was not readily acceptable in the past. However, there are now reports that show ACh can be taken up by brain slices in the presence of other AChE inhibitors and that atropine at certain doses is a competitive inhibitor of the process (20, 25, 27). Small doses of atropine can also cause a release of ACh from isolated cerebral cortex in rats (25). Our data with lower atropine doses can therefore be ex-

TABLE 4. Duration of behavioral excitation during avoidance responding

Injection	Duration of Excitation, min (mean ± SEM)
TBZ	64 ± 11.3 (10)
Atropine (0.1 mg/kg) plus TBZ	101 ± 12.2[a] (10)
Atropine (0.2 mg/kg) plus TBZ	100 ± 13.9[a] (7)
Atropine (0.4 mg/kg) plus TBZ	56 ± 12.1 (8)
Atropine (0.8 mg/kg) plus TBZ	0[b] (7)
Methyl atropine nitrate (0.13 mg/kg) plus TBZ	59 ± 11.4 (8)
Methyl atropine nitrate (1.04 mg/kg) plus TBZ	65 ± 10.8 (5)

Data were obtained after various doses of atropine alkaloid or methyl atropine nitrate administered 1 hr before TBZ (2 mg/kg) and 17 hrs following iproniazid (50 mg/kg) pretreatment. A response rate at least 50% greater than control rates was defined as behavioral excitation. Values in parentheses refer to number of rats in each group. Adapted from Hingtgen and Aprison (18).   [a] $P < 0.05$, compared to TBZ group (t-test).
[b] This 0 value indicates that rats were avoiding shocks at normal rates; hence, no excitation was observed.

plained as follows: 0.1 to 0.2 mg/kg atropine injections produce a biochemical condition in which ACh activates the postsynaptic receptor site for a longer period of time. These psychopharmacological data therefore provide support for the concept that low doses of atropine can cause either a release of ACh or prevent its uptake into the pre- or post-synaptic neurons and/or other cells at the cholinergic synapses.

The time sequence noted to produce an increase in duration of behavioral excitation is interesting. Apparently, a critical period of time exists during which atropine concentrations at key brain synapses must reach a specific level before it can either cause an increased release of ACh or block the uptake of some ACh into the presynaptic nerve endings, postsynaptic neuron, and/or other cells. Following an injection of 0.4 mg/kg atropine the similarity in duration of excitation to the controls suggests that at this dose atropine reaches a level at which it probably can act both pre- and postsynaptically. If it had just blocked the ACh effect, the period of excitation should have decreased, whereas, if it caused a release of ACh or competed with

its uptake, an enhanced duration of effect should have been noted. However, the duration of effect was in the normal range; it therefore appears that both processes may have been occurring.

Several years ago it was shown that lesions in the septal area of the brain result in a 20–37% reduction of total ACh content in the rat brain (24, 29). If neuronal pathways utilizing ACh as a transmitter are involved in the excited responding, noted above, septal lesions should either reduce or eliminate the excitation. When rats with bilateral septal lesions were given the iproniazid-TBZ treatment (19) the duration of behavioral excitation exhibited by these rats was less than 50% of the excitation shown by control rats (septal lesion group: 23.0 ± 5.3 min; control group: 50.4 ± 9.4 min). There was no significant difference in the degree of excitation (an average of 2.5 times greater than baseline rates) between the lesioned rats and the controls, only the length of the period of excitation differed.

Since rats with septal lesions were previously shown to have lowered brain ACh levels, the shorter periods of excitation in the lesioned rats support the suggestion that ACh plays an important role in this type of behavioral excitation. The fact that Pepeu et al. (24) found the lowest levels of ACh in areas of the telencephalon, the area where ACh levels were most closely correlated with increased avoidance responding by our animals, adds further weight to this hypothesis. Since our explanation relates the excitation noted in these rats to the release of ACh at key cholinergic synapses, the apparent smaller amount of ACh available for release in the septal-lesioned rats would account for the shortened periods of behavioral excitation noted.

PROPOSED MECHANISM FOR THE INVOLVEMENT OF ACETYL-CHOLINE IN EXCITATION

In general, we see the model of a cholinergic synapse (Fig. 9) and the model of the serotonergic synapse (reported earlier, see ref 11) as being similar in compartmentation (T_{FB}, T_{LB} and T_F) but different, of course, in the following four ways: a) the biochemical precursors; b) the biosynthetic and degradative enzymes; c) products of the reactions; and d) the location of the degradative enzyme. Since AChE is localized mainly in the postsynaptic membranes, and

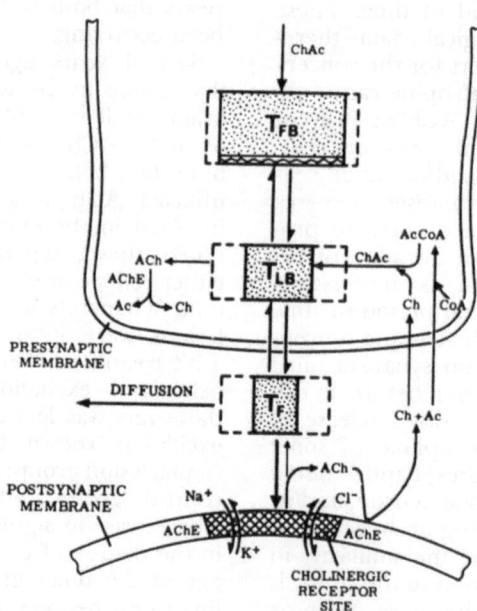

Figure 9. A hypothetical simplified model of a cholinergic synapse. Transmitter pools T_{FB}, T_{LB}, and T_F refer to three pools: a firmly bound pool, a labile bound pool and a free pool, respectively, which are in equilibrium. The dotted lines represent a possible expansion of each pool. The wavy line at the bottom of the T_{FB} pool represents the fact that storage of transmitter in this pool may be affected by drugs. Ch refers to choline, Ac to acetate, ChAc to choline acetyltransferase, AcCoA to acetylcoenzyme A and CoA to coenzyme A. Not shown, is the concept of the physiologically effective pool, T_{PE}, which is to be equated to $T_{LB} + T_F$ (see text and ref 11).

present in the presynaptic membranes in some synapses, the release of ACh during specific behavioral states usually results in a fall of total ACh. Only in the case of inhibiting AChE, with an anticholinesterase drug, does one see an increase in cerebral ACh with changes in behavior (1). However, in each case, one must invoke the mechanism of an increase in released free ACh (T_F) into the synaptic cleft in order to explain the behavioral effect. Thus, this mechanism is similar to the one we have used previously in explaining the role of free 5-HT in key serotonergic synapses in behavioral depression. Although ACh release studies stimilar to those on the release of labeled 5-HT are currently in progress, preliminary evidence of increased "free" ACh are now available from our laboratories (Shea et al., unpublished data).

CONCLUDING COMMENTS

It is our hypothesis that in some functional illnesses, such as cyclic depressions, there may be an abnormal release of two or more neurotransmitters occurring at the same time. This could result in an imbalance of *several* neurotransmitter pathways and might explain why only partial success is obtained by treating patients with this type of depression with only one type of drug. We feel that one approach toward obtaining a better understanding of these functional illnesses is experimentation with animals of the type reported here. Through such research we can implicate certain putative neurotransmitters as being involved in behavioral depression and excitation in animals. If these findings can be extrapolated to man they may provide clues to the possible biochemical basis for these types of functional illnesses. In

this regard, our research program now includes the multiple analyses of a number of putative transmitters in specific areas of the CNS from animals that show depressed or excited activity produced by behavioral manipulations without the use of drugs. Such measurements are now possible because of the development of highly sensitive analytical methods in the picomole range.

REFERENCES

1. APRISON, M. H. *J. Neurochem.* 2: 197, 1958.
2. APRISON, M. H. *Recent Ad. Biol. Psychiatry* 4: 133, 1962.
3. APRISON, M. H., AND C. B. FERSTER. *Experientia* 16: 159, 1960.
4. APRISON, M. H., AND C. B. FERSTER. *J. Neurochem.* 6: 350, 1961.
5. APRISON, M. H., AND C. B. FERSTER. *J. Pharmacol. Exp. Ther.* 131: 100, 1961.
6. APRISON, M. H., AND J. N. HINGTGEN. *J. Neurochem.* 12: 959, 1965.
7. APRISON, M. H., AND J. N. HINGTGEN. *Life Sci.* 5: 1971, 1966.
8. APRISON, M. H., AND J. N. HINGTGEN. *Recent Adv. Biol. Psychiatry* 8: 87, 1966.
9. APRISON, M. H., AND J. N. HINGTGEN. *Biol. Psychiatry* 1: 87, 1969.
10. APRISON, M. H., AND J. N. HINGTGEN. *Int. Rev. Neurobiol.* 13: 325, 1970.
11. APRISON, M. H., AND J. N. HINGTGEN. *Federation Proc.* 31: 121, 1972.
12. APRISON, M. H., T. KARIYA, J. N. HINGTGEN AND M. TORU. *J. Neurochem.* 15: 1131, 1968.
13. APRISON, M. H., M. A. WOLF, G. J. POULOS AND T. L. FOLKERTH. *J. Neurochem.* 9: 575, 1962.
14. ELLIOTT, K. A. C., AND N. HENDERSON. *Am. J. Physiol.* 165: 365, 1951.
15. GESSA, G. L., E. COSTA, R. KUNTZMAN AND B. B. BRODIE. *Life Sci.* 1: 353, 1962.
16. HINGTGEN, J. N., AND M. H. APRISON. *Science* 141: 169, 1963.
17. HINGTGEN, J. N., AND M. H. APRISON. *Life Sci.* 5: 1249, 1966.
18. HINGTGEN, J. N., AND M. H. APRISON. *Neuropharmacology* 9: 419, 1970.
19. HINGTGEN, J. N., M. H. APRISON, W. C. BLACK AND J. C. SLOAN. *Experientia* 29: 74, 1973.
19a HINGTGEN, J. N., AND M. H. APRISON. *Life Sci.* In press.

20. LIANG, C. C., AND J. H. QUASTEL. *Biochem. Pharmacol.* 18: 1187, 1969.

21. MCBRIDE, W. J., AND M. H. APRISON. *Pharmacol. Biochem. Behav.* 1: 587, 1973.

22. MCBRIDE, W. J., M. H. APRISON AND J. N. HINGTGEN. *Neuropharmacology* 12: 769, 1973.

23. MCBRIDE, W. J., M. H. APRISON AND J. N. HINGTGEN. *J. Neurochem.* 23: 385, 1974.

24. PEPEU, G., A. MULAS, A. RUFF AND P. SOTGIU. *Life Sci.* 10: 181, 1971.

25. POLAK, R. L., AND M. W. MEEUWS. *Biochem. Pharmacol.* 15: 989, 1966.

26. PORTER, C. C., J. A. TOTARO AND C. M. LEIBY. *J. Pharmacol. Exp. Ther.* 134: 139, 1961.

27. SCHUBERTH, J., AND A. SUNDWALL. *J. Neurochem.* 14: 807, 1967.

28. SHORE, P. A., D. BUSFIELD AND H. S. ALPERS. *J. Pharmacol. Exp. Ther.* 146: 194, 1964.

29. SORENSEN, J. P., AND J. A. HARVEY. *Physiol. Behav.* 6: 723, 1971.

30. TAKAHASHI, R., AND M. H. APRISON. *J. Neurochem.* 11: 887, 1964.

31. TORU, M., J. N. HINGTGEN AND M. H. APRISON. *Life Sci.* 5: 181, 1966.

32. UDENFRIEND, S., AND P. ZALTZMAN-NIRENBERG. *J. Pharmacol. Exp. Ther.* 138: 194, 1962.

Catecholamines and drug–behavior interactions[1]

L. S. SEIDEN,[2] R. C. MACPHAIL[3] AND M. W. OGLESBY [3]

*Departments of Pharmacological and Physiological Sciences
and Psychiatry The University of Chicago, Chicago, Illinois 60637*

ABSTRACT

The effects of several drugs on schedule-controlled operant behavior depend on the baseline rate of responding and on the nature of the environmental conditions that maintain the behavior. For example, the effects of amphetamine and alpha-methyl-*para*-tyrosine (αMT) on operant performances depend to a large extent on the rate at which organisms respond under nondrug control conditions. A neurochemical mechanism for these rate-dependent effects has not been established. However, several lines of evidence suggest that catecholamines are functionally important in the maintenance of many types of behavior, including operant behavior. The fact that many drugs which exhibit drug–behavior interactions also produce characteristic effects on the metabolism of central nervous system catecholamines suggests that the performance of operant behavior per se modifies brain catecholamine metabolism and thereby the subsequent drug effect. Experiments measuring the depletion of catecholamines following synthesis inhibition with αMT, or changes in the specific activity of norepinephrine after tritium labeling, have shown that operant behavior alters the metabolism of catecholamines. Preliminary evidence is also presented from experiments designed to determine variables associated with the performance-induced changes in catecholamine metabolism. These variables include: rate of response; rate or density of reinforcement; and response-reinforcer contingencies. The results of these experiments suggest a neurochemical mechanism for the rate-dependent effects of amphetamine and αMT. A model is presented that may account for the general phenomenon of drug–behavior interactions in neurochemical terms.—SEIDEN, L. S., R. C. MACPHAIL AND M. W. OGLESBY. Catecholamines and drug–behavior interactions. *Federation Proc.* 34: 1823–1831, 1975.

[1] Supported in part by Public Health Service Grant MH-11, 191 to the University of Chicago.

[2] Career Research Development Award, MH-10, 562.

[3] Supported by NIH Training Grant MH-07, 083.

Abbreviations: αMT, α-methyl-*para*-tyrosine; FR, fixed ratio; FI, fixed interval; DRL, differential reinforcement of low rate.

138   L.S. Seiden et al.

The effects of many drugs on schedule-controlled operant performances have been shown to be a function of several variables, such as the type of drug, the dose, and the baseline rate of responding[4] prior to drug administration (21, 22, 41). It has also become clear that in addition to predrug baseline response rate, other variables in the operant situation play an important role in the determination of a drug effect: these include level of deprivation and magnitude of reinforcement (13, 63, 66), and specific stimulus or schedule variables (44, 45, 49, 53). Many of these behaviorally active drugs affect one or more aspect of the regulation of catecholamine transmitter function in the central nervous system (CNS). The demonstration that several groups of drugs affect both behavior and metabolism[5] of catecholamines raises the possibility of a causal relationship between the behavioral and neurochemical effects of these drugs.

Although the catecholamine-containing neurons comprise less than 5% of the nerve cell population in the CNS (34, 83), there is considerable evidence that the catecholamines are of functional importance in the maintenance of behavior (86). Catecholamines have been implicated in the control of motor activity and food and water intake as well as operant behavior (14, 35, 37, 38, 46, 66). In addition, various types of intense stimulus changes that are aversive in nature also affect the concentrations or metabolism of central nervous system catecholamines (8, 30, 52).

The purpose of this paper is to review studies showing that catecholamines are of functional importance in the maintenance of operant behavior, and conversely, that operant responding or some aspect of the operant situation affects catecholamine metabolism in the CNS (Fig. 1). In the latter part of this paper, the two lines of evidence concerning catecholamine–behavior and behavior–catecholamine interactions will be integrated. Behavioral and biochemical data will be presented that suggest a possible neurochemical mechanism by which drugs, catecholamines and schedule-controlled operant behavior mutually interact.

The baseline rate of responding is an important determinant of the behavioral effects of many drugs. Many effects of the amphetamines as well as other drugs on positively reinforced

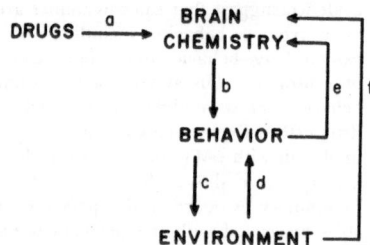

Figure 1. Diagram showing relationships between drugs, behavior, the environment and brain chemistry. Behaviorally active drugs act in some manner to alter brain chemistry (*a*). Alterations in brain chemistry have been shown to affect several conditioned and unconditioned behaviors (*b*). Operant behavior is maintained by its environmental consequences (*c*), and the environment in turn acts on behavior (*d*) by providing the conditions necessary for its occurrence. Operant behavior has recently been shown to alter brain chemistry (*e*). Finally, the environment itself may alter behavior by modifying brain chemistry (*f*). (From Seiden et al. (65).)

---

[4] Baseline rate of responding refers to the performance engendered by schedules of reinforcement. These schedules engender characteristic rates and patterns of responding that are consistent for long periods of time when organisms are tested daily (64).

[5] In this paper, *metabolism* refers to all biochemical and physiological processes that affect a transmitter. These include: synthesis and degradation, accumulation, storage, release, and reuptake.

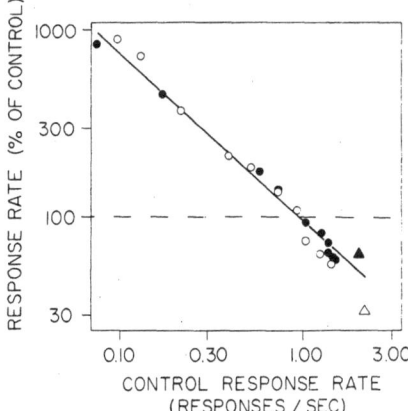

Figure 2. Dependence of effect of d-amphetamine on baseline response rates in a squirrel monkey. Percentage change in rate of responding after 0.3 mg/kg d-amphetamine is plotted as a function of average rate of responding in successive minutes of a 10-min fixed-interval schedule (circles) and under a 30-response fixed-ratio schedule (triangles). Open and filled symbols indicate data from two different sessions. The line through the points was fitted by visual inspection. (From Kelleher and Morse (41).

and negatively reinforced operant performances are characterized as *rate-dependent*. Dews (22) first noted that the effect of methamphetamine on the rate of schedule-controlled responding of pigeons depended on the rate at which the pigeons responded under nondrug baseline conditions. Responding was decreased by methamphetamine when the pigeons ordinarily responded at high rates, but was increased when the pigeons' baseline response rates were low. Rate-dependent effects of amphetamine (Fig. 2), as well as other drugs (e.g., phenothiazines, barbiturates and certain anticholinergic agents) have been subsequently demonstrated in several experiments (23, 50, 51, 53, 54). The baseline rate of responding, however, is not the only

behavioral variable that determines the drug effect. The degree of deprivation (13) and the magnitude of reinforcement are also important variables in drug–behavior interactions. For example, in one experiment (66) groups of rats were trained under a variable-interval schedule using different magnitudes of water reinforcement. Although the baseline response rates among the groups did not differ significantly, after amphetamine treatment there was a significant inverse correlation between response-rate decreases and reinforcement magnitude. While the biochemical and behavioral determinants of these drug–behavior interactions are not entirely clear at present, the basic phenomenon of a behavioral effect of a drug being in part dependent on the baseline behavior and the nature of the conditions that maintain it has been clearly recognized and generally accepted (75).

A number of behaviorally active drugs have been shown to interact with the metabolism of the catecholamines, norepinephrine and dopamine, which are putative neurotransmitters in the CNS. Experimental evidence has suggested several sites of interaction between these drugs and catecholamines in the context of the model of the adrenergic neuron. (See Cooper et al. (19); Andén et al. (3); and Weiner (85)) (Fig. 3). *1*) Drugs that interfere with storage and thereby cause depletion of catecholamines, such as reserpine and tetrabenazine (36, 74). *2*) Drugs that release catecholamines but do not interfere with storage, and therefore do not appreciably change the catecholamine concentration in the CNS. Such drugs include amphetamine (and related substituted phenylethylamines) and methylphenidate (15, 18, 60). *3*) Drugs that block catecholamine reuptake, such as cocaine and imipramine (33). There is evidence that amphetamine and related compounds also block reuptake (59). 4) Drugs that block catecholamine receptors such as chlorpromazine and haloperidol (3, 16). 5) Drugs that directly stimulate catecholamine receptors, such as apomorphine and clonidine,which are believed to stimulate dopamine and norepinephrine receptors, respectively (4,5). 6) Drugs that block enzymes involved in the synthesis and degradation of cate-

cholamines. These include tyrosine hydroxylase inhibitors (α-methyl-*para*-tyrosine), dopa decarboxylase inhibitors (Ro 4-4602 and NSD 1015), dopamine-β-hydroxylase inhibitors (disulfiram and Fla-63) as well as monoamine oxidase inhibitors (pargyline) and inhibitors of catechol-*O*-methyl-transferase (COMT, e.g., pyrogallol) (29, 31, 79).

Many drugs that interact with catecholamines are effective in the treatment of psychiatric disorders (61).

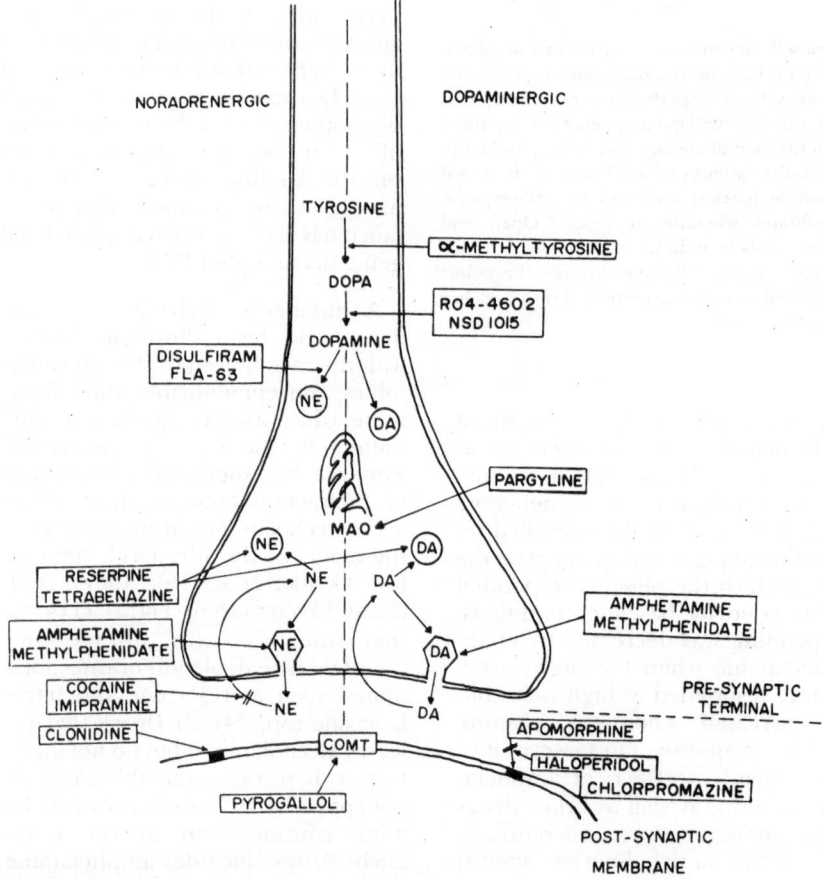

Figure 3. Models of the noradrenergic (left) and dopaminergic (right) neurons indicating possible sites of interaction with drugs. See text for details. NE = norepinephrine: DA = dopamine.

Thus, from the clinical standpoint, the function of the catecholamines in the acquisition and maintenance of behavior is of great interest.

ROLE OF CATECHOLAMINES IN MAINTENANCE OF BEHAVIOR

The analysis of the biochemical mechanisms responsible for drug effects on behavior gained impetus with a series of findings that revealed a close correlation between the sedative effects of reserpine and its neurochemical effects. Among the numerous effects of reserpine, the rapid and prolonged depletion of norepinephrine and dopamine from brain is believed to be causally related to its behavioral effects. As Carlsson et al. (17) and others (7, 77) have shown, reserpine-induced sedation and decreases in spontaneous motor activity can be antagonized with large doses of the catecholamine precursor L-dihydroxyphenylalanine. Further experimentation has demonstrated that the increase in spontaneous motor activity is primarily dependent on the formation of dopamine in the central nervous system (1). In addition, there is substantial clinical evidence that dopamine in the extrapyramidal system in humans plays an important role in the maintenance and coordination of movement and muscle tone; abnormalities of dopamine metabolism result in parkinsonian symptoms that can be at least partly alleviated by administration of L-dopa (38).

There is substantial evidence for the functional importance of brain dopamine and norepinephrine in the maintenance of a discriminated conditioned avoidance response in mice, rats and cats (55, 57, 67–69). In one experiment (72), mice of two different strains were trained to avoid shock in a shuttle-box apparatus. Reserpine pretreatment virtually eliminated avoidance responding while the escape response occurred on approximately 50% of the trials. When L-dopa was administered, the avoidance response was temporarily and partially restored. The time course for conditioned avoidance response restoration differed in the two strains but in each case was roughly correlated with the period of net synthesis of brain dopamine (Fig. 4). The ability of L-dopa to reverse reserpine action was blocked when both cerebral and extracerebral dopadecarboxylase activity was inhibited with Ro 4-4602. These results indicated that a decarboxylated metabolite of L-dopa, and not L-dopa itself, is necessary for reserpine antagonism (72). Furthermore, it has been demonstrated by selective inhibition of extracerebral dopa decarboxylase that the reserpine antagonism is dependent on formation of catecholamines in the CNS (71). In this experiment, peripheral inhibition of dopa decarboxylase with a small dose of Ro 4-4602 resulted in a more prolonged conversion of dopa to dopamine in brain (Fig. 5). Correspondingly, the duration of the reversal of the reserpine effect was increased (Fig. 6). Additional studies (1, 73) have shown that the synthesis of norepinephrine as well as dopamine is necessary for the maintenance of a conditioned avoidance response.

Reduction of catecholamine levels by repeated administration of α-methyl-tyrosine (αMT) disrupts operant behavior maintained by aversive (32) and positive reinforcement contingencies (62). In one experiment by Schoenfeld and Seiden (62), groups of rats were trained to bar press under one of three fixed-ratio schedules (FR 1, FR 5, or FR 10), or one of three fixed-interval schedules (FI-30 sec, FI-60 sec or FI-90 sec) of water reinforcement. Fixed-ratio schedules engender relatively high response rates while FI schedules engender relatively low response rates (26). After αMT treatment 8 and 4 hr before the test session, brain dopamine and norepinephrine concentrations had decreased about 75% and 45% respectively, and responding was decreased for all groups. The decrease in FR responding was a function of the FR requirement and the response rate under baseline conditions (Fig. 7), whereas the decrease in FI responding was similar at each FI value. Moreover, responding under the FR-10 schedule was only slightly decreased early in the session, but was completely eliminated within 20 min (Fig. 8). This steady decline in response rate did not occur in rats performing under the FI schedule; these rats maintained steady but reduced response rates throughout the experimental sessions. It was also demonstrated that the αMT-induced decrease in FR-10 responding could be proportionately antagonized with increasing doses of

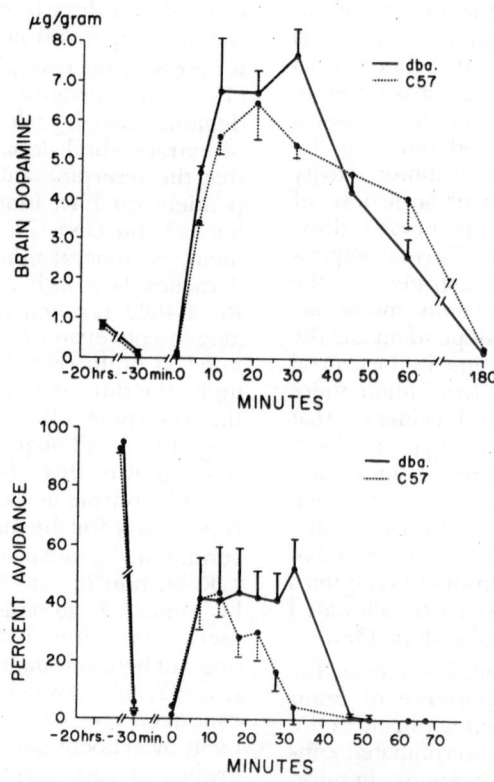

Figure 4. *Upper*: Brain dopamine levels in two strains of mice (dba and C57) treated with reserpine (2.5 mg/kg) followed by L-dopa (400 mg/kg). Vertical lines represent SEM. *Lower*: The effect of reserpine and L-dopa on the conditioned avoidance response; −20 hr represents the level of avoidance responding before any drug treatment. Immediately after this control session the mice were injected with reserpine (2.5 mg/kg) and tested at −30 min. Mice were injected with L-dopa (400 mg/kg) at time 0. (From Seiden and Peterson (72).)

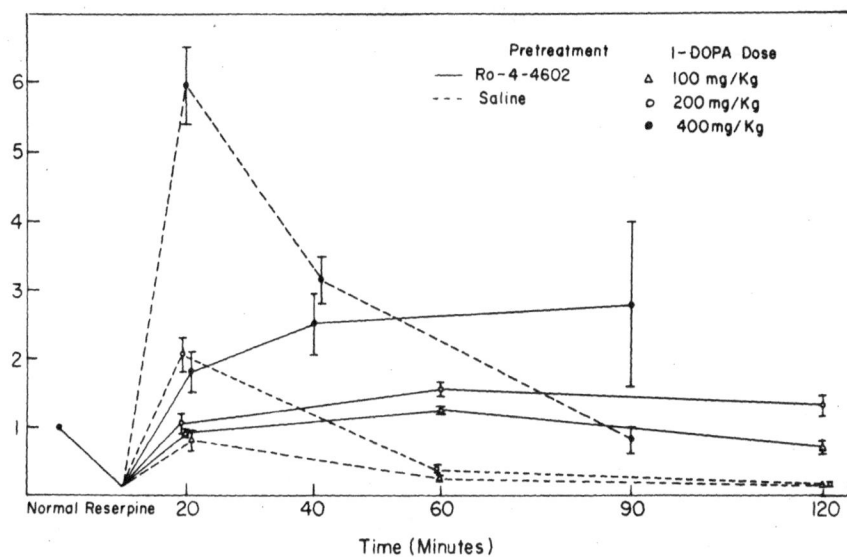

Figure 5. Effects of extracerebral dopa decarboxylase inhibition with Ro 4-4602 on the conversion of systemically administered L-dopa to dopamine in mouse brain. All groups of mice received reserpine (2.5 mg/kg) 20 hr prior to L-dopa. Half the groups received saline and half received Ro 4-4602 (25 mg/kg) 30 min prior to L-dopa. L-Dopa was administered at varying doses and groups of mice were sacrificed at the indicated times for dopamine analysis. Values on the Y-axis represent whole-brain dopamine levels (μg/g). (From Seiden and Martin [71].)

L-dopa, and that the maintenance of FR-10 performance was directly correlated with brain levels of dopamine and norepinephrine (Fig. 9). The results described above suggest that the maintenance of positively reinforced operant behavior depends on the integrity of catecholamines in the central nervous system. In addition, the behavioral effects of αMT are rate-dependent and schedule-dependent. Furthermore, as shown above (Fig. 8), rats lever pressing on a fixed-ratio schedule of reinforcement and treated with αMT show a progressive and sharp decline in the rate of lever pressing during the experimental session; this suggests that a small portion of catecholamines not depleted by αMT at the beginning of the session is utilized as a result of the ongoing behavior, and since the tyrosine hydroxylase is inhibited, catecholamines cannot be replaced through de novo synthesis. From these findings, one might suspect a rate- or schedule-dependent effect of αMT on depletion of catecholamines in the CNS.

ENVIRONMENTAL AND BEHAVIORAL EFFECTS ON CATECHOLAMINE METABOLISM

Early investigations examined the effects of intense aversive, generally referred to as stressful, stimuli on concentrations of brain catecholamines. Although the results of these experiments were at times inconsistent, the majority of the evidence suggests that acute aversive stimulation decreases catecholamine concentrations whereas chronic aversive stimulation

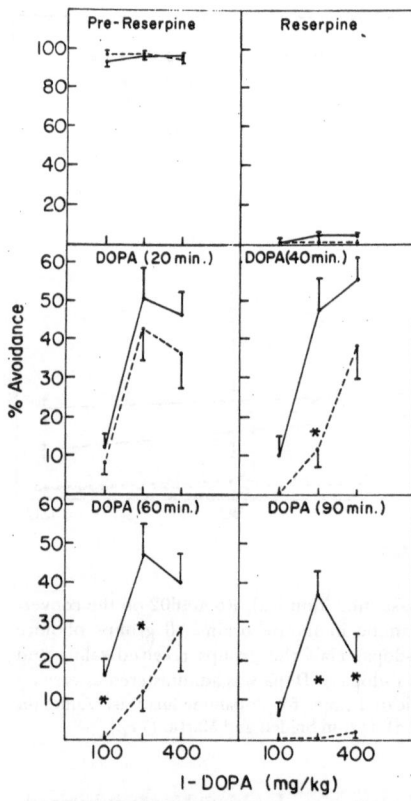

Figure 6. Effect of extracerebral decarboxylase inhibition with Ro 4-4602 on the L-dopa reversal of reserpine induced suppression of avoidance performance. Groups of at least five mice received reserpine (2.5 mg/kg) 20 hr presession and saline (– – –) or Ro 4-4602 (25 mg/kg, ——) 30 min pre-session. L-Dopa was administered at three doses, and avoidance performance was recorded 20, 40, 60, and 90 min later. * indicated *P* less than 0.05. (From Seiden and Martin (71).)

sults have not always been obtained (9). Exposure of rats to a cold environment (−20°C) decreased brain norepinephrine concentrations (52), but others have not obtained these results with less extreme ambient temperatures (9, 30). In contrast to the decline in norepinephrine concentrations after acute treatments, prolonged exposure to electric grid shock (82) and electroconvulsive shock (42) have been reported to increase concentrations of brain norepinephrine.

Marked changes may occur in adrenergic neuronal functioning without concomitant measurable changes in brain concentrations of catecholamines (81). Studies measuring changes in catecholamine metabolism have consistently shown large changes produced by several classes of aversive environmental stimuli. These include: immobilization (20) and forced exercise (30, 80), electro-

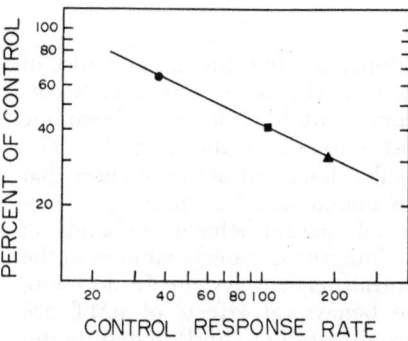

Figure 7. Dependence of the effects of αMT (75 mg/kg at 8 hr and 4 hr presession) on baseline response rates under FR schedules of reinforcement. Six rats were consecutively trained under FR-1 (●), FR-5 (■) and FR-10 (▲) schedules. The effects of αMT on responding are plotted as a function of the baseline response rates (responses per minute) that were obtained with each FR value. Line was fitted by visual inspection. (From data presented in Table 2 of Schoenfeld and Seiden (62).)

increases catecholamine concentrations. For instance, several experiments have demonstrated decreased norepinephrine concentrations in rats and mice after acute exposure to electric grid shock (8, 52, 56). Similarly, forced swimming in rats decreases norepinephrine concentrations (6, 56, 80), although these re-

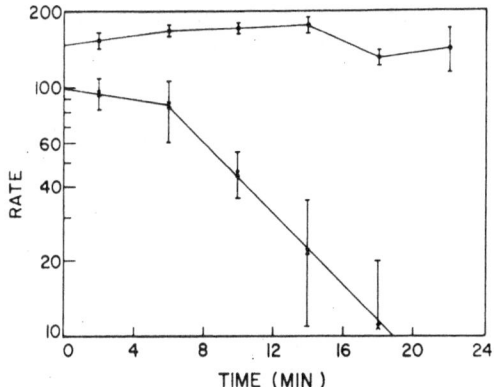

Figure 8. Time course for the effects of αMT (75 mg/kg at 8 hr and 4 hr pre-session) on FR-10 performance in rats. Each value represents the average response (responses per minute) rate of five rats within consecutive 4-min periods. ●, 30-min control session (upper curve); X, 30-min test session (lower curve). From Schoenfeld and Seiden (62).

convulsive shock and electric grid-shock (8, 42) and changes in ambient temperature (76). Moreover, changes in catecholamine metabolism typically occur with less intense stimuli than those that are required to obtain changes in catecholamine tissue concentrations.

Relatively few studies have attempted to correlate changes in brain

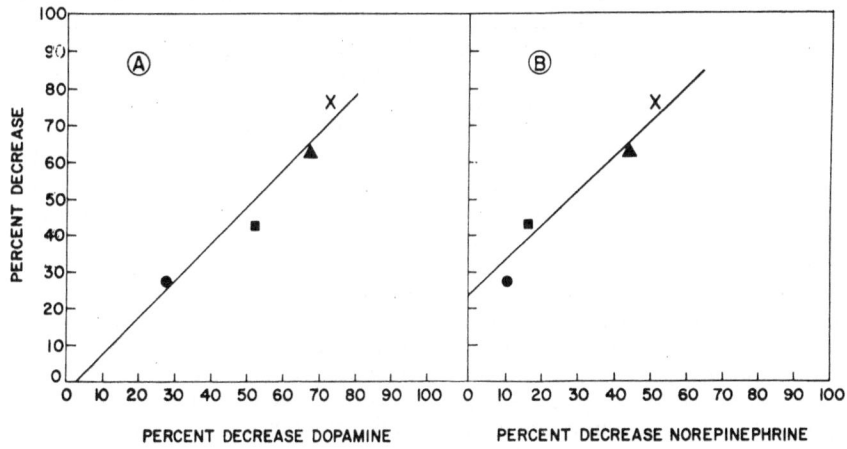

Figure 9. Relationship between brain catecholamine levels and behavioral effects of αMT (75 mg/kg at 8 hr and 4 hr presession) in rats treated with L-dopa (1 hr presession). Values on the ordinate represent response-rate decreases in rats trained under an FR-10 schedule of reinforcement. Values on the abscissa represent decreases in brain dopamine (left) and norepinephrine (right). Each symbol represents the average of five paired observations. X, αMT and L-dopa vehicle; ▲, αMT and 10 mg/kg of L-dopa; ■, αMT and 50 mg/kg of L-dopa; ●, αMT and 100 mg/kg of L-dopa. (From Schoenfeld and Seiden (62).)

catecholamine concentrations or metabolism with the performance of operant behavior. One experiment (27) using histochemical fluorescence techniques showed that in αMT-pretreated rats performing a discriminated lever-press avoidance response, the concentration of brain norepinephrine was decreased further than in untrained αMT-treated controls. As a control for nonspecific effects of electric shock, αMT-treated rats were randomly shocked approximately the same number of times as the trained rats. Compared with nonshocked αMT treated rats, no significant differences in norepinephrine concentrations were obtained. Since the decline of catecholamine concentrations following αMT reflects the rate of catecholamine metabolism (12), the results described above indicate that variables associated with operant avoidance performance lead to increased catecholamine metabolism. Using a dopamine-β-hydroxylase inhibitor (disulfiram), it was found (39) that rats trained to perform a discriminated avoidance response had decreased norepinephrine concentrations and decreased dopamine concentrations when compared with untrained controls. However, because disulfiram affected avoidance performances, these results are open to multiple interpretations. Additional interpretations of the data from these two experiments are suggested by the use of electric shock, which has been shown to alter catecholamine concentrations and metabolism (see above). These experiments emphasize the importance of controlling environmental and behavioral variables in order to evaluate the role of operant performance, as the independent variable, in altering brain catecholamine metabolism.

Changes in catecholamine metabolism associated with positively reinforced operant performances were first demonstrated by Schoenfeld and Seiden (62). Rats treated with αMT and lever pressing on a variable-interval 30-sec schedule of water reinforcement had lower brain concentrations of dopamine and norepinephrine than αMT-treated experimentally naive rats that had free access to water (Fig. 10). In this experiment and in the experiments described below in which behavior was the independent variable, αMT was administered shortly before the test session. The dose and pretreatment were selected to produce inhibition of catecholamine synthesis without depleting catecholamines sufficiently to cause disruption of ongoing behavior. Therefore, the rate decreases observed under FR or FI schedules after repeated administration of αMT (see above) were not observed in this portion of the experiment. The decreases in dopamine and norepinephrine suggest that some aspect of the operant situation increased metabolism of brain catecholamines. These results were extended by Lewy and Seiden (48) using a radioactive tracer technique. Rats were cannulated in the lateral ventricle and randomly assigned to one of three groups: *1*) an ad lib food and water control group; *2*) a water-deprived control group; and *3*) an operant performing group. On the 15th day of training under a variable-interval 30-sec schedule of water reinforcement, all rats were injected intraventricularly with tritiated norepinephrine. Immediately after injection, control rats were returned to their home cages and trained rats were placed in operant test chambers. Two hours later all rats were sacrificed, and subsequently the brainstem-diencephalon from each rat was assayed for tritiated and endogenous norepinephrine. No significant differences were seen in the concentration of endogenous

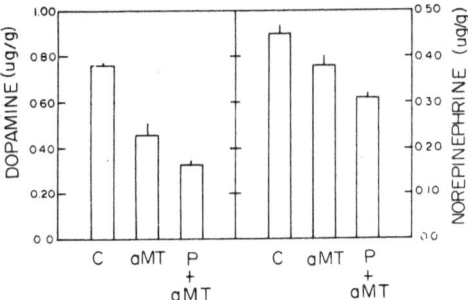

Figure 10. Enhanced depletion of brain catecholamine concentrations after αMT (200 mg/kg, 30 min presession) in rats performing under a variable-interval 30-sec schedule of reinforcement. Each bar represents average concentrations (μg/g ± SEM) of dopamine (left) and norepinephrine (right) in eight rats. C represents untrained and nondeprived control rats; αMT represents controls that received αMT; and P + αMT represents αMT-treated rats that performed under a variable-interval 30-sec schedule for 2 hr before being sacrificed. (From data in Table 4 of Schoenfeld and Seiden (62).)

norepinephrine between the three groups. However, rats performing in the operant test chambers had significantly lower specific activities of norepinephrine than rats in the two control groups that did not differ from each other. Thus, water deprivation by itself is not a sufficient stimulus to alter norepinephrine metabolism. Furthermore, rats trained but not performing on the day of sacrifice did not differ from control rats. In an additional experiment (47), it was found that norepinephrine specific activities and endogenous norepinephrine concentrations did not differ significantly between rats given access (2 hr) to an activity wheel and control rats. These results suggest that neither motor activity nor training of operant performance are important variables in altering norepinephrine metabolism.

The role of the lever-press response in altering catecholamine metabolism has recently been investigated using a yoked-control design. This experiment also compared αMT-induced changes in catecholamine concentrations in discrete brain areas. Each yoked rat was paired with a perform-

ing rat (variable-interval 30-sec schedule of water reinforcement) so that the quantity and temporal distribution of water reinforcement were identical (26). When catecholamine metabolism was compared for a given area, such as caudate or hypothalamus, no differences were obtained between αMT-treated yoked and performing rats. In addition, regional brain concentrations in αMT-treated controls (ad lib water or water-deprived rats) did not differ significantly from each other. However, performing and yoked rats treated with αMT had a greater decrease in dopamine in the caudate and in norepinephrine in the pons-medulla when compared with αMT-treated controls (Table 1). In the cortex, hypothalamus and midbrain, no significant differences between any of the groups were found. These findings indicate that periodic water presentation, water consumption or both are functionally more important in altering catecholamine metabolism than the lever-pressing response.

Alpha-methyl-tyrosine has been shown to antagonize many of the behavioral effects of amphetamine,

TABLE 1. Effect of αMT treatment on dopamine concentrations in the caudate nucleus and norepinephrine concentrations in the pons-medulla of rat brain

Treatment	% of vehicle control concentrations ± SEM	
	Caudate dopamine	Pons-medulla norepinephrine
Control (ad lib water)	61 ± 5	69 ± 6
Water-deprived	63 ± 8	72 ± 4
Performing	41 ± 3	54 ± 4
Yoked	42 ± 3	54 ± 5

Rats (eight per group) were given αMT (100 mg/kg). Thirty minutes later, each performing and yoked animal was placed into an operant test chamber. All animals were killed 2 hr later (2.5 hr after αMT injection). Brains were removed, dissected according to the method of Glowinski and Iversen (28), and subsequently analyzed for catecholamine concentrations. Control values were: caudate dopamine 8.85 ± 0.37 μg/g; pons-medulla norepinephrine 0.46 ± 0.03 μg/g.

including: increased spontaneous motor activity (24); hyperthermic responses (84); rate increases in shuttle-box and nondiscriminated avoidance responding (87); repetitive sniffing and gnawing in rats after large doses (58); and euphoria in humans (40).

We have recently studied the effects of αMT and amphetamine on the performance of rats that were trained under a differential-reinforcement-of-low-rate (DRL) schedule of water reinforcement. Responding under this schedule occurs at a low and steady rate, with occasional short periods of rapid responding (26, 43). Figure 11 shows the effect of amphetamine on the performance of one rat trained under a DRL 17.5-sec schedule. Amphetamine produced dose-related increases in response rate and decreases in the frequency of reinforcement. Analyses of frequency distributions of the time intervals between successive responses (inter-response times) revealed that the increases in responding were associated with shifts in the interresponse time distribution toward shorter values. In addition, pretreatment with αMT blocks the effects of amphetamine on DRL performance (Fig. 12) but αMT alone had no effect on DRL performance.

In contrast to the above, under some conditions αMT pretreatment

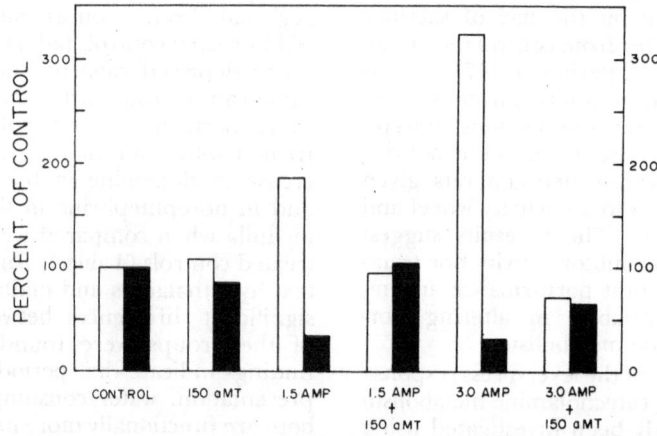

Figure 11. Effects and interactions of d-amphetamine (20 min presession) and αMT (150 mg/kg 60 min presession) on DRL 17.5-sec performance in a rat (R6A). Open bars represent response rate (100% = 3.46 resp. per min) and darkened bars represent session reinforcement frequency (100% = 118 per session). Each dose was administered once. (From Seiden et al. (70).)

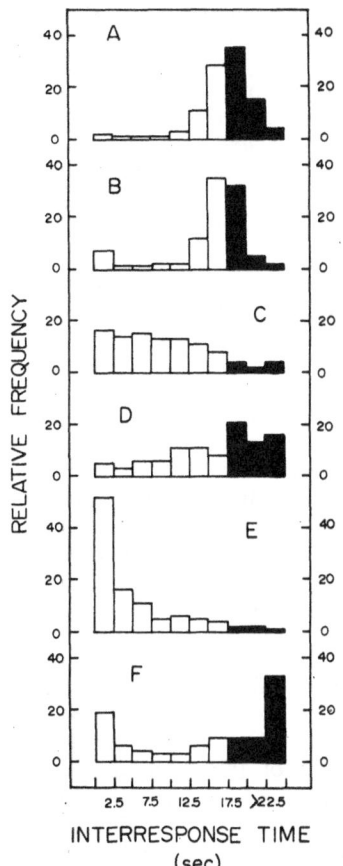

Figure 12. Interresponse time (IRT) distributions for the performance of R6A under DRL 17.5 sec. A, vehicle control performance; B, 150 mg/kg αMT; C, 1.5 mg/kg d-amphetamine; D, 150 mg/kg αMT plus 1.5 mg/kg d-amphetamine; E, 3.0 mg/kg d-amphetamine; F, 150 mg/kg αMT plus 3.0 mg/kg d-amphetamine. αMT and d-amphetamine were administered 60 min and 30 min presession, respectively. Darkened bars represent reinforced IRTs. (From Seiden et al. (70).)

variable-interval component of a multiple variable-interval time-out schedule of food reinforcement. Prior treatment with αMT produced even greater decreases in responding when the rats subsequently received methamphetamine. A similar effect was recently obtained in our laboratory. In this experiment (66), amphetamine decreased the rate of responding of rats trained under a variable-interval 20-sec schedule of water reinforcement. These rate decreases were a direct function of the dose of amphetamine administered and an inverse function of the volume of water reinforcement. Alpha-methyltyrosine given alone produced no noticeable effects on performance, but augmented the rate-decreasing effects of amphetamine when the two drugs were administered together. This experiment attempted to elucidate neurochemical correlates of these behavioral effects by examining changes in catecholamine metabolism as a function of αMT-amphetamine or αMT-behavior interactions. In αMT-treated nonperforming animals, subsequent administration of amphetamine produced a dose-dependent decrease in catecholamines, suggesting a possible biochemical mechanism that can account for the enhanced rate reduction seen with αMT-amphetamine combinations in performing animals. Additional data showed a greater decrease in catecholamine concentrations in αMT-treated performing animals than in αMT-treated nonperforming animals and this greater decrease was inversely correlated with reinforcement magnitude in pons and caudate, and directly correlated with reinforcement magnitude in the hypothalamus. Since the changes in the rate of responding after amphetamine and the changes in catecholamine concentrations in the pons and

does not block and may even augment the behavioral effects of amphetamine. For example, Evans (25) showed that methamphetamine produced dose-related decreases in the rate at which rats responded during the caudate after αMT have the same inverse relationship to the magnitude of the reinforcer, it is possible that changes in catecholamine metabolism in these two areas may be directly related to the rate-decreasing effects of amphetamine.

DRUG–BEHAVIOR–CATECHOLAMINE INTERACTIONS: IMPLICATIONS

Identification of the variables in the operant situation that interact with brain catecholamine metabolism will be of use in predicting the behavioral effects of drugs acting on catecholaminergic mechanisms. Experiments using yoked-control animals suggest that one of the more important variables involved may be the periodic delivery of the reinforcer. This conclusion must be approached with caution, however, because the current neurochemical techniques may not detect minute changes in neuronal function induced by other variables in the operant situation. Additional variables, such as the rate and distribution of responding, the rate and distribution of reinforcement, the magnitude of reinforcement and the introduction of stimuli, need to be more fully investigated.

The results reviewed in this paper have implications for clinical research. These implications are clearly speculative, but they point to an area of research that may result in improved patient responses to psychoactive drugs. For example, catecholamines have been functionally implicated in such clinical disorders as hyperkinesis (78) and depression (2, 61). With information about the functional state of catecholamine nerve cells in these disorders, a more efficient regimen of drug therapy may be prescribed.

At this time the only anatomical areas in which changes in catechol-amine metabolism have been observed are in the lower brainstem (for norepinephrine) and in the caudate nucleus (for dopamine). Dopamine in the caudate nucleus has been strongly implicated in the control of motor activity (38). Recent evidence indicates a functional importance for dopamine in the caudate nucleus in the control of feeding behavior (11, 88). Therefore, dopamine metabolism in this region may reflect motor activity associated with behaviorally-induced changes in responding or an effect of water consumption. Further experiments are necessary to more precisely delineate this effect. Changes in norepinephrine metabolism in the brainstem, an area rich in cell bodies and axons and relatively devoid of norepinephrine nerve endings (34), may reflect alterations in rapid transport of norepinephrine-containing granules to the more rostral brain regions. R.Y. Moore (unpublished results) has shown that axonal transport of [^{3}H]norepinephrine after injection of [^{3}H]dopa into the locus ceruleus of rats occurs at the rate of approximately 55 mm/day. This rapid rate of transport (greater than 2 mm per hour) could account for the decreases in norepinephrine concentrations in brainstem observed 2.5 hr after αMT treatment in the experiments cited above. The changes in specific activity of [^{3}H]norepinephrine in the lower brainstem (48) of performing rats are also consistent with this interpretation. In the peripheral sympathetic nervous system, several rates of rapid transport have been demonstrated (10). Thus, if the rate of transport of norepinephrine seen by Moore in the CNS is subject to modification by drugs and neuronal functioning, the enhanced decline of brainstem norepinephrine in performing animals may be attributable to increased transport of norepinephrine to nerve terminals induced by operant performance. Rostral to the brainstem, we have not observed enhanced depletion of norepinephrine in areas rich in nerve terminals following performance under schedules of water-reinforcement. Increased neural activity would imply increased utilization and release of norepinephrine. However, failure to observe enhanced depletion following performance may indicate that reuptake of norepinephrine keeps pace with enhanced release. Therefore, even in the face of increased utilization of neurotransmitter, one would not expect to find enhanced depletion when reuptake is the major mode of inactivation of the neurotransmitter.

In the first section of this paper the relationships between drug effects and ongoing behavior (i.e., drug–behavior interactions) were briefly reviewed and examples were drawn from the research literature indicating that drug effects are in part a function of the baseline behavior prior to drug administration. The fact that many of these behaviorally active drugs have characteristic biochemical effects on catecholamine metabolism in the CNS suggested a functional relationship between catecholamines and behavior. In this context, evidence was presented that catecholamines are involved in the maintenance of spontaneous motor activity as well as positively reinforced and aversively controlled operant performances. The fact that several drugs that affect catecholamine metabolism in brain also exhibit drug–behavior interactions makes it seem plausible that ongoing behavior itself may affect the metabolism of catecholamines and thereby alter the subsequent drug effect. Evidence was presented that catecholamine metabolism was increased by ongoing operant behavior. Additional preliminary evidence indicated that the effects of ongoing behavior on catecholamine metabolism may account for the drug-behavior interactions seen with αMT (62) and amphetamine (66, 70). Whether or not the general model shown in Fig. 1 can be extended to other drugs and neurochemical systems (e. g., cholinergic, serotonergic) is the subject of future research. Although this model has been supported by relatively little data, it is useful because it can account for the general phenomenon of drug–behavior interactions in neurochemical terms.

REFERENCES

1. AHLENIUS, S., AND J. ENGEL. *Naunyn-Schmeidebergs Arch. Exp. Pathol. Pharmakol.* 270: 349, 1971.
2. AKISKAL, H. S., AND W. T. MCKINNEY. *Science* 182: 20, 1973.
3. ANDÉN, N. E., A. CARLSSON AND J. HÄGGENDAL. *Annu. Rev. Pharmacol.* 9: 119, 1969.
4. ANDÉN, N. E., H. CORRODI, K. FUXE, T. HÖKFELT, C. RYDIN AND T. SVENSON. *Life Sci.* 9: 513, 1970.
5. ANDÉN, N. E., A. RUBENSON, K. FUXE AND T. HÖKFELT. *J. Pharm. Pharmacol.* 19: 627, 1967.
6. BARCHAS, J. D., AND D. X. FREEDMAN. *Biochem. Pharmacol.* 12: 1232, 1963.
7. BLASCHKO, H., AND T. L. CHRUSCIEL. *J. Physiol. London* 151: 272, 1960.
8. BLISS, E. L., J. ALLION AND J. ZWANZIGER. *J. Pharmacol. Exp. Ther.* 164: 122, 1968.
9. BLISS, E. L., AND J. ZWANZIGER. *J. Psychiatr. Res.* 4: 189, 1966.
10. BRADLEY, W. G., D. MURCHISON AND M. DAY. *Brain Res.* 35: 185, 1971.
11. BREESE, G. R., B. R. COOPER AND R. D. SMITH. In: *Frontiers in Catecholamine Research: Third International Catecholamine Symposium,* edited by E. Usdin and S. H. Synder. New York; Pergamon, 1973, p. 701.
12. BRODIE, B. B., E. COSTA, A. DLABAC, N. H. NEFF AND H. H. SMOOKLER. *J. Pharmacol. Exp. Ther.* 154: 493, 1966.
13. BROWN, R. M., AND L. S. SEIDEN. *Federation Proc.* 30: 503, 1971.
14. CAMPBELL, A. B. (Ph.D. Thesis.) Chicago: The University of Chicago, 1973.
15. CARLSSON, A. In: *Amphetamines and Related Compounds,* edited by E. Costa

and S. Garattini. New York: Raven, 1970, p. 289.

16. CARLSSON, A., AND M. LINDQVIST. *Acta Pharmacol. Toxicol.* 20: 140, 1963.

17. CARLSSON, A., M. LINDQVIST AND T. MAGNUSSON. *Nature London* 180: 1200, 1957.

18. CARLSSON, A., E. ROSENGREN, A. BERTLER AND J. NILSSON. In: *Psychotropic Drugs,* edited by S. Garattini and V. Ghetti. Amsterdam: Elsevier, 1957, p. 363.

19. COOPER, J. R., F. E. BLOOM AND R. H. ROTH. *The Biochemical Basis of Neuropharmacology.* New York: Oxford Univ. Press, 1974.

20. CORRODI, H., K. FUXE AND T. HÖKFELT. *Life Sci.* 7: 107, 1968.

21. DEWS, P. B. *J. Pharmacol. Exp. Ther.* 113: 339, 1955.

22. DEWS, P. B. *J. Pharmacol. Exp. Ther.* 122: 137, 1958.

23. DEWS, P. B. *Naunyn-Schmiedebergs Arch. Exp. Pathol. Pharmakol.* 248: 296, 1964.

24. DOMINIC, J. A., AND K. E. MOORE. *Arch. Int. Pharmacodyn. Ther.* 178: 166, 1969.

25. EVANS, H. L. *J. Pharmacol. Exp. Ther.* 176: 244, 1971.

26. FERSTER, C. B., AND B. F. SKINNER. *Schedules of Reinforcement.* New York: Appleton-Century-Crofts, 1957.

27. FUXE, K., AND L. C. F. HANSON. *Psychopharmacologia.* 11: 439, 1967.

28. GLOWINSKI, J., AND L. L. IVERSEN. *J. Neurochem.* 13: 655, 1966.

29. GOLDSTEIN, M., B. ANAGNOSTE, E. LAUBER AND M. R. MCKEREGAN. *Life Sci.* 3: 763, 1964.

30. GORDON, R., S. SPECTOR, A. SJOERDSMA AND S. UDENFRIEND. *J. Pharmacol. Exp. Ther.* 153: 440, 1966.

31. HANSSON, E., R. M. FLEMING AND W. G. CLARK. *Int. J. Neuropharmacol.* 3: 177, 1964.

32. HANSON, L. C. F. *Psychopharmacologia* 8: 100, 1965.

33. HERTTING, G., J. AXELROD AND L. G. WHITBY. *J. Pharmacol. Exp. Ther.* 134: 146, 1961.

34. HILLARP, N. A., K. FUXE AND A. DAHLSTRÖM. *Pharmacol. Rev.* 18: 727, 1966.

35. HOEBEL, B. G. *Annu. Rev. Physiol.* 33: 533, 1971.

36. HOLZBAUER, M., AND M. VOGT. *J. Neurochem.* 1: 8, 1956.

37. HORNYKIEWICZ, O. *Pharmacol. Rev.* 18: 925, 1966.

38. HORNYKIEWICZ, O. *Federation Proc.* 32: 183, 1973.

39. HURWITZ, D. A., S. M. ROBINSON AND I. BAROFSKY. *Neuropharmacology* 10: 477, 1971.

40. JÖNSSON, L. E., E. ÄNGGÅRD AND L. M. GUNNE. *Clin. Pharmacol. Ther.* 12: 889, 1971.

41. KELLEHER, R. T., AND W. H. MORSE. *Ergeb. Physiol. Biol. Chem. Exp. Pharmacol.* 60: 1, 1968.

42. KETY, S. S., F. JAVOY, A. M. THIERRY, L. JULOU AND J. GLOWINSKI. *Proc. Natl. Acad. Sci. U.S.A.* 58: 1249, 1967.

43. KRAMER, T. J., AND M. RILLING. *Psychol. Bulletin* 74: 225, 1970.

44. LATIES. V. G., *J. Pharmacol. Exp. Ther.* 183: 1, 1972.

45. LATIES, V. G., AND B. WEISS. *J. Pharmacol. Exp. Ther.* 152: 388, 1966.

46. LEIBOWITZ, S. F. *Res. Publ. Assoc. Res. Nerv. Ment. Dis.* 50: 327, 1972.

47. LEWY, A. J. (Ph.D. Thesis.) Chicago: The University of Chicago, 1973.

48. LEWY, A. J., AND L. S. SEIDEN. *Science* 175: 454, 1972.

49. MACPHAIL, R. C. *Proc. Am. Psychol. Assoc.* 79: 755, 1971.

50. MACPHAIL, R. C. (Ph.D. Thesis.) College Park: University of Maryland, 1973.

51. MARR, M. J. *J. Exp. Anal. Behav.* 13:291, 1970.

52. MAYNERT, E. W., AND R. LEVI. *J. Pharmacol. Exp. Ther.* 143: 90, 1964.

53. MCKEARNEY, J. W. *J. Exp. Anal. Behav.* 14: 167, 1970.

54. MCKIM, W. A. *Psychopharmacologia* 32: 255, 1973.

55. MOORE, K. E. *Life Sci.* 5: 55, 1966.

56. MOORE, K. E., AND E. W. LARIVIERE. *Biochem. Pharmacol.* 12: 1283, 1963.

57. MOORE, K. E., AND R. H. RECH. *J. Pharm. Pharmacol.* 19: 405, 1967.

58. RANDRUP, A., AND I. MUNKVAD. In: *Amphetamine and Related Compounds,* edited by E. Costa and S. Garattini. New York: Raven, 1970, p. 695.

59. RUTLEDGE, C. O. *J. Pharmacol. Exp. Ther.* 171: 188, 1970.

60. SCHEEL-KRÜGER, J. *Eur. J. Pharmacol.* 14: 47, 1971.

61. SCHILDKRAUT, J. J. *Annu. Rev. Pharmacol.* 13: 427, 1973.

62. SCHOENFELD, R., AND L. S. SEIDEN.. *J. Pharmacol. Exp. Ther.* 167: 319, 1969.

63. SEIDEN, L. S. *Am. Psychol.* 23: 887, 1968.

64. SEIDEN, L. S. In: *Methods of Neurochemistry,* edited by R. Fried. New York: Marcel Dekker, 1973, p. 59.

65. SEIDEN, L. S., R. M. BROWN AND A. J. LEWY In: *Chemical Modulation of Brain Function,* edited by H. C. Sabelli. New York: Raven, 1973, p. 261.

66. SEIDEN, L. S., AND A. B. CAMPBELL. In: *Neuropsychopharmacology of Monoamines and Their Regulatory Enzymes,* edited by E. Usdin. New York: Raven, 1973.
67. SEIDEN, L. S., AND A. CARLSSON. *Psychopharmacologia* 4: 418, 1963.
68. SEIDEN, L. S., AND A. CARLSSON. *Psychopharamacologia* 5: 178, 1964.
69. SEIDEN, L. S., AND L. C. F. HANSON. *Psychopharmacologia* 6: 239, 1964.
70. SEIDEN, L. S., R. C. MACPHAIL AND J. ANDRESEN. In preparation.
71. SEIDEN, L. S., AND T. W. MARTIN. *Physiol. Behav.* 6: 453, 1971.
72. SEIDEN, L. S., AND D. D. PETERSON. *J. Pharmacol. Exp. Ther.* 159: 422, 1968.
73. SEIDEN, L. S., AND D. D. PETERSON. *J. Pharmacol. Exp. Ther.* 163: 84, 1968.
74. SHORE, P. A., S. L. SILVER AND B. B. BRODIE. *Science* 122: 284, 1955.
75. SIDMAN, M. *Ann. N.Y. Acad. Sci.* 65: 282, 1956.
76. SIMMONDS, M. A. *J. Physiol., London* 203: 199, 1969.
77. SMITH, C. B., AND P. B. DEWS. *Psychopharmacologia* 3: 55, 1962.
78. SNYDER, S., AND J. L. MEYERHOFF. *Ann. N.Y. Acad. Sci.* 205: 310, 1973.
79. SPECTOR, S., A. SJOERDSMA AND S. UDENFRIEND. *J. Pharmacol. Exp. Ther.* 147: 86, 1965.
80. STONE, E. A. *Psychosom. Med.* 32: 51, 1970.
81. SULSER, F., AND E. SANDERS-BUSH. *Annu. Rev. Pharmacol.* 11: 209, 1971.
82. THIERRY, A. M., F. JAVOY, J. GLOWINSKI AND S. S. KETY. *J. Pharmacol. Exp. Ther.* 163: 163, 1968.
83. UNGERSTEDT, U. *Acta Physiol. Scand. Suppl.* 367: 1, 1971.
84. VALZELLI, L., E. DOLFINI, M. TANSELLA AND L. GARATTINI. *J. Pharm. Pharmacol.* 20: 595, 1968.
85. WEINER, N. *Annu. Rev. Pharmacol.* 10: 273, 1970.
86. WEISS, B., AND V. G. LATIES. *Annu. Rev. Pharmacol.* 9: 297, 1969.
87. WEISSMAN, A., B. K. KOE AND S. S. TENEN. *J. Pharmacol. Exp. Ther.* 151: 339, 1966.
88. ZIGMOND, M. J., AND E. M. STRICKER. *Science* 177: 1211, 1973.

Behavioral toxicology:
A developing discipline[1]

NANCY K. MELLO

Alcohol and Drug Abuse Research Center, McLean Hospital
Harvard Medical School, Belmont, Massachusetts 02178

Exposure to toxic chemicals may adversely affect a number of organ systems and hepatic, cardiac, pulmonary and central nervous system function may be severely compromised. In some instances, the concomitant behavioral changes may be unremarkable, as in the case of asbestos exposure which may eventually result in the development of a rapidly fatal mesothelioma, as long as 25 to 30 years after initial contact (16). In contrast, other toxic agents such as lead and methylmercury may produce early and direct behavioral effects which range from interference with sensory and motor integration to profound intellectual deficits and emotional disturbances (3, 13).

In experimental behavioral toxicology, the task is not only to detect and measure the behavioral effects of known pollutants, but also to devise sensitive techniques to evaluate the possible consequences of exposure to new and untested chemicals. The emergence of a subspecialty of behavioral toxicology is a relatively recent development and received little formal recognition

prior to 1969 (22). It is now increasingly acknowledged that the behavioral analysis of toxicological effects is critical for providing a sound empirical basis for legislative regulations concerning the maximally acceptable concentrations of pollutants in the environment and industry (23, 24).

Traditionally, toxicology has been a subspecialty of pathology and "toxic effects" have been defined in terms of tissue changes which can be seen on gross or microscopic examination. It is obvious that if a toxic chemical produces gross pathological changes in the central nervous system, it may also produce discernible behavioral effects. However, this issue is complicated by the variable concordance between central nervous system (CNS) lesions and behavioral deficits. It is the common experience of neurologists and neuropathologists that gross CNS insults, revealed on postmortem examination, may have been associated with surprisingly few

[1] Supported by grant DA-4R6010 from the National Institute on Drug Abuse.

155

neurological signs and symptoms or behavioral effects during the patient's life. The functional capacity of the central nervous system to compensate for focal cortical damage is one of the most frequent general findings of several generations of neurologists and neuropsychologists. Conversely, gross behavioral impairments may not be accompanied by discernible CNS lesions on postmortem examination. The role of abnormalities in neurotransmitter release and uptake in behavioral problems is unknown.

The argument for experimental behavioral studies is not only that toxic effects of chemicals on the central nervous system may produce early discernible effects on behavioral function, but more importantly that the behavior of the organism is the end point of the functional integration of the nervous system encompassing sensory, motor and cognitive aspects. The functional capacity of the central nervous system cannot be determined by histological or even physiological studies independent of behavior analysis. Examination of both learned and unlearned (reflex and phylogenetically specialized) behaviors may reveal subtle deficits in CNS function, which may or may not be accompanied by demonstrable tissue pathology.

BEHAVIORAL TOXICOLOGY AND PHARMACOLOGY: SOME COMPARISONS

The inclusion of a session on behavioral toxicology in this symposium on the current status of behavioral pharmacology implies the several commonalities that are shared by both disciplines. However, there are also major differences between behavioral toxicology and behavioral pharmacology which dictate some differ-

ences in methodology between the two fields. In behavioral pharmacology, a psychoactive agent is known or suspected to have short-term *transitory* effects on the central nervous system which are reflected in behavioral function. The drug can be administered in a range of doses and the effects of the drug on performance in a variety of dimensions can be determined.

In behavioral toxicology the primary problem is *not* to devise a behavioral characterization of a transitory drug effect but rather to detect the effect of an agent that may cumulate very slowly through time and produce discernible behavioral effects only after it reaches a critical concentration. Prolonged exposure to the toxic agent may permanently compromise neurobehavioral function. Some familiar examples are the effects of methylmercury poisoning and lead poisoning.

Many of the most powerful techniques used in behavioral pharmacology have been developed from Skinner's experimental analysis of behavior (8, 17). Operant techniques are equally applicable to the study of behavioral toxicology and have provided the basic models for behavioral analysis. Accurate psychophysical techniques have been developed to assess the sensitivity of sensory systems (20). However, particularly in the study of developing organisms, a number of very simple tests of basic motor function have also proved sensitive to the toxic effects of chemicals (see 18). As the work of Spyker (18, 19) so clearly illustrates, a central nervous system deficit may become evident only upon a specific kind of behavioral challenge. These data have obvious implications for clinical evaluation insofar as a compromised central nervous system may be evident only upon presentation of additional challenge such as aging, stress, physical illness, and so on.

Exposure to toxic pollutants may occur through environmental pollution or through occupational exposure. Industrial pollutants often create widespread environmental hazards. The far-reaching effects of industrial and environmental exposure to asbestos have recently been dramatically documented (1). The individual who is exposed in an industrial or environmental situation may be quite unaware of his exposure, the effects, or in instances such as asbestos poisoning, the potentially lethal consequence of exposure until it is much too late. The deleterious effects of nonchemical pollutants such as noise and radiation have also been well documented (13).

In contrast to behavioral pharmacology, the chemical agents that usually concern behavioral toxicologists are rarely deliberately self-administered (e.g., instances of pica in children with consequent lead poisoning). However, there is increasing evidence that drugs of abuse may produce pathological toxic effects not unlike those of industrial and environmental pollutants. Some recent findings concerning amphetamine toxicity are discussed in this symposium (15). The several adverse, and often irreversible, medical consequences of alcohol abuse are well known (9, 12). However, it has only recently been shown that chronic alcohol abuse prior to and during pregnancy may severely compromise the neurobehavioral development of the progeny in man (7). Growth retardation has also been reported in rodents when females have been given high doses of morphine prior to conception (4). The importance of longitudinal research in behavioral teratology to evaluate potentially toxic agents is discussed in this symposium (19).

CLINICAL ASPECTS OF BEHAVIORAL TOXICOLOGY

A second methodological problem that confronts the behavioral toxicologist is that, in addition to the insidious onset of behavioral effects, the *clinical* symptom picture usually consists of primarily subjective complaints, few of which are unique, in the sense of pathognomonic of exposure to a particular toxic chemical. Some behavioral signs and symptoms that have been associated with chronic exposure to toxic chemicals and pollutants are summarized in Table 1. A cursory examination of this partial listing shows that many are not easily amenable to quantitative evaluation. The assessment of subjective states (e.g., mood, affect, pain, fatigue) is always limited by the subject's ability and willingness to provide accurate self-reports. Moreover, in view of the debilitating physical and medical concomitants of exposure to toxic chemicals, it would be difficult (and perhaps irrelevant) to establish whether depression, for example, was a primary symptom or a secondary reaction to progressive incapacity.

The frequency of occurrence of gross behavioral derangements such as disorientation and confusion, delusions, and homicidal behavior can be assessed with more objectivity. However, long before such behavioral problems were evident, some significant and perhaps irreversible central nervous system insult must have already occurred.

Even within the realm of sensory (discrimination) capacities, perception and memory, for which precise and reliable behavior measures are available, the early detection of a slight change might fall within the error of measurement. Gross changes would probably follow induction of permanent CNS damage. Moreover, even the most powerful operant dis-

TABLE 1. Some behavioral and neurological consequences of representative pollutants (5, 10, 13, 14, 16, 21)

	Carbon disulfide	Asbestos	Mercury	Lead	Carbon monoxide	Noise
Tremor			√			
Muscle weakness	√		√		*	
Headache			√		*	
Dizziness			√			
Seizures				√		
Sensory loss			√		*	√
Mental retardation			√	√		
Memory defects	√		√			
Severe insomnia	√					√
Fatigue			√		*	√
Drowsiness			√		*	
Irritability	√		√	√		√
Uncontrolled anger	√		√	√		
Homicide attempts	√					
Sexual impotence	√					
Crying spells			√			
Depression	√					
Delusions	√					
Hallucinations	√					
Confusion	√				√	
Disorientation					√	
Autism					√	
Suicide	√					
Death—direct medical consequence		√			√	

* Effects of acute intoxiction.   √ Chronic effects.

crimination procedures are open to some ambiguities of interpretation. A performance decrement, i.e., a reduction in the stimulus control of behavior, may be a function of changes in attention, in sensory capacity, in motivation, or in any combination of these several factors (see 11).

These several considerations seem to lead to the conclusion that the early detection and measurement of the behavioral effects of pollutants in man involve a number of methodological problems for which behavioral science can offer no immediate or definitive solutions. The early detection of subclinical changes in human behavior as a function of exposure to chemicals probably can-

not be accomplished with any degree of accuracy or reliability.

In fact most endogenous toxic states that are frequently heralded by subtle behavior changes, such as unusual fatigue or lassitude, are difficult to quantify objectively. Slowly developing and insidious hepatic and renal disease, with eventually fatal termination, may first be evidenced by such behavioral complaints long before good evidence of organ dysfunction can be determined with existing diagnostic techniques (6).

Similar problems in evaluation have also been encountered by investigators concerned with qualitative variations in ability in children; variations which are usually termed "minimal brain dysfunction." An exhaustive catalogue of interdisciplinary diagnostic evaluation procedures has been used in an effort to identify minimal brain dysfunction. These range from standard IQ and visual motor performance tests to the examination of discrimination capacities in all sensory modalities; screening for specific neurological signs and disorders of motor function, as well as examination of reasoning, memory, concept formation, speech and communication disorders; sleep disturbances; emotional and social capacities and academic achievements (2).

It is apparent from the foregoing that experimental animal studies of pollutant effects on behavior hold most promise for effective *prediction* whereas clinical studies are of necessity limited to after-the-fact *descriptions*. The nature of the clinical concomitants of exposure to toxic chemicals, as well as the possibility of irreversible CNS damage antecedent to symptom expression, severely reduces the potential utility of a *predictive* model in man.

CONCLUSIONS

The importance of devising techniques to detect the early consequences of exposure to toxic chemicals is a challenge of particular urgency. Unlike most intoxicating drugs of abuse, by the time the toxic agent has had a central and behavioral effect these effects may be irreversible. Without the existence of accurate data concerning both the behavioral and medical consequences of pollutant exposure, the basis for setting standards for environmental and occupational safety will be at best imprecise and at worst self-serving for the polluter (see 1).

Ideally behavioral toxicological studies will permit predictions of the deleterious effects of pollutants be-

fore such pollutants are released for contact with man. The ultimate public health consequences of a major effort to predict toxicity are difficult to estimate. However, on the basis of clinical studies of individuals who have been chronically exposed to chemical pollutants (5, 10, 13, 16, 21), it is apparent that a major investment in experimental predictive efforts is potentially a more constructive approach to these problems than elaborating a post-exposure index of toxicity and lethality. The talents and attention of behavioral pharmacologists are badly needed in this emerging area of behavioral analysis.

REFERENCES

1. BRODEUR, P. Annals of industry; casualties of the work place. (I-V), *New Yorker,* October 29–November 26, 1973.
2. CHALFANT, J. C., AND M. A. SCHEFFELIN. *Central processing dysfunction in children: A review of research.* NINDS Monograph No. 9. Washington, D. C.: U.S. Govt. Printing Office, 1969.
3. EVANS, H. L., V. G. LATIES AND B. WEISS. Behavioral effects of mercury and methylmercury. *Federation Proc.* 34: 1858, 1975.
4. FRIEDLER, G., AND J. COCHIN. Growth retardation in offspring of female rats treated with morphine prior to conception. *Science* 175: 654, 1972.
5. GARLAND, H., AND J. PEARCE. Neurological complications of carbon monoxide poisoning. *Q. J. Med.* 36: 445, 1967.
6. *Harrison's Principles of Internal Medicine* (6th Edition), edited by M. M. Wintrobe, R. O. Adams, I. L. Bennett, E. Braunwala, K. Isselbacher, R. Petersdorf and G. W. Thorne. New York: McGraw-Hill, 1970.
7. JONES, K. L., D. W. SMITH, A. P. STREISSGUTH AND N. C. MYRIANTHOPOULOS. Outcome in offspring of chronic alcoholic women. *Lancet* 1: 1076, 1974.
8. KELLEHER, R. T., AND W. H. MORSE. Determinants of the specificity of behavioral effects of drugs. *Ergeb. Physiol. Biol. Chem. Exp. Pharmakol.* 60: 1, 1968.

9. LIEBER, C. S., E. RUBIN AND L. M. DECARLI. Chronic and acute effects of ethanol on hepatic metabolism of ethanol, lipids and drugs: Correlation with ultrastructural changes. In: *Recent Advances in Studies of Alcoholism*, edited by N. K. Mello and J. H. Mendelson. Washington, D. C.: U. S. Govt. Printing Office, Pub. No. (HSM) 71-9045, 1971, p. 3.

10. MANCUSO, T. F., AND B. Z. LOCKE. Carbon disulphide as a cause of suicide: Epidemiological study of viscose rayon workers. *J. Occup. Med.* 14: 595, 1972.

11. MELLO, N. K. Behavioral Studies of Alcoholism. In: *The Biology of Alcoholism: Vol. II, Physiology and Behavior*, edited by B. Kissin and H. Begleiter. New York: Plenum, 1972, p. 219.

12. MENDELSON, J. H. Biological concomitants of alcoholism. *N. Engl. J. Med.* 283: 24 and 71, 1970.

13. *Pollution: Its impact on mental health.* DHEW Publication No. (HSM) 72-9135. Washington, D. C.: U. S. Govt. Printing Office, 1972.

14. SCHULTE, J. H. Effects of mild carbon monoxide intoxication. *Arch. Environ. Health* 7: 524, 1963.

15. SCHUSTER, C. R., AND M. W. FISCHMAN. Amphetamine toxicity: Behavioral and neuropathological indexes. *Federation Proc.* 34: 1845, 1975.

16. SHERRILL, R. Asbestos, the saver of lives, has a deadly side. *N. Y. Times Magazine* Jan. 21, 1973.

17. SKINNER, B. F. *The Behavior of Organisms, An Experimental Analysis.* New York: Appleton-Century-Crofts, 1938.

18. SPYKER, J. M., S. B. SPARBER AND A. M. GOLDBERG. Subtle consequences of methylmercury exposure: Behavioral deviations in offspring of treated mothers. *Science* 177: 621, 1972.

19. SPYKER, J. M. Assessing the impact of low level chemicals on development: Behavioral and latent effects. *Federation Proc.* 34: 1835, 1975.

20. STEBBINS, W. C. *Animal Psychophysics: The Design and Conduct of Sensory Experiments.* New York: Appleton-Century-Crofts, 1970.

21. WARKANY, J., AND D. M. HUBBARD. Adverse mercurial reactions in the form of acrodynia and related conditions. *Am. J. Dis. Child.* 81: 335, 1951.

22. WEISS, B., AND V. G. LATIES. Behavioral pharmacology and toxicology. *Ann. Rev. Pharmacol.* 9: 297, 1969.

23. WEISS, B., AND V.G. LATIES. *Behavioral Toxicology.* New York: Plenum, 1975.

24. XINTARAS, C., B.L. JOHNSON, AND I. DEGROOT. *Behavioral Toxicology: Early Detection of Occupational Hazards.* DHEW Publication No. 74-126. Washington, D.C.: U.S. Govt. Printing Office, 1974.

Assessing the impact of low level chemicals on development: behavioral and latent effects[1]

JOAN M. SPYKER

University of Virginia Medical School, Department of Anatomy
Charlottesville, Virginia 22901

ABSTRACT

There is growing evidence that nervous tissue, especially the brain, is more sensitive to many foreign chemical substances than has previously been suspected, and that toxic effects may be manifested as subtle disturbances of behavior long before any classical symptoms of poisoning become apparent. Early detection of an insidious toxic process (behavioral toxicology) may enable the prevention or attenuation of harm to humans and other organisms. Adding to both the sensitivity and complexity of behavioral toxicologic testing is the increasing evidence that individuals are more vulnerable to adverse factors during the period of development (conception → puberty) than at any other time in life. Subtle functional disturbances in organisms exposed while immature (behavioral teratology) may be one of the most sensitive indicators of chemical toxicity. Furthermore, defects in a developmental process may have only delayed effects. A morphological or biochemical lesion can be dormant and not manifest itself until later in life as a behavioral disorder, mental deficiency, or overt functional impairment. Longitudinal evaluation is required to detect long-term or delayed effects of a particular developmental influence on biological and behavioral functions. Examples from research on the subtle and latent consequences of prenatal and early postnatal exposure to methylmercury that illustrate the above principles are presented. It is concluded that behavioral and long-term evaluation of organisms exposed during development are essential for a thorough assessment of the impact of certain low level chemicals on human health.— SPYKER, J. M. Assessing the impact of low level chemicals on development: behavioral and latent effects. *Federation Proc.* 34: 1835–1844, 1975.

Industrial wastes, pesticides, food and fuel additives, drugs, herbicides, fungicides, and numerous environmental pollutants represent chemicals to which humans are routinely exposed. Methylmercury and lead are familiar examples of contaminants that have accumulated slowly through time, reached critical levels, and unexpectedly caused permanent, deleterious effects in humans and other

[1] Supported in part by a grant from The National Foundation/March of Dimes and Public Health Service Grant FR05431.

161

organisms. There is a pressing need to find out what effects the substances accumulating in our environment may be having on us.

Detection of the insidious onset of toxically induced pathological processes presents a major challenge. Modern toxicology has the capability of predicting, attenuating and even preventing deleterious effects that may result from the ubiquitous chemicals found in man's environment. Behavioral teratology and toxicology play key roles here.

BEHAVIORAL TOXICOLOGY

Behavioral changes may serve as the earliest indicators that some, as yet covert, toxic action is occurring—perhaps at a time when the process can still be reversed. There is growing evidence that nervous tissue, especially the brain, is more sensitive to many foreign chemical substances than has previously been suspected, and that toxic effects may be manifested as subtle disturbances of behavior long before any classical symptoms of poisoning become apparent (4, 24, 25, 33).

The detection of an insidious toxic process and its cumulative effects through time may be greatly facilitated by sensitive and reliable behavioral evaluation procedures. Repeated behavioral samples can measure the extent of reversibility of toxic effects and reveal delayed and progressive impairments. Thus, changes in either isolated or functionally related behaviors can serve as early warning indicators of potential damage to organisms and their environment.

Behavioral processes are also important in themselves. Deficits in intellectual processes, sensory function, motor control (especially coordination and skilled performance), emotional responses and so forth, may be exceedingly disadvantageous to an

organism even though morbidity and mortality may remain unaffected.

VULNERABILITY OF DEVELOPING ORGANISMS

Adding to both the sensitivity and complexity of behavioral toxicologic testing is increasing evidence that exposure to chemicals while immature is more likely to produce toxic effects than exposure as an adult. It is now fairly well accepted that an individual is more vulnerable to certain adverse factors during the period of development than at any other time in life. Distinguishing features that contribute to the developing organism being more vulnerable to chemical insult than the mature organism include differences in metabolizing enzymes (12), excretory capacity (2), degree of development of protective systems like the blood–brain barrier (20), binding capacities of the serum and tissue proteins (14), proportion and distribution of various tissues (3), and differences in tissue concentrations of the chemical (20).

An organism continues to develop both prenatally (embryo → fetus) and postnatally (neonate → infant → child → adolescent) until puberty is reached. Thus, reference will be made to developing organisms as encompassing all immature stages. Although prepubertal individuals (because they are incompletely developed) are still at more risk than adults, most of our evidence for the vulnerability of developing organisms comes from humans and experimental animals exposed during the prenatal period.

It was previously believed that a placental barrier protected the fetus. Although the maternal organism may alter a chemical or at least reduce its concentration, the function of the placenta as a barrier is limited and molecules of most substances can cross the "barrier" either by simple. diffusion or by some type of active transport system. Consequently, many chemicals entering the pregnant woman ultimately will be found in the fetus. Furthermore, the immature organism does not have the same capacities as the adult to metabolize and detoxify noxious substances. It has been shown that the fetus and newborn have not yet developed the mechanisms to detoxify and excrete a variety of drugs and environmental chemicals (8, 17). Perhaps nowhere is the vulnerability of the unborn more dramatically evident than in the thousands of congenital malformations and severe functional deficits resulting from prenatal exposure to certain drugs, radiation, industrial wastes, and other chemicals in our environment (9, 15, 18, 21). Almost without exception, the mothers were unaffected.

Testing of new drugs in pregnant experimental animals has been required in the United States ever since the thalidomide tragedy of the 1960's —which revealed for the first time that a drug, given with the best intentions for the benefit of the mother, could have disastrous consequences for the human fetus. These teratological testing procedures are primarily designed to uncover the potential of a substance, when given to the mother, to cause death, structural abnormalities, growth retardation, or overt functional impairment in the young. However, it is unlikely that prenatal exposure to chemicals at levels routinely encountered in our environment will result in clinically evident birth defects. The problem is that, in the absence of obvious impairment, subclinical damage may still exist and may be overtly expressed with age.

BEHAVIORAL TERATOLOGY

Assessment of the subtle functional consequences of an insult (e.g., exposure to a toxic substance) during development has been termed "Behavioral Teratology." The underlying theme is that teratogens may have special affinities for particular developing fetal brain centers, that the developing brain is very vulnerable to insult, and that alterations in neurodevelopment become manifest as alterations in behavior. The developmental deviation is thought to be of a neuroanatomical (perhaps seen only at the ultrastructural level) or neurochemical nature. However, since behavior represents an integrated response of the organism, an impairment in the functioning of systems other than the nervous system may also be reflected as a behavioral change.

Teratology can be defined as the study of the adverse effects of the environment (i.e., everything outside the organism) on developing systems. A more comprehensive definition is that teratology is the science dealing with the causes, mechanisms, and manifestations of developmental deviations of either structural or functional nature (37). Behavior is at least as susceptible to teratogenic influence as other developing systems. However, unlike structural birth defects, subtle behavioral abnormalities are not readily evident and may be revealed only by special tests during postnatal life. Particularly at low levels, teratogens may cause behavioral changes in the absence of gross functional or structural defects (1, 7, 11, 30, 31, 34, 35, 38).

In behavioral evaluation of subjects exposed to a toxin during development, the factors affecting both teratology and behavioral toxicology testing should be taken into account. Figure 1 schematically illustrates the areas of responsibility and overlap in the emerging science of behavioral teratology. Basic considerations in teratology and behavioral toxicology, respectively, are briefly summarized in the following two paragraphs;

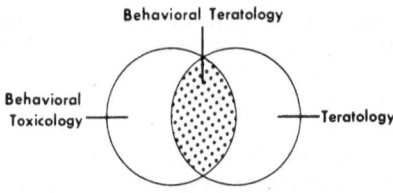

Figure 1. Schematic representation of the areas of responsibility and overlap in behavioral teratology; the principles of teratology as well as of behavioral toxicology must be considered.

these factors are discussed in detail in the references cited.

Although many questions remain to be answered regarding the behavioral consequences of teratogens, it appears that those factors that determine the type and extent of structural abnormality also influence the type and extent of behavioral abnormality. These five, basic factors can be outlined as follows: a given inherited *genetic predisposition*, linked together with a *particular teratogen*, administered at a certain *dosage level*[2], during a specific *stage of development*[2] results in the observed abnormality. Finally, since the maternal organism provides the environment for the developing organism, *maternal/fetal-offspring interaction* is also a factor that must be considered (13, 36).

Results of behavioral tests are likewise influenced by a number of factors. The most important influences include the housing and testing environments that constrain and shape the organism's behavior (16), the consequences of various responses that

organisms perform during the tests (22), past experience in laboratory and natural settings (10), and the adaptive significance and evolutionary history of the behaviors chosen for observation (6). All of these factors must be considered when evaluating the significance of behavioral test results.

LATENT EFFECTS

Defects in a developmental process may not become evident for years. A morphological or biochemical lesion can be dormant and not manifest itself until later in life as a behavioral disorder, mental deficiency, or overt functional impairment (19, 26, 32). Perhaps there are compensatory mechanisms that initially mask the defect, but are not adequate as aging, repeated exposure to stress, and cell death occur. Delayed effects can be uncovered by long-term evaluation.

Long-term evaluation of animals chronically exposed as adults to a particular drug or substance is part of standard toxicologic testing procedure. Long-term assessment of subjects exposed while immature is seldom done, yet may be a more revealing indicator of the potential toxicity of the substance in question. This approach is especially warranted if children, adolescents and women of child-bearing age are anticipated to be in the "exposed population."

Determination of long-term or delayed effects of a particular developmental influence on biological or behavioral functions requires the use of a "longitudinal"[3] research design

---

[2] In many behavioral teratologic tests to stimulate the real-life situation, the dosage level employed is well below the "teratogenic range," per se, (i.e., dosage sufficient to interfere with specific developmental events without destroying the whole embryo) and the organism is exposed during all stages of development (i.e., the mother is treated throughout pregnancy).

[3] Most developmental research uses the "cross-sectional" (versus longitudinal) method, which involves studying several age periods concurrently, with the assumption that antecedent events would have been held constant. This method cannot address those questions concerned with long-term changes.

Periods in Life-span→

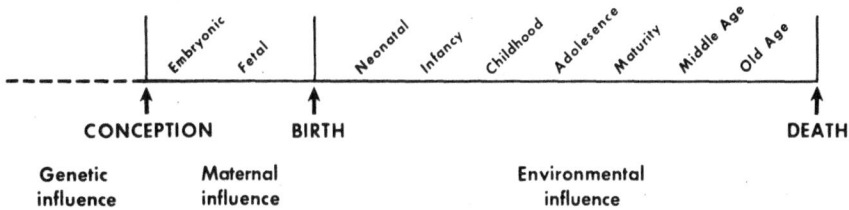

Figure 2. A longitudinal research design must be employed to detect delayed and long-term consequences of an insult during development. This involves following specific individuals, controlling or monitoring genetic background, controlling or monitoring prenatal and postnatal influences, and periodically assaying biological and behavioral functions during the subject's life-span.

(Fig. 2). This involves following specific individuals from birth through maturity, satisfactorily controlling or monitoring genetic background and environmental experiences, and periodically assaying biological and behavioral functions.

Since changes may be seen at one period in life and not at another, it is important to assay functions at each major stage of the subject's life-span. Tests in animal subjects with a short life-span permit the complete study of developmental effects from conception to death in a relatively brief period (e.g., the life-span in mice is approximately 2 years).

MATURATIONAL AND BEHAVIORAL EVALUATION

One form of behavior may be affected and not another following interference with a particular developmental process. Since it is extremely difficult to predict what types of subtle or delayed effects will be seen, or when, it is best to utilize a variety of maturational and behavioral measures at various periods of the life-span.

The discriminative power of any behavioral analysis increases as a function of the range of behavioral end

points examined. The following categories are frequently included in postnatal functional evaluation:

1) Morphological and physical characteristics: e.g., congenital defects; sex; general appearance, such as posture, fur condition; objective signs, such as cyanosis, ptosis.

2) Maturational landmarks: e.g., age of ear opening, eye opening, piliation, rearing, mating; delay in these parameters often implies that other processes will be retarded or otherwise affected.

3) Growth: e.g., weight gain at normal rates; growth is frequently used as the best index of general health.

4) Specific reflexes, responses and sensory-motor capacities: e.g., righting reflex, corneal reflex, grasping, orienting, response to nose and tail pinch; test especially during the first few weeks of life to assess maturation.

5) Activity levels: e.g., measures of hyper- or hypoactivity at various ages, such as spontaneous activity in home cage, activity wheel, open-field; distinguish type of activity change, such as tremor or locomotion.

6) Neuromuscular ability: e.g., tests of coordination, strength, speed, endurance, agility; gait evaluation.

7) Sensory functions: e.g., tests of the "intactness" of various sensory processes, such as vision, audition, olfaction, somesthesia.

8) Learning ability: e.g., measures of learning, reasoning and retention; vary difficulty from simple classical conditioning to complex operant behavior.

9) Emotionality: e.g., measures to assess role of autonomic nervous system, such as response to a foreign environment; techniques may be parallel to evaluating role of personality variables in man.

10) Sexual parameters: e.g., sexual development, sex role, reproductive efficiency, mating behavior, maternal behavior, fertility rate.

The categories listed above are relatively exhaustive. Even if money, time and expertise were no object, it is impractical—as well as impossible—to test every biological and behavioral function. The results of cost:risk:benefit analysis of the chemical in question, knowledge of its distribution and mechanism of action in the pregnant organism, as well as results from an initial dose-response evaluation will help guide the investigator in choice and number of test procedures to be employed. In addition, careful, periodic observation of an experimental colony in a well-controlled environment can often provide good clues as to where to look. In fact, one of the most useful applications of this type of research design is to *predict*— that is, to give clues of what to look for and monitor in exposed human populations in order to pick up deficits early—hopefully at a time when they are still reversible.

The evaluation should be done during the early stages of the animals' life-span, i.e., in the mouse, birth through puberty (3–4 wks) and into young adulthood (2–6 months). If no deviations are detected during the maturational screen, some offspring should be maintained for longitudinal testing and biological and behavioral

functions periodically assayed to determine if delayed effects can arise from prenatal or early postnatal exposure to the chemical in question. Initially, a dose-response function for readily observable central nervous system effects should be obtained.

CROSS-FOSTERING PROCEDURES

Longitudinal research in behavioral teratology is fraught with potentially confounding variables that may obscure or contribute to real effects. One of the most subtle such influences is that of postnatal effects induced by the mother. In mammals, any experimental maternal treatment producing prenatal effects (i.e., chemical effects on the fetus directly via placental transfer or indirectly by interfering with placental function) must also be considered capable of affecting offspring postnatally. Maternal residual (postnatal) chemical effects may be mediated directly via the milk of the nursing mother, or indirectly through maternal neglect of offspring and other early experience factors (e.g., aberrant maternal retrieving, grooming and activity).

In order to separate prenatal from postnatal influences on subsequent maturation and development, fostering (exchanging offspring with similarly treated mothers) and cross-fostering (exchanging treated progeny with control mothers and vice versa) procedures should be employed prior to nursing. In addition, a proportion of both control and experimental offspring should be raised by the biological mother to control for the fostering variable itself (Fig. 3). (Described in Spyker (25)).

EXAMPLES FROM RESEARCH

Generally speaking, the objective of research in our laboratory is to investi-

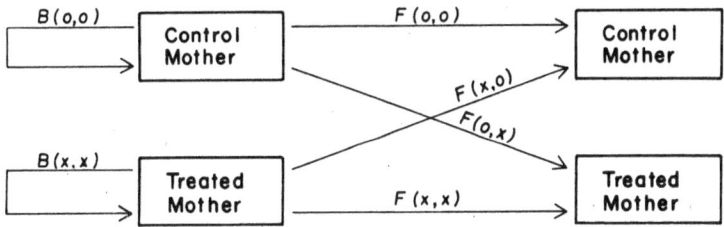

Figure 3. Schema of experimental groups generated according to maternal treatment and rearing-mother type, where: B = litters remained with their *biological* mother, F = litters were transferred to a *foster* mother. The figures in parentheses represent dosage (e.g., x = 2 mg methylmercury/kg mouse): first figure = dosage to biological mother, second figure = dosage to rearing mother. These procedures are employed at birth (before progeny nurse) in order to discriminate prenatal from postnatal influences in subsequent maturation and development.

gate the effects of environmental pollutants on immature organisms. Of special interest is the evaluation of subtle and long-term effects resulting from low-level exposure. In addition to detecting and describing effects at all stages of the life-span, attempts are being made to identify sites and mechanisms of action.

Longitudinal research design

Figure 4 schematically represents our experimental design and procedures for long-term evaluation of mice exposed to a chemical during development. The broken line represents the possible period of exposure either prenatally (via placental transfer) or postnatally (via mother's milk or treatment) or both.

At birth, the neonates are physically examined (P) and suitable proportions are fostered and cross-fostered (x-f). Various maturational parameters are evaluated between birth and puberty, at which time the offspring are weaned (21 days). At four major stages in the animals' life-span, i.e., adolescence (1 mo), maturity (6 mo → 1 yr), middle age (1 → 1.5 yr), and old age (1.5 → 2 yr), mice are examined for physical signs and

symptoms (P) and then functionally evaluated (F).

When maturational or behavioral deficits are detected, neurochemical, neuroanatomical, or other indicated procedures (e.g., immunological; see below) are done to complement functional findings.

The longitudinal research design illustrated in Fig. 4 is now being used in our laboratory to assess the subtle and delayed consequences of exposure to methylmercury[4] during development. This protocol for long-term evaluation was arrived at after my co-workers and I carried out developmental, behavioral, biochemical and morphological studies over the lifetime of mice from mothers exposed to methylmercury (MeHg) at different stages of gestation (5, 23–30).

In these studies, the pregnant female was treated on specific days during gestation in order to determine

---

[4] Methylmercury is a cumulative environmental contaminant, is currently affecting a large number of people, crosses blood–brain and placental barriers, concentrates in the fetus, and has primarily neurotoxic effects that may be reversible to some extent if detected early enough.

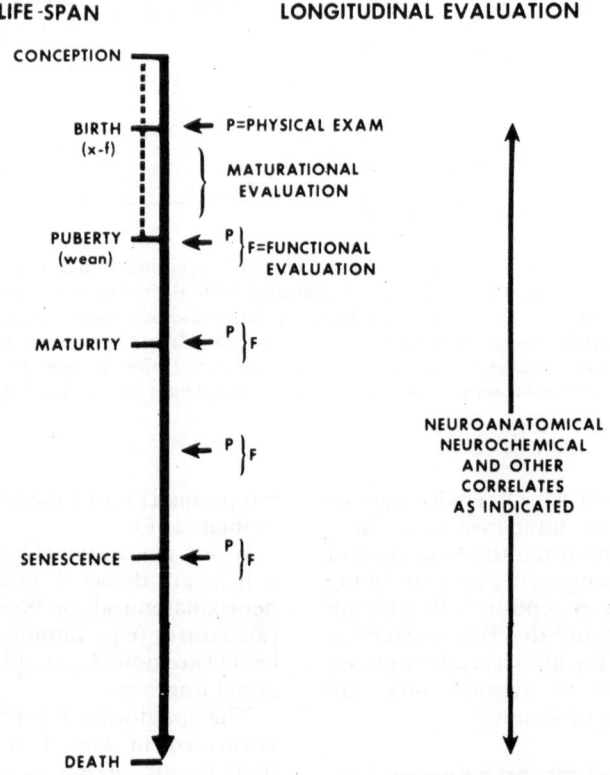

Figure 4. Schematic representation of the experimental design and procedures for long-term evaluation of mice exposed to a chemical during development. The broken line represents the possible period of exposure either prenatally (via placental transfer) or postnatally (via mother's milk or treatment) or both. (x-f) = cross-fostering procedures employed at birth.

the vulnerability of the embryo or fetus at different ages. In addition, some offspring were exposed to the chemical postnatally via mother's milk; others were cross-fostered to a control dam and thus were only exposed prenatally. (An alternate treatment approach is to chronically expose the mother to low doses prior to conception and throughout gestation. This method probably simulates the real-life situation better; however, it cannot answer the question of when in development the organism is most susceptible to adverse effects or what type of functional

deficit is associated with the stage of development when exposed).

Methods

On one of days 7, 9, 12 or 13 of pregnancy, nulliparous mice (strain 129/SvSl) received a single intraperitoneal (i.p.) injection of 0, 0.5, 1, 2, 4, or 8 mg/kg MeHg dicyandiamide (Panogen)[5] freshly dissolved in 0.5 ml

[5] Donated by NOR-AM Agricultural Products, Inc., Woodstock, Ill. The highest dose used (8 mg/kg) corresponds to approximately one-fourth the LD_{50} (median lethal dose) for the nonpregnant adult mouse.

of 0.9% NaCl/100 g body weight.[6] Females receiving only saline served as controls. One day prior to term (day 18), the females were put in separate cages and allowed to deliver. Sixty-two treated females produced 372 neonates that were apparently normal at birth. Subsequent developmental, behavioral, biochemical and morphological evaluation of these "normal" offspring during their lifespans produced some interesting and unexpected results. A summary of some of these findings are presented below. The detailed methodology and results are described separately (25).

Abnormal development

Although, on close inspection, treated and control neonates were indistinguishable at birth, MeHg retarded general growth and development in a significant number of offspring by 1 mo of age. Many of these animals never attained normal size. Not only did a larger number of prenatally exposed offspring, overtly normal at birth, grow less rapidly than controls, but a significant number died before weaning. As one might expect, both survival time and weight-gain data were dose-dependent. Functionally, neither control nor MeHg young exhibited any discernible neuromuscular deficits (such as retarded righting response, tremor, ataxic gait, leg dragging, hind-leg crossing) during maturation.

Behavioral deviations

The majority of offspring in the experimental colony appeared unaf-

[6] This experimental design (i.e., days of injection, doses of MeHg, strain of mouse) was employed because in a previous study to determine the effects of MeHg on prenatal development, using this design, 490 of 498 viable fetuses appeared normal when recovered at term (28). This surprising result prompted the postnatal study to see if deviations were detectable after birth in apparently unaffected offspring.

fected at the time of weaning, even though a significant number either were under-developed or did not survive. In an attempt to further elucidate how maternal MeHg exposure may affect postnatal development of apparently normal offspring, behavior was assessed. Our objective was to determine if behavioral deviations were detectable in prenatally and postnatally exposed animals that were grossly indistinguishable from controls. Treated and control mothers were also observed for possible differences in behavioral tests; no differences were found. Summaries of two behavioral tests are included here.

Open-field test: The open-field is an effective, widely used instrument for observing and quantifying basic behavior. This testing device is frequently used to measure the response of an organism to an unfamiliar environment. When 30-day-old, overtly normal offspring were tested in the open-field on 2 consecutive days, there was a significant difference between MeHg and control animals with respect to four of eight parameters evaluated (Table 1). When control animals were placed in the center of the field they proceeded forward and began exploratory activity. In addition to differences in the two indexes used most frequently to assess emotionality (defecation and urination), offspring from MeHg-treated mothers took a significantly longer time to begin exploration and when they did, a significant number took three or more backward steps initially or during the test period (30).

Swimming evaluation: Following the last period in the open-field, mice were placed in a glass tank filled with water at room temperature and their swimming behavior was observed for a period of 10 min. Video tape recordings were made for subsequent evaluation and confirmation of findings. In spite of the fact that all treated animals were grossly indistinguishable from controls at time of testing, behavioral differences were observed in a simple swimming apparatus. Figure 5 illustrates some of the deviant swimming behavior characteristic of the treated group (30).

Neuromuscular deficits

As the offspring reached adulthood (6 mo–1 yr), those exposed to the highest dose began showing signs of overt neurological impairment (e.g., tremor, incoordination, ataxia, difficulty righting). However, the majority of the colony was unremarkable upon routine observation. This observation prompted a behavioral study to assess motor ability in the apparently unaffected group.

From the colony of 12 to 15 month-old mice that appeared normal when observed in home cages, 10 animals were randomly selected from the following three treatment groups: 1) controls (born and reared by a saline mother), 2) prenatally exposed (born of a MeHg mother; reared by a saline mother), and 3) postnatally exposed (born of a saline mother; reared by a MeHg-mother). In each case the

MeHg-mother had received the 4 mg/kg dose. These 30 offspring were evaluated in each of the following test situations: 1) horizontal surface, 2) inclined plane, and 3) vertical grid (described in Spyker (25)). Table 2 shows the number of animals, according to treatment, exhibiting neuromuscular deficits when evaluated in the three test situations listed above. Evaluation of these older animals on the vertical grid was the most discriminating of the tests whereas assessment of motor ability on the inclined plane was the least sensitive. As indicated in Table 2, for each test situation more offspring were found abnormal in the group exposed to MeHg in utero than in the group that was exposed postnatally via the treated, rearing mother.

Immuno-deficiencies

Among the unexpected findings during long-term postnatal evaluation was an increased incidence of infection (usually eye infections) in older treated animals. Infections were seldom found in controls or younger treated animals. On laboratory evaluation, the afflicted animals proved

TABLE 1. The effect of prenatal exposure to methylmercury dicyandiamide on open-field performance

Parameter	Treatment[a]		Sex[a]	
	Saline[b]	MeHg[c]	Female	Male
Center latency, sec	5.5	7.1[e]	6.1	6.5
Center squares entered[d]	4.1	2.9	3.7	3.3
Peripheral squares entered	34.9	34.1	35.4	33.6
Defecation	2.9	1.7[e]	2.9	1.7[f]
Urination	0.89	0.52[f]	0.92	0.50[f]
Rearing	3.6	3.3	3.4	3.6
Grooming	1.3	0.65	0.58	1.3
Backing	0.03	0.37[e]	0.18	0.22

[a] Mean number of responses from 2-min trials on 2 consecutive days beginning at 30 days of age. [b] No. = 19 (8 female, 11 male). [c] No. = 20 (11 female, 9 male). Mother of test animal injected with 0.16 mg MeHg dicyandiamide/20 g body weight. [d] Entering is defined as placing all four legs into any square. [e] $= P < 0.05$, [f] $= P < 0.01$, 2-way least squares analysis of variance, no interactions significant.

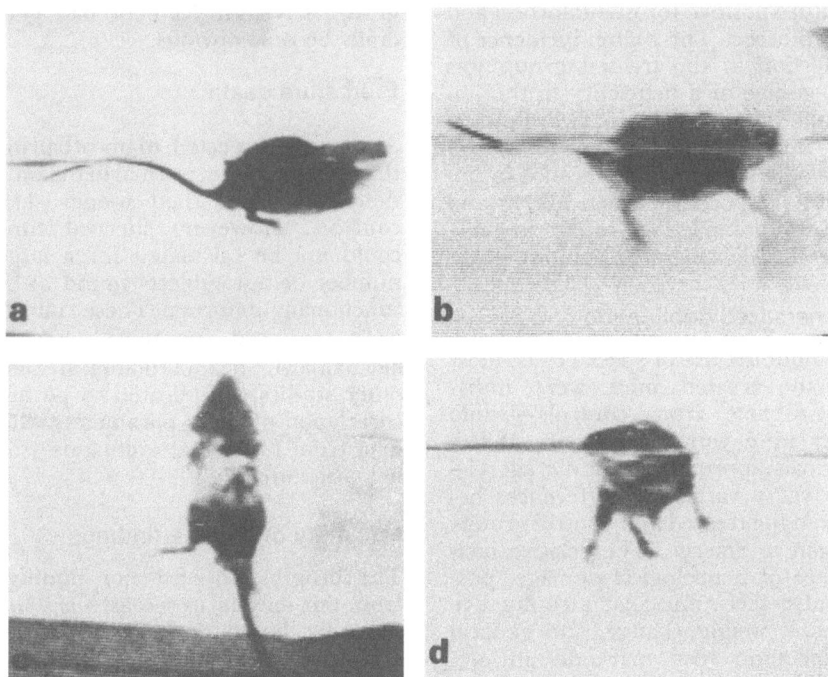

Figure 5. A significant number of apparently normal offspring from MeHg-treated mothers exhibited deviant behavior during swimming. Controls swam with *a*) front legs tucked, hind legs alternately kicking to turn or propel, and tail under water for balance and propulsion (note general posture with respect to waterline). Characteristic of the treated group were frequent episodes of incoordination and impaired swimming ability such as *b*) "freezing" in the water with all legs extended for periods up to 2 min (compare posture to that of control in Fig. 5a); *c*) floating suspended in a vertical position with only head above water; and *d*) swimming with legs askew and inability to maintain normal orientation in water. (From (30). Copyright 1972 by the American Association for the Advancement of Science.)

TABLE 2. Results of neuromuscular evaluation of mature mice exposed to methylmercury during prenatal or postnatal development

Treatment	Number tested	Number abnormal in test situation		
		Inclined plane	Horizontal surface	Vertical grid
Control	10	0	0	0 ⎤ a
Prenatally exposed	10	3	4	5 ⎬
Postnatally exposed	10	1	2	2 ⎦

a $P < 0.05$.

culture positive for pneumococci and streptococci. This higher incidence of infection in the treated group was suggestive of a deficiency in the immune system. Preliminary evaluation of immunological competence revealed a significant difference in immune response between MeHg and control animals (See under *Immunological evaluation*, next section).

Generalized debilitation

Without the use of special tests, most of the treated mice were indistinguishable from controls—while they were young. However, as the animals approached middle age (1– 1.5 yr), a variety of differences between the treated and control groups began to emerge. The relative incidence of neurological damage, postural defects, muscular atrophy, eye lesions, weight change, and general debilitation rose markedly in offspring exposed to methylmercury prenatally, via mothers' milk, or both. Some of these obvious differences are illustrated in Fig. 6. Subtle behavioral tests were no longer needed to significantly discriminate the treated

group; CNS involvement had generally become obvious.

Premature death

As might be expected, many offspring that deviated from normal in the ways described above died sooner than controls. However, survival time could not be calculated for a large number of test subjects found to be functionally abnormal. These animals were sacrificed for morphological, biochemical, histopathological and other studies as indicated to permit correlation of these parameters with data from functional evaluation (see next section).

Summary of lifetime findings

The progression of major findings from this long-term research is schematically illustrated in Fig. 7 in the order in which they were detected. In summary, offspring from treated mothers, although apparently normal at birth, responded differently from controls when tested for maturational and behavioral deviations at various stages throughout postnatal develop-

Figure 6. Figure illustrates representative abnormalities found during long-term evaluation of mice prenatally exposed to methylmercury. *a*) Purulent exudate apparent in right eye of offspring at 14 months of age. *b*) Same animal as in Fig. 6*a* three months later. Cornea of right eye is now dense, dull and opaque. Left eye appears unaffected. *c*) Same animal as in Fig. 6*a* and *b* when 2 yr old. Right eye has completely atrophied leaving only a sunken orbit filled with connective tissue. Left eye is now also undergoing atrophy, although no exudate was present. *d*) Kyphosis, an abnormally increased convexity of the thoracic spine, occurred frequently in the treated group after 1 yr of age. *e*) After kyphosis had developed, most of the deformed animals displayed muscular atrophy in the hind legs (left). Control (right) *f*) Treated offspring (1.5 yr old) which became obese within a 2-week period. *g*) Control (above) compared to obese animal from treated group (below). *h*) Generally debilitated female with prolapsed uterus photographed just before dying at 15 mo of age. *i*) Many offspring that behaved abnormally when young had severe neuromuscular deficits when older. Animal shown here is unable to right itself when placed on its side.

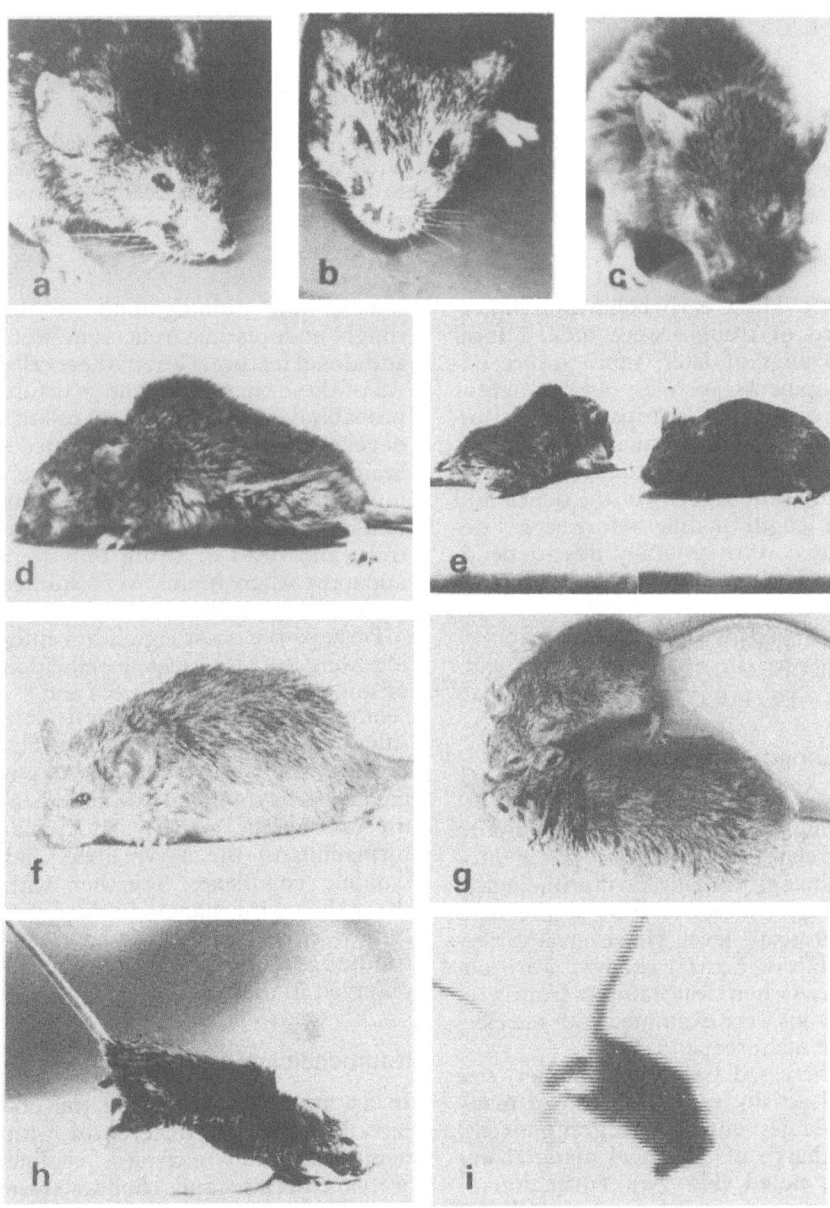

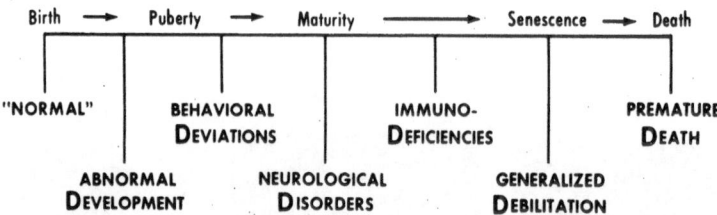

LIFETIME FINDINGS - *The "Six D's"*

Birth → Puberty → Maturity → Senescence → Death

"NORMAL"   BEHAVIORAL   IMMUNO-   PREMATURE
                  DEVIATIONS   DEFICIENCIES   DEATH

         ABNORMAL   NEUROLOGICAL   GENERALIZED
         DEVELOPMENT   DISORDERS   DEBILITATION

Figure 7. Schematic summary of six major ways animals prenatally exposed to methylmercury were found to be significantly different from controls as they grew older. All 372 offspring evaluated postnatally were apparently normal at birth. None of these deviations from normal would have been detected without long-term evaluation.

ment. These early behavioral indications of trouble were indeed forewarnings of later, more severe developments such as obvious motor impairment, neuropathology, postural problems, immuno-deficiencies, generalized debilitation, and early aging. The severity of the deficit and the length of time before it was detected were generally dose-dependent. The mothers were not affected.

MORPHOLOGICAL AND BIOCHEMICAL CORRELATES OF BEHAVIORAL DEFICITS

Neuroanatomical assessment

Brains of behaviorally abnormal offspring were studied in an attempt to correlate subtle deficits with neuropathology. Initially, no morphological changes were observed at the light microscope level. However, a variety of ultrastructural changes were observed when tissue samples from cerebellums were examined with the electron microscope (5, 26).

Increased lysosomal number, size and activity were observed in many Purkinje neurons and granule cells; discharge of lysosomal material was also noted (Fig. 8a). Formation of giant sized lysosomes or autophago-

somes, and disintegration of the rough endoplasmic reticulum were additional features seen in these cells. All of these changes in fine structure probably denote early signs of cellular degeneration of neurons. Indeed, brains from *older*, more severely involved animals in the treated group weighed significantly less than controls, and dead or dying cells were apparent when brains were studied under the light microscope.

Perhaps the most significant findings were the incomplete myelination of some nerve fibers (Fig. 8b) and absent or diminutive post-synaptic densities in many synaptic junctions (Fig. 8c and d). The size of the post-synaptic membrane is thought to be an indicator of synaptic activity. Such malformations of the nerve fibers and synaptic complexes, together with the pathological changes in the nerve cells described above, may have contributed to the behavioral deficiencies observed in these animals.

Neurochemical analysis

In an attempt to correlate the observed behavioral differences with neurotransmitter enzymes, choline acetyltransferase and cholinesterase determinations were done on the brains of 24 randomly chosen mice (equally distributed between treated and controls, males and females). A two-way (MeHg by sex) analysis of variance was done for each of the following parameters: diencephalic-telencephalic weight (everything above the superior colliculus); cerebellar weight; total milligrams protein in whole brain; and activity of choline acetyltransferase and cholinesterase (in μmoles/gram protein per hour or μmoles/brain per hour. The only significant ($P < 0.05$) effect was a sex difference in cerebellar weight (males > females). No significant alteration was found in any parameter between MeHg and saline offspring when evaluated at 1 mo of age (30). Perhaps neurochemical differences would have been detected if we had made determinations by brain region rather than by whole brain, or if we had assessed the animals later in their life-span.

Immunological evaluation

Awareness of an increased incidence of bacterial infection in older animals that had been exposed to MeHg in utero prompted a preliminary evaluation of the dual-natured immune system. Sheep red blood cells were used to selectively challenge the T-cell immune system (responsible for the expression of cellular immunity) and *Brucella abortus* antigen was used to challenge the B-cell system (responsible for the expression of humoral immunity). Ten young (4 mo) and 10 mature (14 mo) randomly chosen female offspring from both treated and control groups were immunized with the two antigens and antisera were collected to measure hemagglutinin titers (primary antibody producing capacity).

A highly significant low antibody response was detected in treated animals in the older group in-jected with *Brucella* antigen (Fig. 9). No other significant differences were found (27). These preliminary results indicate that the thymus-dependent (T cell) immune function was left intact, whereas thymus-independent (B cell) immune function, which resists infections caused by bacteria, was impaired in mature offspring from MeHg-treated mothers. The results may, in part, explain the increased incidence of bacterial infections in the older animals. This humoral deficiency may in turn relate to a plasma cell dysfunction as suggested by a high percentage of abnormal plasma cells found in the spleens of these animals (Chang and Spyker, unpublished observations).

Although this is a preliminary study, impaired immune function may indeed be a delayed effect of prenatal exposure to MeHg. Like other systems, the immune system is differentiating during fetal development. Furthermore, the two-compartment immune system can be distinguished quite early. Therefore, it is possible that insult during embryogenesis could affect one immune system and not the other.

I would like to emphasize here that this delayed effect, as well as many others, could not have been detected without the use of a longitudinal research design. This apparent dysfunction of the immune system may also be an example of how impairment of a system other than the nervous system can affect behavior. On the other hand, decreased immunological competence may be altering nervous system function and thus indirectly affecting behavior.

SUMMARY AND CONCLUSIONS

Behavioral toxicology

Behavioral assessment is an important component in the evaluation of ad-

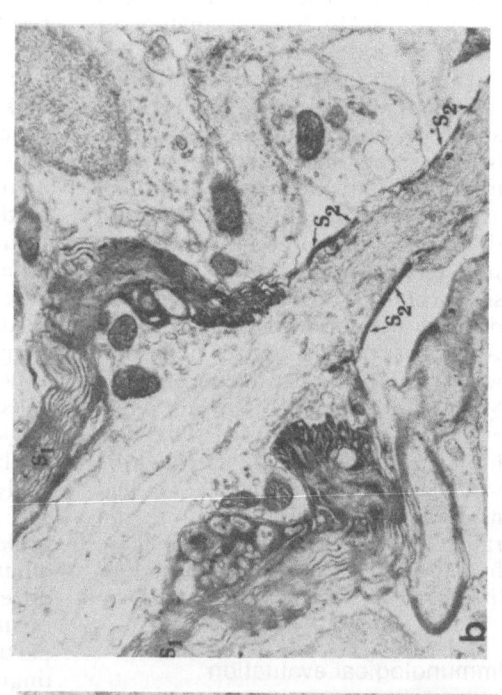

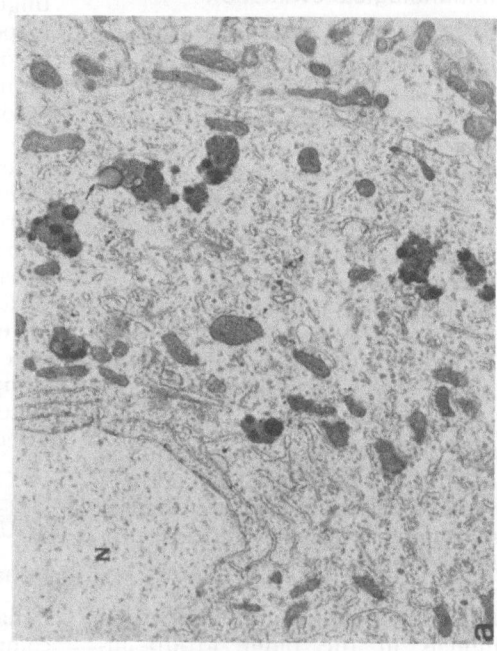

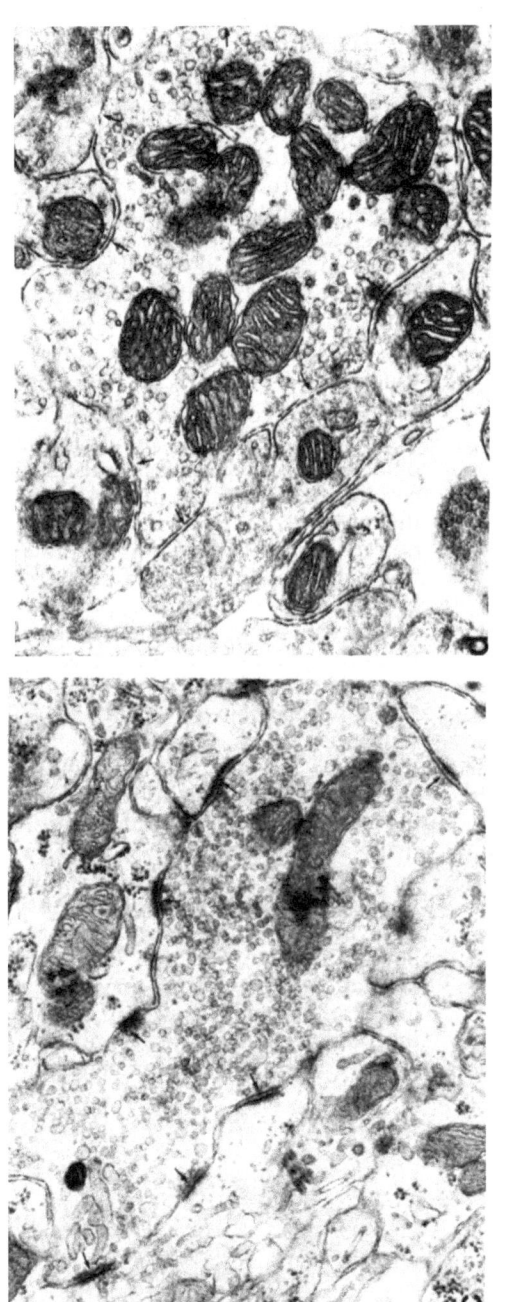

Figure 8. Electron micrographs showing alterations in ultrastructure of cerebellar cells from mice prenatally exposed to low levels of methylmercury. a) Purkinje neuron, treated. Note increased lysosomal activity. Release of lysosomal enzymes into the cytoplasmic matrix (→) is also apparent. ×17,000 b) Longitudinal section through a large nerve fiber at the node of Ranvier, treated. Reduction in myelination in one segment of the nerve fiber (S₂) as compared to the neighboring segment (S₁) of the same fiber is shown. Degenerative changes are also evident at the node of Ranvier. ×18,000. c) Mossy fiber terminal, control. Note the regularly arranged densities (→) and normal synaptic pattern. ×48,000. d) Mossy fiber terminal, treated. Note the reduction and absence of synaptic densities (→) in contrast to control in Fig. 8c.

PRIMARY IMMUNE RESPONSE

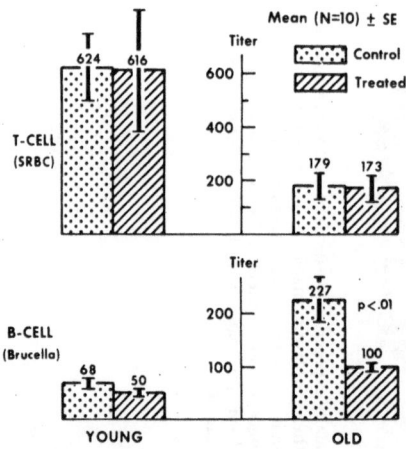

Figure 9. Impaired immune function may be a delayed effect of prenatal exposure to methylmercury. Sheep red blood cells (SRBC) and *Brucella abortus* antigen were used to selectively challenge the T-cell and B-cell immune systems, respectively. A highly significant low antibody response was detected in *older* treated animals injected with *Brucella* antigen.

verse effects for two main reasons: *1*) Subtle, behavioral changes may serve as early indicators or predictors of later, more severe, consequences. Early detection of toxic effects may enable the prevention or attenuation of harm to humans and other organisms; *2*) Behavioral deficits may be extremely disadvantageous to an organism and, therefore, are important in their own right.

Developing organisms

Individuals are more vulnerable to adverse factors during the period of development (conception → puberty) than at any other time in life. Thus, the fetus and child are at greater risk from toxic effects than the adult. Subtle functional disturbances in offspring from exposed mothers (behavioral teratology) may be one of the

most sensitive indicators of chemical toxicity.

Latent effects

Defects in a developmental process may be dormant and not manifest themselves until later in life as a behavioral disorder, mental deficiency, or overt functional impairment. Longitudinal evaluation is required to detect long-term or delayed effects of a particular developmental influence on biological and behavioral functions.

Safety standards

Exposure to certain substances during development can have subtle and/or delayed consequences that almost certainly would remain undetected by the test procedures currently required. There is need for continual review of the tests required for safety evaluations — and of the protocols for carrying out these tests — both from the point of view of protecting the public health and of assuring efficient utilization of scientific resources.

Due to the nature of behavioral changes, the vulnerability of developing organisms, the fact that we normally are exposed to low concentrations of chemicals, and finally, to the implications of the research already done in Behavioral Teratology, I believe that behavioral and long-term evaluation of organisms exposed during development are essential for a thorough assessment of the impact of low level chemicals on human health.

REFERENCES

1. ARMITAGE, S. G. Effects of barbiturates on behavior of rat offspring as measured in learning and reasoning situations. *J. Comp. Physiol. Psychol.* 45: 146, 1952.

2. BERNSTEIN, J. Postnatal development of the kidney. *Am. J. Pathol.* 66: 16a, 1972.

3. BOYD, E., AND C. KRIJNEN. Tolerated doses of phenacetin in relation to body weight and organ weights. *Jpn. J. Pharmacol.* 19: 386, 1969.

4. BRYCE-SMITH, D. Behavioral effects of lead and other heavy metal pollutants. *Chem. Br.* 8: 240, 1972.

5. CHANG, L. W., AND J. M. SPYKER. Ultrastructural changes in the nervous system after in utero exposure to low doses of methylmercury. *Acta Neuropath.* In press.

6. EIBL-EIBESFELDT, I. *Ethology, the Biology of Behavior.* New York: Holt, Rinehart & Winston, 1970.

7. FURCHTGOTT, E. Behavioral effects of ionizing radiations: 1955–1961. *Psychol. Bull.* 60: 157, 1963.

8. HAGERMAN, D., AND C. A. VILLEE. Transport functions of the placenta. *Physiol. Rev.* 40: 313, 1960.

9. HICKS, S. P., C. J. D'AMATO AND M. J. LOWE. The development of the mammalian nervous system. *J. Comp. Neurol.* 113: 435, 1959.

10. HINDE, R. A. *Animal Behavior: A Synthesis of Ethology and Comparative Psychology.* 2nd ed. New York: McGraw-Hill, 1970.

11. HOFFIELD, D. R., AND R. L. WEBSTER. Effect of injections of tranquilizing drugs during pregnancy on offspring. *Nature* 205: 1070, 1965.

12. JONDORF, W., R. MAICKEL AND B. BRODIE. Inability of newborn mice and guinea pigs to metabolize drugs. *Biochem. Pharmacol.* 1: 352, 1959.

13. KALTER, H. *Teratology of the Central Nervous System.* Chicago: University of Chicago Press, 1968.

14. KOBYLETZKI, D. Basis of prenatal medications: maternal-fetal distribution, peripartal elimination. In: *Prenatal Infections*, edited by O. Thalhammer. Stuttgart: Georg Thieme-Verlag, 1971.

15. LENZ, W. Thalidomide and congenital abnormalities. *Lancet* 1: 45, 1962.

16. MARLER, P., AND W. J. HAMILTON III. *Mechanisms of Animal Behavior.* New York: Wiley, 1966.

17. MOYA, F., AND B. E. SMITH. Distribution and placental transport of drugs and anesthetics. *Anesthesiology* 26: 45, 1965.

18. MURAKAMI, U. Embryo-fetoxic effect of some organic mercury compounds. *Annu. Rep. Res. Inst. Environ. Med., Nagoya Univ.* 18: 33, 1971.

19. NAIR, V., AND K. P. DuBOIS. Prenatal and early postnatal exposure to environmental contaminants. *Chicago Med. Sch. Q.* 27: 75, 1968.

20. NYHAN, W. Toxicity of drugs in the neonatal period. *J. Pediatr.* 59: 1, 1961.

21. RUGH, R., AND M. WOHLFROMM. Previous reproductive history and the susceptibility to X-ray induced congenital anomalies. *Nature* 210: 969, 1966.

22. SKINNER, B. F. *Contingencies of Reinforcement: A Theoretical Analysis.* New York: Appleton-Century-Crofts, 1969.

23. SPYKER, J. M. Methylmercury, mice and men. (Ph.D. Thesis), Minneapolis: Univ. of Minnesota, 1971 (summarized in *Diss. Abstr. Int.* 32: 8, 1972).

24. SPYKER, J. M. Subtle consequences of methylmercury exposure. *Teratology* 5: 267, 1972.

25. SPYKER, J. M. Behavioral teratology and toxicology. In: *Behavioral Toxicology*, edited by B. Weiss and V. G. Laties. New York: Plenum, 1975, p. 311–344.

26. SPYKER, J. M., AND L. W. CHANG. Delayed effects of prenatal exposure to methylmercury: brain ultrastructure and behavior. *Teratology* 9: A37, 1974.

27. SPYKER, J. M., AND G. FERNANDES. Impaired immune function in offspring from methylmercury-treated mice. *Teratology* 7: 28, 1973.

28. SPYKER, J. M., AND M. SMITHBERG. Effects of methylmercury on prenatal development in mice. *Teratology* 5: 181, 1972.

29. SPYKER, J. M., AND S. B. SPARBER. Behavioral teratology of methylmercury in the mouse. *Pharmacologist* 13: 275, 1971.

30. SPYKER, J. M., S. B. SPARBER AND A. M. GOLDBERG. Subtle consequences of methylmercury exposure: Behavioral deviations in offspring from treated mothers. *Science* 177: 621, 1972.

31. VAN GELDER, G. A., T. L. CARSON AND W. B. BUCK. Slowed learning in lambs prenatally exposed to lead. *Toxicol. Appl. Pharmacol.* 25: 466, 1973.

32. VORSTER, D. W. Psychiatric drugs and treatment in pregnancy. *Br. J. Psychol.* 3: 431, 1965.

33. WEISS, B., J. BROZEK, H. HANSON, R. C. LEAF, N. K. MELLO AND J. M. SPYKER. Effects on behavior, Chap. X. In: *The Evaluation of Chemicals for Societal Use*, edited by N. Nelson. Washington, D.C.: National Academy of Sciences, 1974.

34. WEISS, B., AND J. M. SPYKER. Behavioral implications of prenatal and early post-

natal exposure to chemical pollutants. *Pediatrics* 53: 851, 1974.

35. WERBOFF, J. Developmental psychopharmacology. In: *Principles of Psychopharmacology*, edited by W. G. Clark and J. del Guidice. New York: Academic, 1970.

36. WILSON, J. G. Embryological considerations in teratology. In: *Teratology:*

Principles and Techniques, edited by J. G. Wilson and J. Warkany. Chicago: University of Chicago Press, 1965.

37. WILSON, J. G. *Environment and Birth Defects.* New York: Academic, 1973.

38. YOUNG, R. G. Developmental psychopharmacology: a beginning. *Psychol. Bull.* 67: 73, 1967.

Amphetamine toxicity: behavioral and neuropathological indexes[1]

CHARLES R. SCHUSTER[2] AND MARIAN W. FISCHMAN

Departments of Psychiatry and Pharmacological and Physiological Sciences, University of Chicago Pritzker School of Medicine, Chicago, Illinois 60637

Toxicology, a science that "deals with poisons and their effects" (43), has traditionally emphasized the identification and analysis of chemicals that interfere with life processes. As techniques became more sophisticated, evaluation of morphological lesions came within the province of toxicologists. Only recently, however, has the possibility of correlating function and morphology been suggested to the student in toxicology (e.g., 42); he has been instructed to critically evaluate chemically induced physiological, biochemical and morphological changes. The indexes of toxicity have usually been alterations in blood and urine chemistry, body temperature, and the like. Many of these physiological changes may be the effect of irreversible morphological pathology, and therefore, in fact, indicative of life-threatening toxicity. These irreversible changes represent an end point in the development of overt indications of toxicity. It is of the utmost importance to be able to detect less obvious, but still destructive damage. What is needed are a series of sensitive measures that might, together or separately, indicate the presence of a potentially poisonous substance.

The category of substances that toxicologists study is drugs. The evaluation of drug toxicity employs a variety of different techniques, the usefulness of each being dependent on its relevance to the type of drug being studied. In this context, one aspect of function that has been relatively ignored is behavior, although there are many sensitive measuring devices available. Efficient and reliable methods exist for assessing small behavioral changes. It remains to the creative investigator to apply them appropriately. The

---

[1]Supported by Public Health Service Grant DA-00085 (W. Richter, Principal Investigator).
[2]C. R. Schuster is the recipient of Career Award MH-11,042.

181

proper design of a behavioral test could enable investigators to detect these small changes, and thus allow for short exposure to the toxin at relatively low doses. These tests could be used as a microscope to magnify any possible pathologies, pointing perhaps to functional changes that would otherwise go undetected. Initially, the organism would be given high doses of the drug, perhaps over an extended period of time, during which a wide variety of gross measures would be made. If the substance is toxic, this approach would ensure a substantial effect. It should then be possible to utilize the behavioral, biochemical and morphological data thus obtained to design more sensitive behavioral tests which allow for a fine-grained analysis of functional changes at much lower doses of the chemical in question. These doses would reflect more accurately the usual level of human exposure.

It is the purpose of this paper to outline an animal model for the assessment of a drug's toxicity. In this model are included behavioral assays, the results of which can be correlated with other functional and morphological changes occuring simultaneously in the experimental organism. The use of infrahuman organisms in toxicological research allows for analysis of some of the basic mechanisms involved in the toxicity and the range of experimental manipulations ethically possible is far greater than would be possible with man.

For the past several years our laboratory has been engaged in research investigating the behavioral changes induced by psychomotor stimulant drugs. We have been particularly interested in the amphetamines because of the numerous accounts of its abuse in many parts of the world over the last 10–15 years (2, 22, 27, 30, 32). First synthesized by Edeleano in 1887, the amphetamines were introduced into the medical pharmacopoeia in 1932 as a decongestant with the hope that they would prove therapeutically useful in the treatment of bronchial asthma. Although this promise was not fulfilled, the central nervous system stimulant effect of this class of drugs has become the basis for their therapeutic as well as their nonmedical use. These drugs produce an arousal state, which is demonstrable with either electroencephalographic (3) or behavioral (9) techniques as well as marked appetite suppression (41). These effects have been the basis for their therapeutic application in the treatment of obesity, mild depression, narcolepsy, counteracting the toxic effects of narcotics and sedatives, and in alleviating fatigue, as well as in their "paradoxical" reduction of childhood hyperkinesis. The subjective effects of these drugs have usually been perceived as a sense of increased energy, decisiveness and euphoria. It is probably for these reasons that they are also abused.

When administered in therapeutic doses, the amphetamines do not appear to be particularly toxic. It is only when these drugs are taken in doses exceeding those used therapeutically, usually on a somewhat chronic basis, that medical complications occur. During the past few years, clinical and experimental reports have appeared suggesting evidence of morphological and behavioral toxicity associated with amphetamine intake (6, 15, 29, 35). Research by Citron et al. (6) has led to the speculation that abuse of the amphetamines could lead to irreversible functional and morphological pathology. These authors described the appearance of a necrotizing angiitis in a series of polydrug abusers and suggested that methamphetamine was the agent responsible for this change. Clinical reports such as this are unfortunately confounded by variables such as the ingestion of other drugs, infections, and the predrug status of the patient. Laboratory studies, although more controlled, have not been sufficiently extensive to warrant any firm conclusions. Rumbaugh et al. (35), studying rhesus monkeys given 1.5 mg/kg intravenous methamphetamine every other day for 2 weeks, found extensive neuropathological changes including petechial hemorrhages and extensive cerebral edema. Kasirsky et al. (29) described similar vascular changes in rabbits given 300 mg/kg oral methamphetamine for 120 days. They also reported neuronal ganglion cell degeneration, fatty degeneration of the liver, and necrosis or degeneration of the renal tubules in these chronically dosed animals. Amarose, Schuster and Muller (1), on the other hand, found comparable doses of methamphetamine to be lethal in the rabbit. Intravenous doses of 2.5 mg/kg administered daily for 5 days produced no morphological (personal communication) or chromosomal damage in these animals. The hemorrhagic changes reported by Kasirsky et al. (29) are common terminal anoxic changes in the rabbit, and the other lesions can occur spontaneously in that organism. Therefore, it is possible that the drug or experimental procedure may have exacerbated a preexisting condition. In cats, intravenous amphetamine administration for 2 weeks has been reported to cause catecholamine depletion in neurons of the reticular system of the medulla (13) and chromatolysis in association with this change (15).

The implications of these studies for human amphetamine abusers are unclear. In each laboratory experiment in which methamphetamine was delivered on a chronic basis, only one dose of the drug was chosen, usually a high dose that causes immediate observable toxic effects in that species.

Under these circumstances, it is not surprising that some severe physiological changes might occur. Ellinwood (12) has pointed out that the toxic fatal or subfatal syndromes are seen, not in the chronic high-dose users, but in the relative neophyte user who overdoses. The most life-threatening aspect of stimulant overdose is the hyperthermia and convulsions that precede cardiovascular collapse. Ellinwood has suggested that if these symptoms are dealt with, the remaining amphetamine effects are not overwhelming. Harrison, Ambrus and Ambrus (23) have shown that tolerance to the temperature-increasing effects of the amphetamines does develop in rats over as short a period as 6 days. If this is a major mechanism of its toxicity, then, chronic use in gradually increasing doses could provide some protection against its lethal effects. Thus, there appear to be two separate problems when the clinical complications accompanying amphetamine abuse are considered. The first of these problems is seen in the death or near-death from overdose, reflected in the animal literature where only one large dose of the drug is administered to the organism and pathological changes are assessed after some period of time. In this case, the hyperthermia, possible convulsions, hypertension, and so on might very well be expected to produce irreversible morphological changes. On the other hand, human amphetamine abusers rarely begin their drug taking at these toxic doses. Rather, they generally start with relatively low oral doses, taken 3–4 times daily, switch to intravenous use to improve the quality of the drug experience, and then increase their dose as tolerance builds up, injecting drug from 4 to 10 times daily (32). The other aspect of clinical complications after amphetamine misuse is reflected in the psychosis associated with the chronic stimulant intake and the secondary accompaniments of the life style (e.g., malnutrition, infection, and the like). In addition, in the Rumbaugh et al. study (35), and frequently on the streets, ground methamphetamine tablets were dissolved and injected intravenously; the contribution of the fillers and binders in the tablets to the observed pathological changes remains to be determined.

The behavioral effects of chronically administered amphetamines can be assessed either while the animal is being maintained on drug, or after the drug has been discontinued. When drug is withdrawn, the animal can be observed for withdrawal symptoms and other short-term effects. In addition, longer term, possibly irreversible, behavioral changes might be measured several weeks to months later, after the animal has had time to readapt to the nondrug environment.

While animals are being maintained on chronic amphetamines, they show many of the same symptoms seen in the human drug abuse pattern, including anorexia and weight loss, open sores, hyperactivity, stereotyped activity and some evidence of visual hallucinations (11, 21, 33). In fact, in a recent paper, Ellinwood and co-workers (14) have drawn a parallel between the development of amphetamine psychosis in humans and the motoric changes seen in animals after chronic amphetamine intoxication. The stereotypies and perseveration of specific behaviors characterize chronic amphetamine intoxication in both man and infrahuman organisms. Tolerance, or the disappearance of most of this behavioral syndrome, does develop in monkeys and in man as the chronic regimen is continued (19, 32).

The effects of withdrawing methamphetamine after chronic adminis-

tration are unclear. Kalant (28) has concluded that physical dependence does develop, but the slow elimination of methamphetamine from the body may make it difficult to detect severe withdrawal symptoms. Both Feinberg and Irwin (16), in cats, and Herman (24), in rats, reported a decrease in activity when chronically administered amphetamine was abruptly withdrawn. "Crashing" or "coming down" from amphetamine has been described in the clinical literature (7, 40) and Smith (39) has described "classic withdrawal reactions" lasting 2–4 days in people. However, both Seevers, Ganz and Deneau (personal communication) and Fischman and Schuster (19) have reported no observable withdrawal symptoms after long-term high dose chronic administration of methamphetamine in the rhesus monkey.

When the amphetamines are taken chronically in therapeutic doses, there appears to be little or no behavioral malfunction. Unfortunately, tolerance soon develops to the euphoric (as well as other) effects of the drug, and the abuser continuously increases the dose (32). Although therapeutic doses of amphetamines are generally no higher than 15–60 mg/day, Kramer and his colleagues (32) have interviewed "speed freaks" who claim to inject 1–2 grams of methamphetamine 2–10 times daily. Kalant (28), casting doubt on these reports, has quoted Gibbons, who analyzed samples of "street drug" and concluded that total daily dose of the high intake stimulant abusers varies, but remains under 40 mg/kg or 3 grams/day. Even this more conservative estimate indicates phenomenal tolerance to both the behavioral and physiological actions of the drug. Until recently, no comparable data from the laboratory existed. Kosman and Unna (31), in a comprehensive review of the literature, pointed out that all of the data on the chronic behavioral effects of amphetamine had been obtained with subcutaneous or oral administration of the drug, not intravenous, as is usual with the human abuser. In an attempt to replicate and better understand the mechanisms underlying tolerance development, as well as to investigate some of the behavioral effects associated with long-term methamphetamine intake and their possible correlation with morphological and biochemical changes, Fischman and Schuster (19) undertook a series of experiments in which rhesus monkeys were maintained on a chronic methamphetamine regimen over a 3–10 month period.

BEHAVIORAL EFFECTS OF CHRONIC METHAMPHETAMINE

Prior to beginning the study of the chronic toxicity of methamphetamine, it was necessary to determine its acute intravenous lethal dose in the rhesus monkey. Nineteen male rhesus monkeys individually housed in wire mesh primate cages were the subjects. Each animal was removed from his cage, given a single, 1 ml intravenous injection of d-methamphetamine HCl[3] over a 10-sec time period, and immediately returned to his cage, where he was observed until the drug effect had dissipated or the animal died. Drug was administered in doses of 9.0, 5.0, 3.0, 2.5, 2.0 or 1.0 mg/kg. Animals receiving either 9.0 or 5.0 mg/kg methamphetamine collapsed almost immediately and all four of them died of respiratory arrest within 30 min following the injection. None of these animals showed any of the

usual sympathomimetic or stimulant effect usually associated with methamphetamine intake. The five monkeys tested with 2.5 and 3.0 mg/kg of methamphetamine also died. However, at these doses, sympathomimetic effects (piloerection, pupillary dilation, salivation), as well as gross tremors in the extremities, were observed. Death, associated with respiratory paralysis, took significantly longer to occur at these intermediate doses than at the higher doses. At the lower doses of 1.0 and 2.0 mg/kg, all of the animals tested survived. It is thus possible to roughly estimate that the acute intravenous LD_{50} dose of methamphetamine for the rhesus monkey housed under the environmental conditions of this experiment is between 2.0 and 3.0 mg/kg. Subsequent attempts to replicate these results using nine rhesus monkeys chronically maintained in primate chairs were unsuccesful. Doses of intravenous methamphetamine ranging from 2.0 mg/kg to 10.0 mg/kg were not lethal to the animals under those experimental conditions. A dose of 20.0 mg/kg methamphetamine was required to produce death. Clearly, the toxicity of this drug varies considerably with environmental conditions, a fact that has been pointed out previously. Höhn and Lasagna (25) for example have reported that the LD_{50} of amphetamine is 10 times as great for mice tested individually as compared with mice tested under aggregated conditions. In addition, such environmental variables as size of the cage, temperature of the test room, experimental history, and the like have effects on the LD_{50} of amphetamine (4, 5).

One of the initial decisions concerning the design of further experiments was whether to allow the experimental animals to self-adminis-

---

[3]d-Methamphetamine HCl was graciously supplied by Abbott Laboratories, North Chicago, Illinois.

ter the drug or to give them noncontingent infusions. (For a description of the self-administration procedures, see Schuster and Johanson, 37). Rhesus monkeys, given 24 hour/day access to intravenous injections of methamphetamine contingent upon a lever press response, will overdose and die in less than 15 days (Johanson, Balster and Bonese, in preparation). It was obvious that under these circumstances the toxicity of the amphetamines occurred before tolerance could develop. It was decided, therefore, not to allow the animals to self-administer the drug, but rather to program injections so that the dose administered could be controlled.

Rhesus monkeys, surgically prepared with intravenous catheters, were housed individually in wooden cubicles. Each monkey wore a stainless steel harness connected to a spring arm that was attached to the back wall of the cubicle. The spring arm had the additional function of protecting the intravenous catheter which was connected to the drug bottle via a peristaltic pump. A complete description of this apparatus can be found in Fischman and Schuster (20). Animals received one intravenous infusion of methamphetamine every 3 hours, eight times daily in doses starting at 0.0625 mg/kg per infusion (0.5 mg/kg per day) and gradually increasing to doses as high as 6.5 mg/kg per infusion (52 mg/kg per day). The effects of the chronic drug regimen on responding maintained by a fixed-ratio 10 schedule of food reinforcement were investigated. The major finding of this study (20) was the development of behavioral and physiological tolerance to doses of methamphetamine that would have been lethal to a drug-naive monkey. The initial severe interference with food-reinforced fixed-ratio responding disappeared and the monkeys

were eating and drinking normal amounts within a relatively short period of time. The behavioral disruption was also manifested by continuously repetitive picking and grooming behavior, most evident after doses of drug ranging from 4.0 to 10.0 mg/kg per day. This abnormal behavior continued during the 22-hr time-out from the experimental session long after the animals exhibited tolerance to the anorexic and response suppressant effects of the drug, reminiscent of the selective tolerance to the effects of amphetamine reported by Tormey and Lasagna (41), Schuster and Zimmerman (38), and Schuster, Dockens and Woods (36). Behavioral tolerance developed first to those aspects of the animals' repertoire that were under contingency control. It took considerably longer for tolerance to develop when the aberrant behavior did not interfere with meeting the reinforcement requirements. Terminally the animals were sacrificed for morphological[4] and biochemical analysis. These data will be discussed in a later section.

Although tolerance developed to the schedule controlled disrupting effects of methamphetamine on behavior maintained by a fixed-ratio schedule of food reinforcement, the animals did continue to show signs of toxicity that were grossly observable, e.g., stereotypical behavior. It therefore appeared that the measure of responding for food on a FR 10 schedule of reinforcement was not a sensitive indicator of drug toxicity. In fact, as tolerance devloped to the response suppressant effects of the drug, other drug effects such as hy-

---

[4] Morphological analysis was conducted by Ward R. Richter, D. V. M., Associate Professor of Pathology, University of Chicago, Pritzker School of Medicine.

DRL 40"

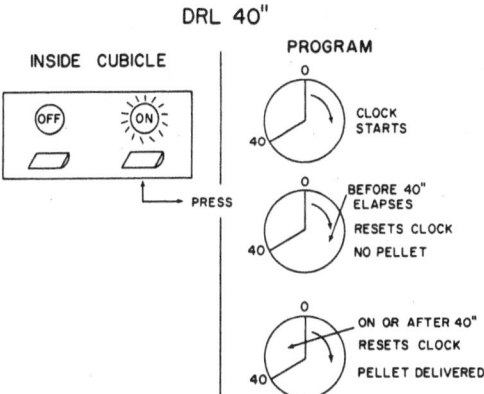

Figure 1. Schematic diagram of the differential reinforcement of low rate (DRL) 40" schedule of reinforcement. One lever is operable. Lever presses spaced by 40 sec or more are rein-forced with a food pellet. Lever presses spaced by less than 40 sec reset the continuously running clock and have no other programmed consequence.

peractivity acted to increase response rate to predrug levels and above. In view of these rate increases, it was decided that a task involving the slow patterning of responses might provide a sensitive behavioral assay of the effects of chronic methamphetamine intoxication. Therefore, a second experiment was performed in which food-deprived rhesus monkeys were trained to space lever-press responses by at least 40 sec. As diagrammed in Fig. 1, one lever in the cubicle is operable, and the programming clock runs continuously. Each lever press resets the clock to zero. Lever presses spaced by at least 40 sec are reinforced with a food pellet in this differential reinforcement of low rate schedule (DRL; Ferster and Skinner (17)).

An acute drug regimen was established in which one intravenous infusion of methamphetamine was administered 5 min prior to an experimental session every 3 or 4 days. In this way, dose–response relationships for number of responses and number of reinforcements per session

could be determined prior to exposure to the chronic methamphetamine regimen. The doses tested were: 0.0625, 0.125, 0.25, 0.5, 0.75 and 1.0 mg/kg. In Fig. 2, data from the acute regimen are shown for a representative animal responding on the DRL schedule of reinforcement. Interresponse time (IRT) distributions are presented showing the percent of total number of responses made within a session that were spaced by 1–10 sec, 11–20 sec, 21–30 sec, and so on. Theoretically, an efficient animal responding on a DRL 40" schedule would have a curve that peaked sharply at 40 sec and fell off rapidly on either side. The left hand column of the figure shows results obtained prior to the chronic drug regimen. For this animal there were decreases in response rate and number of reinforcements with increases in dose of methamphetamine. A dose of 1.0 mg/kg methamphetamine suppressed all responding. The monkeys were then maintained on the chronic methamphetamine regimen starting at 0.0625 mg/kg per infusion every 3 hours (0.5 mg/kg per day) and increasing gradually to 2.0 mg/kg per infusion (16.0 mg/kg per day). Figure 3 shows interresponse time distributions for the first and last days at selected doses for the same animal. Predictably, responding maintained by this schedule of food reinforcement was disrupted by methamphetamine administration. With increased exposure to a specific dose, however, tolerance developed, as was indicated by an increase in the number of reinforcements obtained per session. For animal A043 (Fig. 3), increases in drug dose at or above 1.0 mg/kg per in-

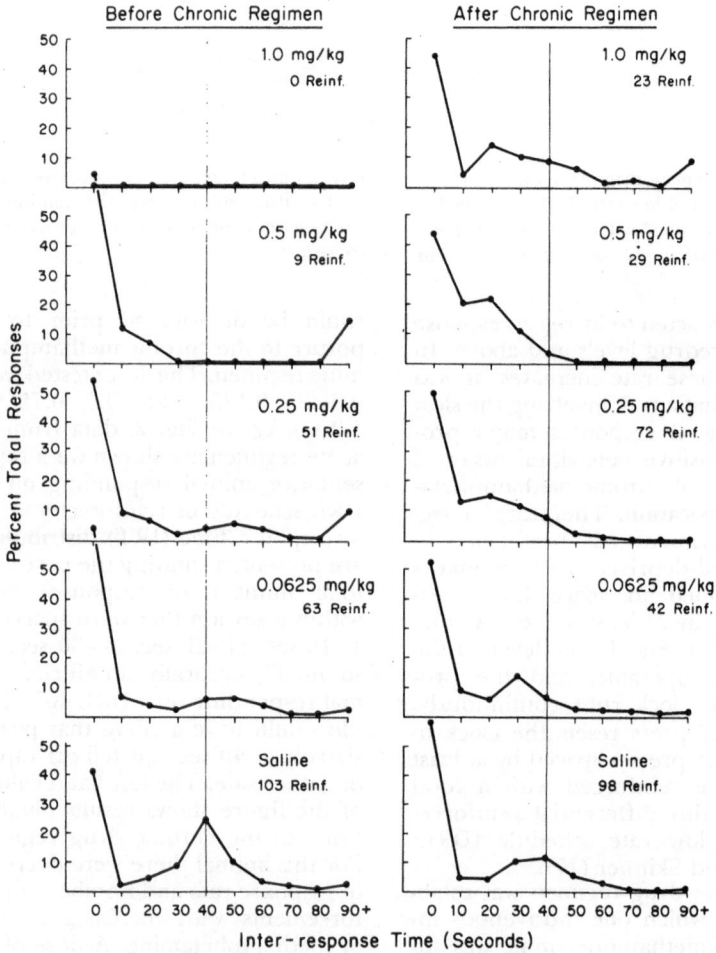

Figure 2. Interresponse time distributions for animal A043, working on the DRL 40″ schedule of reinforcement while on the acute methamphetamine regimen. Data are shown after selected acute doses of drug before (*left column*) and after (*right column*) the chronic drug regimen. In each case, the saline data represent a mean for the 5 days just prior to the acute drug regimen. The number of reinforcements per session is listed next to each interresponse time distribution.

fusion did not have a severely disruptive effect on responding if number of reinforcements obtained is used as the criterion. In addition, tolerance to the suppressant effects of methamphetamine is clearly present. This is apparent if the IRT distribution after an acute dose of 1.0 mg/kg methamphetamine prior to the chronic regimen (Fig. 2) is compared with the IRT distribution obtained on the first day that 1.0 mg/kg methamphetamine was administered during the chronic regimen (Fig. 3). On the latter day, this animal received 83 reinforcements and made 290 responses as opposed to 0 responses at that dose prior to the chronic regimen.

Following cessation of drug administration after 3–6 months on the chronic regimen, no gross behavioral or physiological signs of withdrawal were observed in any of the monkeys. No suppression of responding was seen, and, in fact, the IRT distributions tended to approximate the predrug distribution more closely than the distributions obtained under the chronic regimen. When acute dose–response curves were again determined 30–45 days after cessation of the chronic regimen, residual tolerance was seen. Figure 2 (*left column*) shows the lack of a response suppressant effect on responding for food at the doses of methamphetamine used. One mg/kg methamphetamine totally suppressed responding prior to the chronic administration of that drug, while after the chronic regimen, animals lever-pressed, and in some cases received as much as half the control number of reinforcements. In addition to the tolerance seen, an acute dose of 0.5 mg/kg after the chronic regimen had a distinctly excitatory effect. For example, A043 made approximately 450 responses per session during the postchronic saline

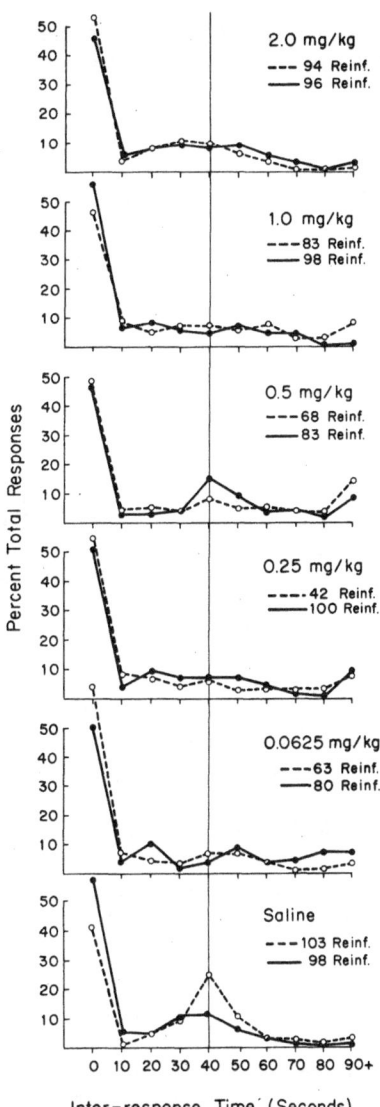

Figure 3. Interresponse time distributions for animal A043 working on the DRL 40″ schedule of reinforcement at selected doses on the chronic methamphetamine regimen. The first day at each drug dose is indicated by a dashed line; the last day by a solid line. The number of reinforcements for each session is also indicated.

control sessions. Responses per session were unchanged after 0.0625 mg/kg drug, increased to 662 after 0.25 mg/kg drug, and to 1004, more than twice the control level, after an acute dose of 0.5 mg/kg methamphetamine. Finally, at 1.0 mg/kg of drug, given acutely, responses per session decreased to 79. The excitatory effect was not seen in drug-naive animals. The large increase in response rate after methamphetamine is reminiscent of the increased responding seen on the FR 10 schedule for food reinforcement after monkeys chronically maintained on methamphetamine developed tolerance to the anorexic and response suppressant effects of the drug (19). The major effect seen in monkeys chronically maintained on methamphetamine for 3 to 6 months and then maintained drug free for 1–3 months is a residual tolerance to the response suppressant effects of the drug. This long-term behavioral change is possibly an irreversible effect of chronic maintenance on methamphetamine.

MORPHOLOGICAL AND BIOCHEMICAL CONSEQUENCES OF CHRONIC METHAMPHETAMINE

Despite the high dose of methamphetamine administered to the monkeys in both of the chronic studies and the concomitant behavioral changes seen, no gross neuropathological changes were observed during the complete necropsy performed after sacrifice. In addition, light microscopic examination of tissue collected during necropsy failed to reveal any abnormal changes. These findings are in contrast to the vascular changes that have been reported to occur in rabbits (29), monkeys (13, 35), and humans (6) after varying periods of time on methamphetamine. It is pos-

sible that transient lesions might occur and are repaired without any residual evidence in the animals receiving the drug chronically. However, the uninterrupted chronic regimen followed, in which dose was raised only after tolerance development, might have afforded some protection and avoided the morphological changes observed in the other studies.

Meltzer (unpublished) has found that the monkeys used in the present study, which were chronically maintained on intravenous methamphetamine in doses ranging from 2.0 to 6.5 mg/kg, showed skeletal muscle pathology. In frozen sections, the most common pathological findings were scattered atrophic fibers and abnormalities of mitochondrial staining, chiefly a complete lack of staining. Although variable, at least 5% of the fibers were pathological in one or more of the muscles studied. In muscle embedded in plastic, more subtle focal areas of myofibrillar degeneration and Z-band streaming were seen. Generally the level of pathology of this type was twice the amount found in saline-treated or predrug control samples.

Biochemical analyses of the brains of these animals showed no change in norepinephrine levels compared to those obtained from control animals. Dopamine, however, was considerably depleted in the caudate nucleus (Seiden, personal communication).

The assumption of irreversible pathology after chronic amphetamine use has been based on a paucity of experimental data. The lack of pathological changes in the brain of rhesus monkeys chronically maintained on high doses of methamphetamine for 3–6 mo (20) makes necessary a reassessment of the data on cerebral vascular pathologies reported by Rumbaugh et al. (35) in rhesus monkeys. There are, however, several rather basic procedural differences that might account for these inconsistencies. For example, the petechial hemorrhages reported after methamphetamine administered every other day for 2 weeks could very well be a consequence of the sudden and abrupt changes in blood pressure seen after a single moderately high dose of amphetamine. This abrupt rise in blood pressure might not occur when the drug dose is very gradually raised, since it is possible that tolerance did develop to the cardiac and pressor effects of the drug. Eble and Rudzik (10) have reported that, in the dog, there is a decrease in the systolic blood pressure response to amphetamine after repeated doses. In addition, Rosenberg et al. (34) have demonstrated tolerance in man to the systolic blood pressure and temperature elevation effects of d-amphetamine (0.6 mg/kg) after 14 days of chronic drug administration.

In an attempt to evaluate the relationship between amphetamine and cerebral vascular pathology, we have replicated the Rumbaugh study, excluding, however, the angiographic studies. Five adult male rhesus monkeys were intravenously injected with 1.5 mg/kg methamphetamine (pulverized Desoxyn tablets in physiological saline) every other day for 2 weeks. Control monkeys received only saline. At necropsy, after sacrifice, all animals failed to show any significant gross pathological lesions. In addition, histological examination of tissue samples failed to show any abnormal pathology previously reported with amphetamine administration. Since this experiment exactly duplicated that reported by Rumbaugh and colleagues (35), with the exception of the angiography, we are now exploring the possibility that the angiography or its interactions with the effects of methamphetamine might be responsible for the pathological changes previously reported.

IMPLICATIONS FOR FUTURE RESEARCH

The finding that large behavioral changes did not occur after drug was withdrawn is not surprising. It should be clear that the functional deficits evidenced by drug-induced behavioral changes that disappear with the development of tolerance cannot be expected to be present when drug is withdrawn. It is entirely possible that the initial behavioral disruption seen with chronic methamphetamine may have been correlated with central nervous system damage, but if so, it was not detectable using our techniques. Other areas of the brain may have taken over the necessary functions to enable the monkeys to return to behavior patterns generally similar to those seen prior to drug. The general functional flexibility of the nervous system is a problem that has plagued physiological psychologists for years. Time and time again, using lesion techniques to localize specific functions, they have discovered that deficits are short-lived and the behavior in question returns to prelesion levels (e.g., Fischman and Meikle (18)).

The preliminary findings reported here have redirected our research in several ways. As pointed out in an early section of this paper, it should be possible to demonstrate that functional changes can be shown to occur with shorter exposure or with exposure to lower doses of the toxic substance than are necessary to produce morphological changes. Therefore, we are determining whether the changes in muscle morphology and depletion of dopamine in the caudate can be produced in a 14-day regimen of rapidly escalated doses of methamphetamine. Preliminary evidence indicates that the muscle pathology and depletion of dopamine in the caudate can be produced by this shorter exposure to the drug. Finally, as was the case with the longer regimen, no gross signs of vascular pathology or cellular changes as seen under the light microscope are observed with this shorter regimen. It would therefore appear that we can produce the same morphological and biochemical changes in a 14-day exposure period that were produced by many months of exposure to methamphetamine. It remains to be seen, however, whether tolerance to the behaviorally disrupting effects of amphetamines is also seen with the shorter drug regimen.

In addition, because of the observed changes in muscle morphology and chemistry, we are developing several procedures to determine their functional significance. Animals are being trained in three separate tasks; one measuring gross strength of the arm; one measuring fine motor coordination; and one measuring the ability to perform complex eye tracking movements. The fact that these procedures were prompted by the observed muscle pathology illustrates the general strategy that has evolved from our interdisciplinary approach to the problem of measuring the toxicity of chronically administered intravenous methamphetamine. It is not sufficient to simply select any behavior that is sensitive to the acute actions of the drug in question. We are now attempting to utilize the information supplied by the neurochemical and morphological changes seen after chronic methamphetamine to design behavioral tests specific and sensitive enough to measure correlated functional changes. It will then be possible to determine whether behavioral measures show disruptions at lower doses of the drug than those

necessary to produce morphological changes. Clearly, if behavioral measures are to have any importance in the field of toxicology we must demonstrate their utility and predictive accuracy. Only further research will determine the efficacy of our strategy in the development of sensitive and predictive behavioral measures.

The authors would like to thank Mr. Gary Reynolds for his help in running experiments and maintaining the animals, and Dr. Chris Johanson for critical review of the manuscript.

REFERENCES

1. AMAROSE, A. P., C. R. SCHUSTER AND T. P. MULLER. An animal model for the evaluation of drug-induced chromosome damage. *Oncology* 27: 550–562, 1973.
2. ANGRIST, B. M., AND S. GERSHON. Amphetamine abuse in New York City— 1966–1968. *Semin. Psychiatry* 1: 195–207, 1969.
3. BRADLEY, P. M., AND B. J. KEY. The effect of drugs on the arousal responses produced by electrical stimulation of the reticular formation of the brain. *Electroencephalogr. Clin. Neurophysiol.* 10: 97–110, 1958.
4. CHANCE, M. Aggregation as a factor influencing the toxicity of sympathomimetic amines in mice. *J. Pharmacol. Exp. Ther.* 87: 214–219, 1946.
5. CHANCE, M. Factors influencing the toxicity of sympathomimetic amines to solitary mice. *J. Pharmacol. Exp. Ther.* 89: 289–296, 1947.
6. CITRON, B. P., M. HALPERN, M. McCARRON, G. D. LUNDBERG, R. McCORMICK, I. J., PINCUS, D. TATTER AND B. J. HAVERBACK. Necrotizing angiitis associated with drug abuse. *N. Engl. J. Med.* 283: 1003–1011, 1970.
7. COHN, S. Abuse of centrally stimulating agents among juveniles in California. In: *Symposium on the Abuse of Central Stimulants*, edited by F. Sjoquist and M. Tottie. New York: Raven, 1968.
8. COLE, J. N., AND A. D. RUDZIK. Interaction between amphetamine and sympathomimetic agents on the cardiovascular system. In: *Amphetamines and Related Compounds*, edited by E. Costa

and S. Garattini. New York: Raven, 1970, p. 513–530.

9. COLE, S. O. Experimental effects of amphetamine: supplementary report. *Percept. Mot. Skills* 31: 223–232, 1970.

10. EBLE, J. N., AND A. D. RUDZIK. Interaction between amphetamine and sympathomimetic agents on the cardiovascular system. In: *Amphetamines and Related Compounds*, edited by E. Costa and S. Garattini. New York: Raven, 1970, p. 513–530.

11. ELLINWOOD, E. H. Amphetamine psychosis: A multidimensional process. *Semin. Psychiatry* 1: 208–226, 1969.

12. ELLINWOOD, E. H. Emergency treatment of acute reactions to CNS stimulants. *J. Psychedelic Drugs* 5: 147–151, 1972.

13. ELLINWOOD, E. H., AND O. D. ESCALANTE. Behavior and histopathological findings during chronic methedrine intoxication. *J. Soc. Biol. Psychiatry* 2: 27–39, 1970.

14. ELLINWOOD, E. H.; A. SUDILOVSKY AND L. M. NELSON. Evolving behavior in the clinical and experimental amphetamine (model) psychosis. *Am. J. Psychiatry* 130: 1089–1093, 1973.

15. ESCALANTE, O. D., AND E. H. ELLINWOOD. Central nervous system cytopathological changes in cats with chronic methedrine intoxication. *Brain Res.* 21: 555, 1970.

16. FEINBERG, G., AND S. IRWIN. Effects of chronic methamphetamine administration in the cat. *Federation Proc.* 20: 396, 1961.

17. FERSTER, C. B., AND B. F. SKINNER. *Schedules of Reinforcement.* New York: Appleton-Century-Crofts, 1957.

18. FISCHMAN, M. W., AND T. H. MEIKLE, JR. Visual intensity discrimination in cats after serial tectal and cortical lesions. *J. Comp. Physiol. Psychol.* 59: 193–201, 1965.

19. FISCHMAN, M. W., AND C. R. SCHUSTER. Tolerance development to chronic metamphetamine intoxication in the rhesus monkey. *Pharmacol. Biochem. Behav.* 2: 503–508, 1974.

20. FISCHMAN, M. W., AND C. R. SCHUSTER. Behavioral, biochemical and morphological effects of methamphetamine in the rhesus monkey. In: *Behavioral Toxicity*, edited by B. Weiss and V. Laties. New York: Plenum, 1975, p. 375–394.

21. FITZ-GERALD, F. Effects of d-amphetamine upon behavior of young chimpanzees reared under different conditions. In: *Neuropsychopharmacology*, edited by H. Brill and J. Cole. Amsterdam: Elsevier, 1967, vol. 5, p. 1226.

22. GRIFFITH, J. D., J. CAVANAUGH AND J. OATES. Schizophreni form psychosis induced by large-dose administration of d-amphetamine. *J. Psychedelic Drugs* 2: 25–27, 1969.

23. HARRISON, J. W. E., C. M. AMBRUS AND J. L. AMBRUS. Tolerance of rats toward amphetamine and methamphetamine. *J. Am. Pharm. Assoc.* 41: 539–541, 1952.

24. HERMAN, Z. S. The influence of prolonged amphetamine treatment and amphetamine withdrawal on brain biogenic amine content and behavior in the rat. *Psychopharmacologia* 21: 74, 1971.

25. HÖHN, R., AND L. LASAGNA. Effects of aggregation and temperature on amphetamine toxicity in mice. *Psychopharmacologia* 1: 210, 1960.

26. JOHANSON, C. E., R. L. BALSTER AND K. F. BONESE. Self-administration of psychomotor stimulant drugs: The effects of unlimited access. *Pharmacol. Biochem. Behav.* Submitted.

27. JÖNSSON, L. E., AND L. M. GUNNE. Clinical studies of amphetamine psychosis. In: *International Symposium on Amphetamines and Related Compounds*, edited by E. Costa and S. Garattini. New York: Raven, 1970, p. 929–936.

28. KALANT, O. J. *The amphetamines: Toxicity and addiction.* Springfield, Ill.: Thomas, 1966.

29. KASIRSKY, G., I. H. ZAIDI AND M. F. TANSY. LD 50 and pathogic effects of acute and chronic administration of methamphetamine HCl in rabbits. *Res. Commun. Chem. Pathol. Pharmacol.* 3: 215–231, 1972.

30. KILOH, L. G., AND S. BRANDON. Habituation and addiction to amphetamines. *Br. Med. J.* 2: 40–43, 1962.

31. KOSMAN, M. E., AND K. R. UNNA. Effects of chronic administration of the amphetamines and other stimulants on behavior. *J. Clin. Pharmacol. Ther.* 9: 240–254, 1968.

32. KRAMER, J. C., V. S. FISCHMAN AND D. C. LITTLEFIELD. Amphetamine abuse. *J. Am. Med. Assoc.* 201: 305–309, 1967.

33. PICKENS, R., AND W. HARRIS. Self-administration of d-amphetamine by rats. *Psychopharmacologia* 12: 158–163, 1968.

34. ROSENBERG, D. E., A. B. WOLBACH, E. J. MINER AND H. ISBELL. Observations on direct and cross tolerance with LSD and d-amphetamine in man. *Psychopharmacologia* 5: 1–15, 1963.

35. RUMBAUGH, C. L., R. T. BERGERON, R. L. SCANLAN, J. S. TEAL, H. D. SEGALL, H. C. H. FANG AND R. McCORMICK. Cerebral vascular changes

secondary to amphetamine abuse in the experimental animal. *Radiology* 101: 345–351, 1971.

36. SCHUSTER, C., W. DOCKENS AND J. WOODS. Behavioral variables affecting the development of amphetamine tolerance. *Psychopharmacologia* 9: 170–182, 1966.

37. SCHUSTER, C. R., AND C. E. JOHANSON. The use of animal models for the study of drug dependence. In: *Research Advances in Alcohol and Drug Problems*, edited by R. J. Gibbons. New York: Wiley, 1974, p. 1–32.

38. SCHUSTER, C. R., AND J. ZIMMERMAN. Timing behavior during prolonged treatment with *d,l*-amphetamine. *J. Exp. Anal. Behav.* 4: 327–330, 1961.

39. SMITH, D. E. The characteristics of dependence in high-dose methamphetamine abuse. *Int. J. Addict.* 4: 453–459, 1969.

40. SMITH, D. E., AND C. M. FISCHER. An analysis of 310 cases of acute high-dose methamphetamine toxicity in Haight-Ashbury. *Clin. Toxicol.* 3: 117–124, 1970.

41. TORMEY, J., AND L. LASAGNA. Relation of thyroid function to acute and chronic effects of amphetamine in the rat. *J. Pharmacol. Exp. Ther.* 128: 201–209, 1960.

42. STEWART, C. P., AND A. STOLMAN. The toxicologist and his work. In: *Toxicology: Mechanisms and Analytical Methods*, edited by C. P. Stewart and A. Stolman. New York: Academic, 1960, pp. 1–25.

43. *Webster's Seventh New Collegiate Dictionary.* Springfield, Mass.: Merriam, 1967, 936.

Psychophysical evaluation of toxic effects on sensory systems

HARLEY M. HANSON

Merck Institute for Therapeutic Research
West Point, Pennsylvania 19486

ABSTRACT

Toxic effects on sensory systems have rarely been evaluated by psychophysical methods. As examples of possible applications four studies are described. Sodium salicylate and kanamycin, both reported to produce hearing deficits in man, have also been demonstrated to affect auditory thresholds in monkeys. With the latter drug the deficits measured were found to be correlted with specific loss of receptor cells in the cochlea. Pheniprazine, known to induce red–green color blindness, was found to disrupt a wavelength discrimination in pigeons. *Trans* 11-amino-10,11-dihydro-5-(3-dimethylaminopropyl)-5,10-epoxy-5*H*-dibenzo[*a,d*]-cycloheptene dihydrochloride, which was found to bleach the tapidum lucidum in dogs when given subacutely, was found to decrease sensitivity to light. The loss in sensitivity measured by behavioral techniques was correlated with the loss of coloration of the tapidum. Monkeys, not having a tapidum, did not show a similar effect.—HANSON, H. M. Psychophysical evaluation of toxic effects on sensory systems. *Federation Proc.* 34: 1753, 1852–1857, 1975.

There is a sizable literature dealing with the toxic effects of certain chemical agents on sensory systems in man. The effects of methanol on the visual system are well-known and part of the general folklore. Other types of effects on sensory systems are less well-known to the general public but are of potentially greater concern to the medical community. For example, aspirin, some antibiotics like streptomycin, and certain diuretics such as ethacrynic acid are capable of affecting the auditory system. In spite of the dire nature of such effects, there have been few laboratory studies of even these compounds using psychophysical techniques; and the evaluation of compounds for such toxicity before they are administered to man has seldom been done. Basic laboratory data are scant, although this deficiency is slowly being rectified.

This paper will present data collected using behavioral methods for four compounds affecting either the visual or auditory system. In the first three examples the toxic effects were first noticed in man and the laboratory studies were merely confirmatory.

The administration of large doses of aspirin is not unusual in the medical treatment of arthritis, and lay people often overdose with this readily available compound. Clinical observations of hearing loss, tinnitus, and vertigo following high dosage levels of aspirin are by no means uncommon. The fact that these side effects are readily reversible probably explains why the use of this agent has not been restricted.

Using a shock avoidance test situation Myers and Bernstein (4) trained squirrel monkeys to jump a barrier in the presence of a tone. In testing, if jumping occurred, the intensity of the tone was reduced with each additional trial until jumping ceased (the threshold value). Ten animals were injected with sodium salicylate subcutaneously (s.c.), 500–600 mg/kg. Twenty-four hours later testing was done over a frequency range of 0.125–8 kHz. Every animal showed hearing loss ranging from 17 to 36 dB. Plasma salicylate levels were found to correlate well with the degree of hearing loss. Complete recovery was noted 24–48 hours after the last pharmacological signs were noted. No changes were found by either light or electron microscopy in the inner ears of animals sacrificed at the time of maximal hearing loss.

Stebbins et al. (6), using larger monkeys, *Macaca irus* and *Macaca nemestrina*, developed an operant conditioning procedure capable of yielding detailed audiograms over a wide range of frequencies. The animals were tested in restraining chairs with the head fixed insuring delivery of tones directly to the ear via a specially adapted earphone. Testing was done in a sound-treated room. The test animals after careful training pressed one of two available levers which occasionally (about four times a min) resulted in the delivery of a pure tone to the earphones. Pressing the other lever during a tone, and only during a tone, resulted in the delivery of a food pellet. Responses to the second lever in the absence of a tone were followed by a 3 sec time-out from the experiment. After extensive training auditory thresholds were determined by decreasing the intensity of a tone until the monkey failed to respond on the second lever. Five intensities at 10 dB intervals around this value were then presented randomly (method of constant stimuli). Two different frequencies were tested each day, and final values were collected when for two successive sessions the thresholds were less than 5 dB apart for a given frequency. Data collected with four monkeys are shown in Fig. 1. The threshold values are those estimated intensities eliciting a response on the second lever for 50% of presentations.

The audiograms are very similar in appearance to those constructed for human data but show an extended sensitivity to very high frequencies. The agreement between the functions for the four animals is excellent.

One of the animals tested in this situation was injected intramuscularly with doses of 250 and 500 mg/kg of sodium salicylate preceding test sessions. Threshold values at 4,000 Hz were determined at 1, 3, 6 and 24 hours following injection. The data collected are shown in Fig. 2. The threshold of hearing was markedly elevated following administration of drug and had returned to normal within 24 hours. These data elegantly confirm those reported by Myers and Bernstein (4). The more rapid recovery of normal hearing in this experiment could well be attributed to species differences.

Kanamycin and some similar antibiotic compounds are known to produce hearing deficits in man. Goure-

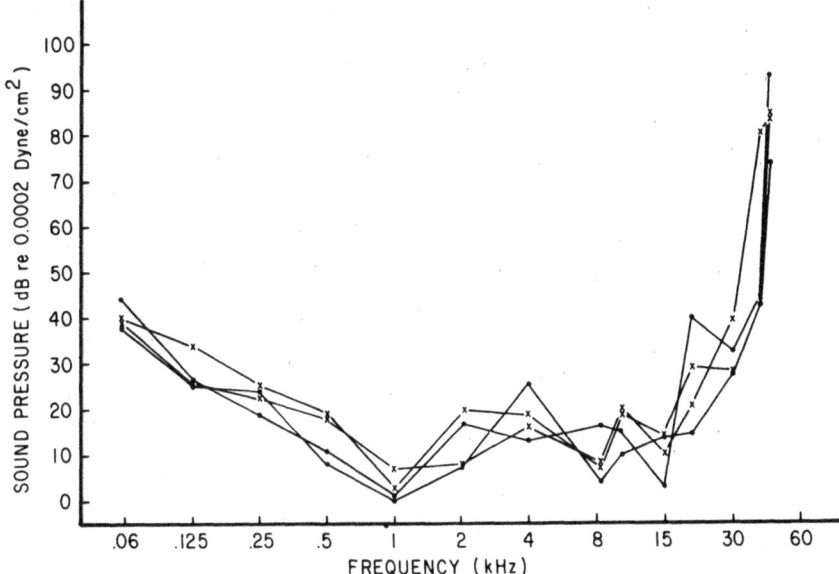

Figure 1. Auditory threshold functions collected in a sound-treated chamber for three *Macaca irus* and a *Macaca nemestrina*. The monkeys were trained to press a lever which occasionally produced a 3-sec duration pure tone. Pressing a second lever during the tone resulted in the delivery of a food pellet. The data points represent the response to five intensities, 5 dB apart, around the estimated threshold randomly presented at each frequency 10 to 20 times during a test session. Two frequencies were tested each day. (Figure redrawn from Stebbins et al. (6).)

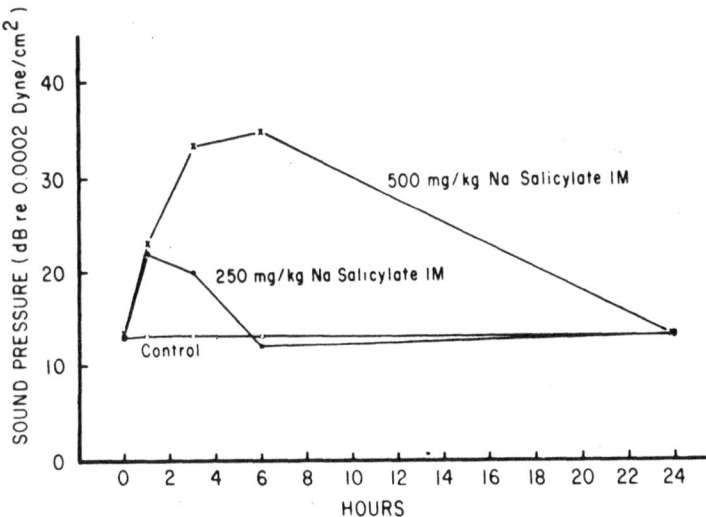

Figure 2. Effects of sodium salicylate on auditory thresholds at 4 kHz in a *Macaca irus*. The drug was administered at 0 hr. The test method used was the same as that described in Fig. 1. (Figure redrawn from Stebbins et al. (7).)

vitch et al. (2) using an operant test situation with rats found an increase in the auditory threshold at 2,000 Hz after dosing daily with kanamycin, 100–400 mg/kg. The effect was seen after 20–30 days of dosing. The largest increases in threshold occurred following dosing with 400 mg/kg per day. One rat at this dose showed an apparent complete loss of hearing.

Stebbins (5) reported data collected in a macaque during dosing with kanamycin. The test method was that described earlier; the course of the drug effect was followed for more than 200 days. Part of the data collected are shown in Fig. 3. The hearing loss produced by kanamycin was progressive in nature beginning with the highest frequency, 15 kHz, followed by losses at 11, 8 and 4 kHz with essentially no loss of sensitivity at the lowest frequency tested, 2 kHz. In this experiment kanamycin, 100 mg/kg, was administered for 180 days, essentially the entire period of testing. This dramatic and polished documentation of kanamycin's ototoxic effects was found to correspond well with histologic findings in this and other similarly treated monkeys. High frequency hearing loss was correlated with loss of receptor cells in the basal turn of the cochlea and as hearing loss progressed to the lower frequencies, receptor loss in the upper turns occurred.

The stability of the threshold values over extended periods of time, the

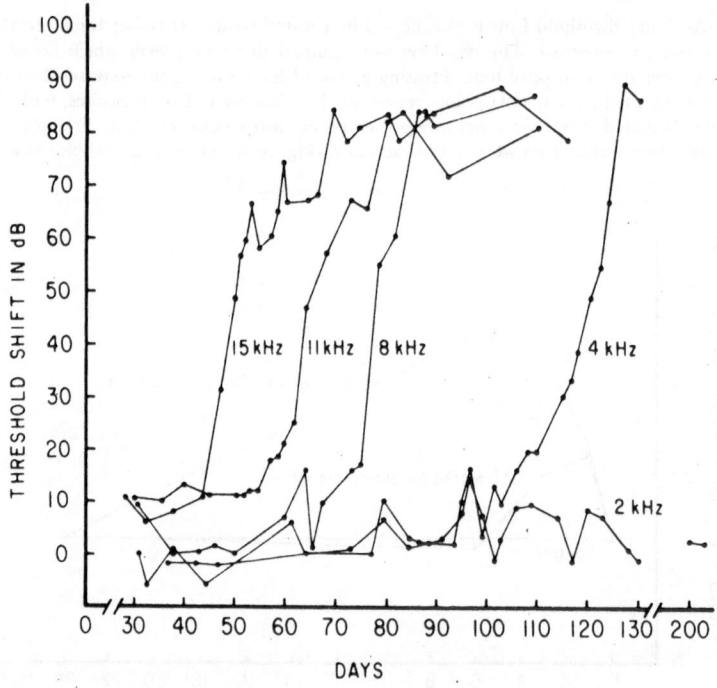

Figure 3. Changes in auditory thresholds at five frequencies for a macaque resulting from intramuscular administration of kanamycin, 100 mg/kg per day. The data collected during the first 30 days of dosing are not shown. The test method used was the same as that described in Fig. 1. (Data redrawn from Stebbins (5).)

face validity of the audiograms collected, as well as the dramatic correlation of the behavioral data with the histological findings suggest this methodology could well be used for the evaluation and prediction of ototoxicity.

A challenging laboratory problem was posed in the late 1950's when papers began to appear in the literature reporting that pheniprazine was found to induce red–green color blindness. This compound, 1-phenyl-2-hydrazinopropane hydrochloride, a potent inhibitor of monoamine oxidase, was found useful for the treatment of depression. Defective red–green color vision is an unusual side effect and there were no laboratory techniques then available to predict its ocurrence (there are still none in general use). A study using pigeons for subjects was undertaken in my laboratory to see if a laboratory analog of this phenomenon could be developed (3). The pigeon was selected because of its known sensitivity to wavelengths in the visual spectrum and because of the wealth of specifically developed behavioral techniques that were available for this animal.

Six pigeons were trained in a wavelength discrimination in a specially designed operant conditioning chamber by selectively reinforcing pecking when lights of 5,700 Å and 6,000 Å (yellow and reddish-orange) were on the key and witholding all reinforcement when lights of 5,500, 5,900, or 6,300 Å (yellowish-green, orange and red) illuminated the key. Presentation of the various stimuli was for short intervals and was random; reinforcers were delivered only occasionally for key pecks according to a variable ratio schedule (VR 160). All animals learned to discriminate between the five stimuli; responding to the nonreinforced stimuli seldom occurred. After stable performance was obtained, daily dosing with pheniprazine (9 or 18 mg/kg orally) was started. Selected performances of four pigeons that showed a disruption of the wavelength discrimination after 12 to 76 days of treatment are shown in Fig. 4. The average numbers of responses per minute emitted to each of the wavelengths are shown as histograms. The control data were collected before the first dose of pheniprazine; the second row of histograms shows the performance gathered during the first test session after dosing with pheniprazine was discontinued. The final set of data was collected some 40–80 days later when the discrimination had been recovered. A definite loss of the almost perfect discrimination oc̀curred as a result of dosing with pheniprazine. The rate of key pecking immediately after drug administration was lower than the control rate, but was high enough to give assurance as to the ability of the birds to respond. Two of the birds did not show disruption even though extremely high daily doses of pheniprazine were given. One bird (highest daily dose 24 mg) after 179 days of dosing showed toxic effects but never lost its ability to discriminate. That the deficits in ability to discriminate were due to pheniprazine is supported by the data collected after dosing had been stopped, i.e., the recovery of the original performance. Also supporting the implication of pheniprazine as the causative agent is the similarity of the recovery period to that reported for man (23–54 days).

Drug-induced defects in color vision are an unusual side effect and, to my knowledge, pheniprazine is the only compound that has been reported to produce this effect. The technique described is a complicated one and one that is not readily available to toxicologists, although it or some similar method would seem to be the only way to assay suspected compounds for their effects on differential sensitivity to wavelengths in the visual spectrum.

The next experiment dealing with vision that I wish to describe resulted from the unexpected observation by toxicologists concerned with its evaluation that *trans* 11-amino-10,11-dihydro - 5 - (3 - dimethylaminopropyl)-5,10 - epoxy - 5*H* - dibenzo[*a,d*] cycloheptene dihydrochloride resulted in the bleaching of the tapidum lucidum, that light reflective structure in the eye of the dog. There appeared to be no other effects on vision in the affected animals as far as one could

tell from simple observation. I undertook the study of this compound in animals that had been specially trained in an operant situation yielding an estimate of the absolute threshold of vision, as well as a measure similar to the "dark-adaptation" curve as collected with human subjects. The techniques used were similar to those used with humans.

Rhesus monkeys and beagle dogs were trained to respond to the presence of a dim light in an otherwise dark chamber by a lever response. Nonresponding to a light presentation resulted in painful electric shock, as did responding in the absence of a

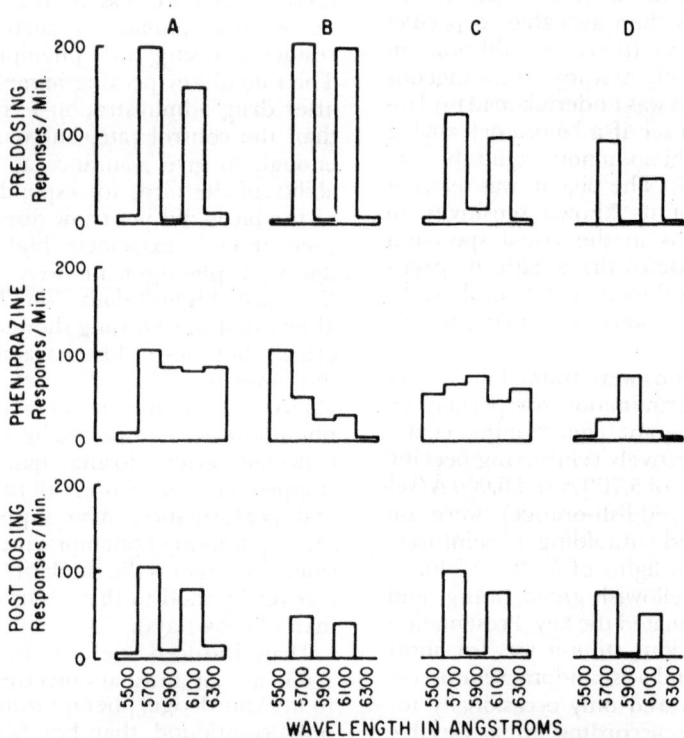

Figure 4. Effects of administration of pheniprazine on a visual wavelength discrimination in pigeons. The birds were reinforced only for key pecking to lights of 5,700 and 6,100 Å. Loss of the wavelength discrimination as a result of pheniprazine is shown in the second row of histograms, the third row shows performance after drug withdrawal. (Redrawn from Hanson et al. (3).)

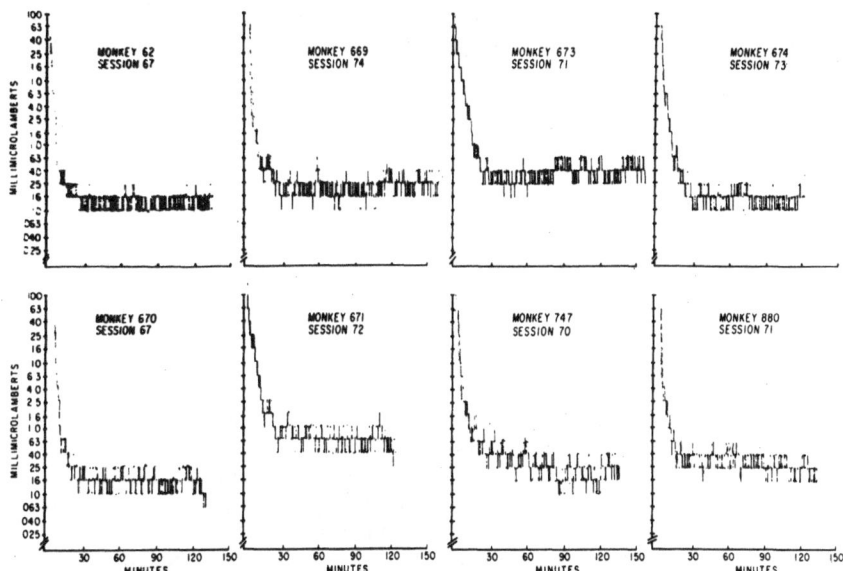

Figure 5. Dark adaptation records of eight rhesus monkeys. The luminance of the light source, the only light in the otherwise dark test chamber, is shown in millimicrolamberts. Lever responses by the experimental animal during a stimulus presentation decreased the luminance of the stimulus and avoided electric shocks. Lever responses in the absence of the stimulus resulted in electric shocks. Programmed pulses which occurred randomly averaging once/minute increased the luminance in the absence of responding.

light stimulus. Light presentations occurred randomly, averaging one per minute, and were 10 sec in duration; a lever response during a signal immediately extinguished it. The light source was a calibrated electroluminescent panel (1) that allowed the level of luminance to be controlled by simply changing voltage. Since the panel had only one phosphor its dominant wavelength was 5,500 Å with a relatively narrow bandwidth.

The final behavior, as programmed and recorded by means of the usual relay control devices, arranged that the test animals' behavior controlled the brightness of the signal light. If the experimental animal responded to a light stimulus the next presentation was decreased in brightness and vice versa, a laboratory analog of the

"Method of Adjustment." All animals were "light adapted" by exposure to paired 100 W bulbs in the chamber preceding the start of any test session. Figures 5 and 6 show representative data collected with rhesus monkeys and dogs in this situation. The adaptation curves collected were highly individual for each animal and were consistent from day to day and over extended periods of testing. Although it is not pronounced at this wavelength, there is a "rod–cone" break apparent in the curves collected with the monkeys that is clearly absent in the curves gathered with the dogs, an effect probably attributable to basic anatomical differences in the eyes of two species.

Figure 7 shows data collected with four dogs given daily doses of *trans* 11 - amino - 10,11 - dihydro - 5 - (3-

dimethylaminopropyl) - 5,10 - epoxy-5H - dibenzo[a,d]cycloheptene dihydrochloride. Although data comparable to those shown in Fig. 6 were collected during each test session, for reductive purposes the data points in the figure are the average luminances programmed during the last 30 min of each test session (a measure estimating the absolute threshold). The dogs were dosed daily immediately after the test session. Three of the dogs showed an immediate and dramatic loss of sensitivity as a result of administration of *trans* 11 - amino-10,11 - dihydro - 5 - (3 - dimethylaminopropyl) - 5,10 - epoxy - 5H - dibenzo[a,d]cycloheptene dihydrochloride. In one dog (#201), it was possible to dose for a short period and then allow recovery to occur before starting another course of treatment.

Correlated with the losses in sensitivity seen were the slow but dramatic changes in the coloration of the tapidum. Figure 8 shows retinal photographs taken of the left eye of dog #201 preceding treatment, at the time of maximal effect and after recovery. As can be seen, a dramatic loss of coloration was in evidence at the time of minimal sensitivity. A similar effect was seen in the other two dogs affected. The curvature of the dark adaptation curves was not appreciably modified nor was the variability of the controlled luminance at threshold affected — both indicators that the effects seen were not due to a general disability of the test dogs, or due to some lack of the ability to discriminate.

Although this experiment helped very little in explaining the occur-

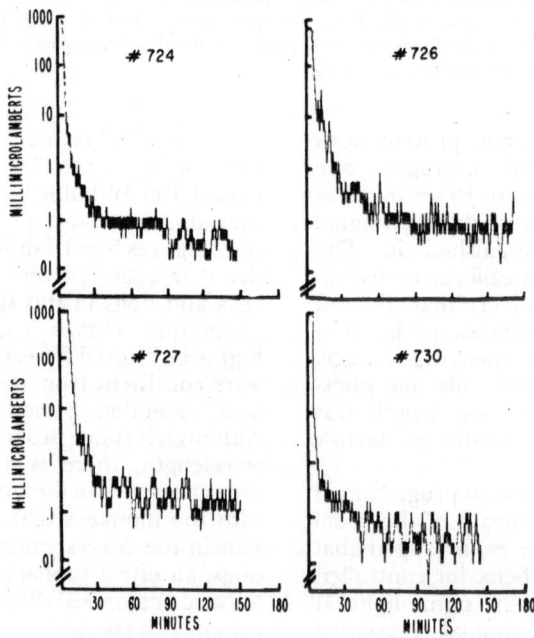

Figure 6. Dark adaptation of four beagle dogs. The conditions of the experiment were the same as in Fig. 5, except the dogs lifted a lever with their noses to respond to a stimulus presentation.

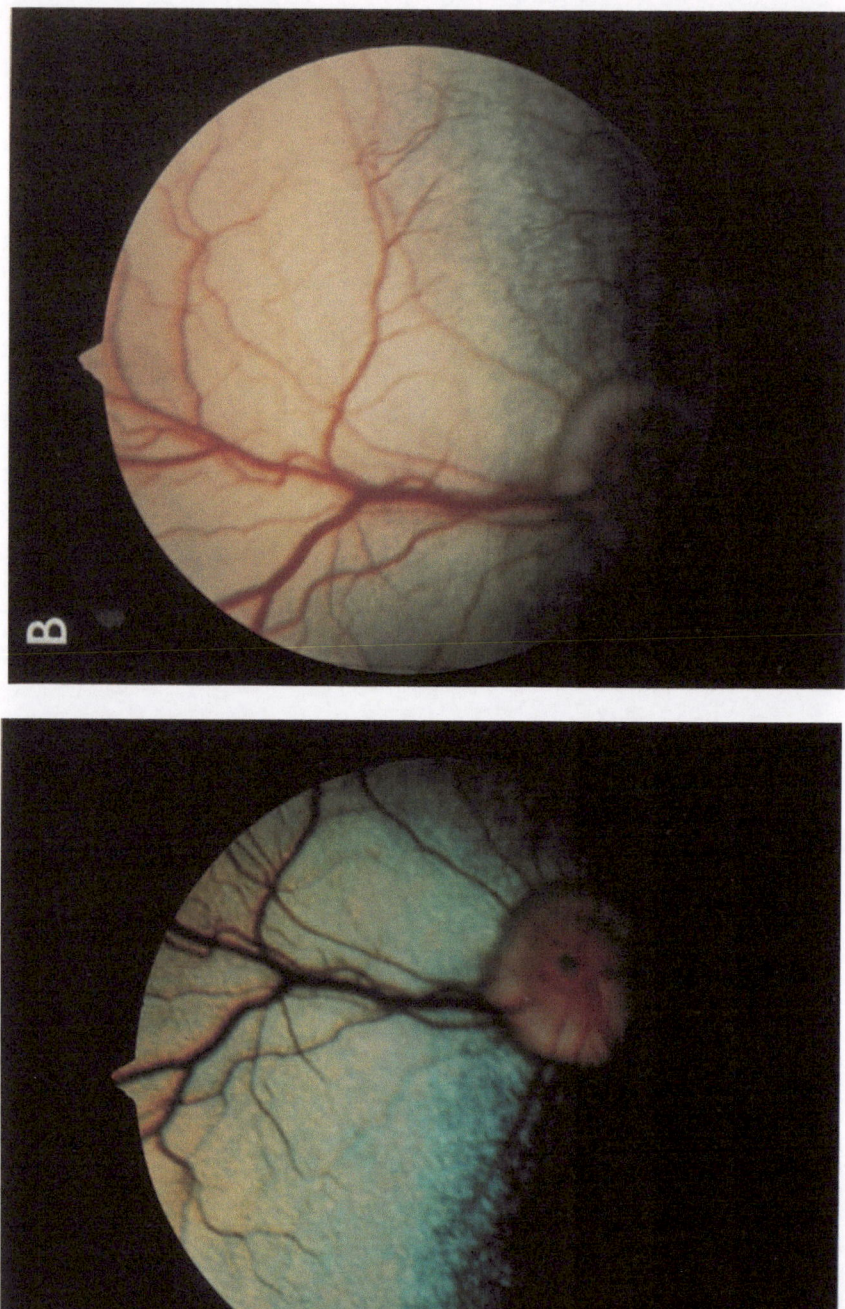

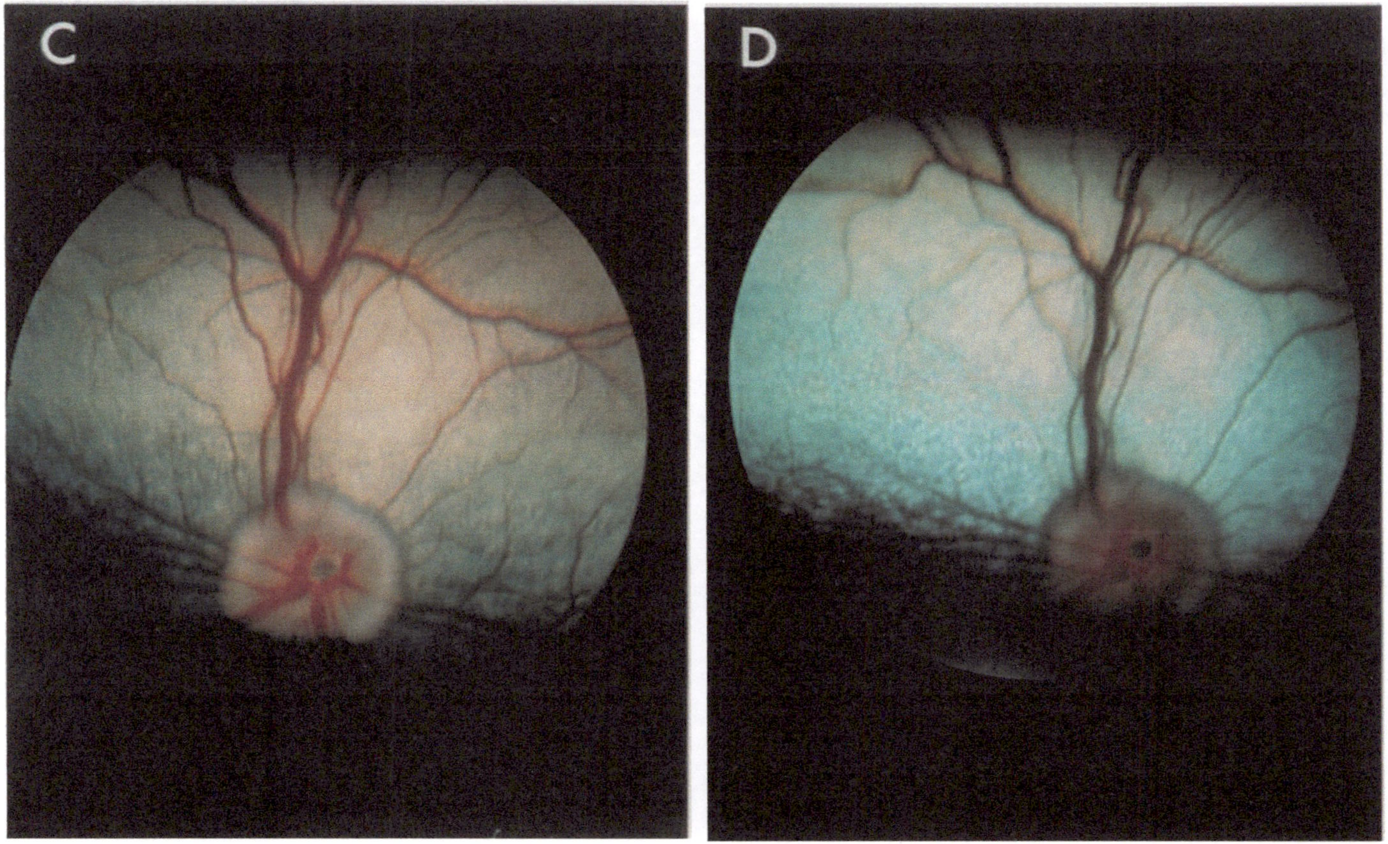

Figure 8. Retinal photograph of the left eye of dog #201 shown in Fig. 7 . Photographs *A* and *D* show the normal-appearing eye. Photo *A* was taken preceding testing. Photo *B* was taken on experimental day 36 (as shown in Fig. 7) after dosing with *trans*-11-amino-10,11-dihydro-5-(3-dimethylaminopropyl)-5,10-epoxy-5*H*-dibenzo[*a,d*]-cycloheptane. Photo *C* was taken on experimental day 57 also after dosing. Photo *D* was taken on experimental day 71 when recovery of the original threshold values was nearly complete.

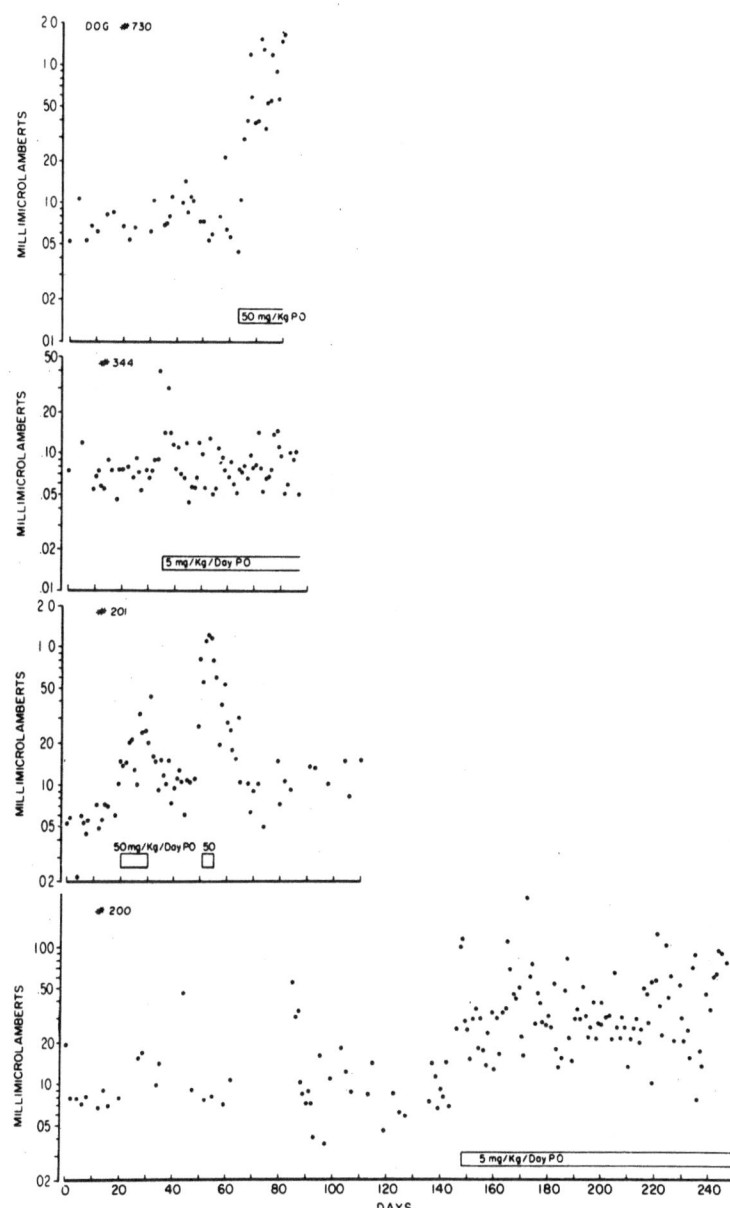

Figure 7. Effects of dosing with an agent which resulted in bleaching of the tapidum lucidum in the dog on visual thresholds. The data points represent the average luminance values programmed during the last 30 min of a 120-min test session. Dog #201 was dosed with 50 mg/kg orally per day for a 10-day period and then for a 5-day period.

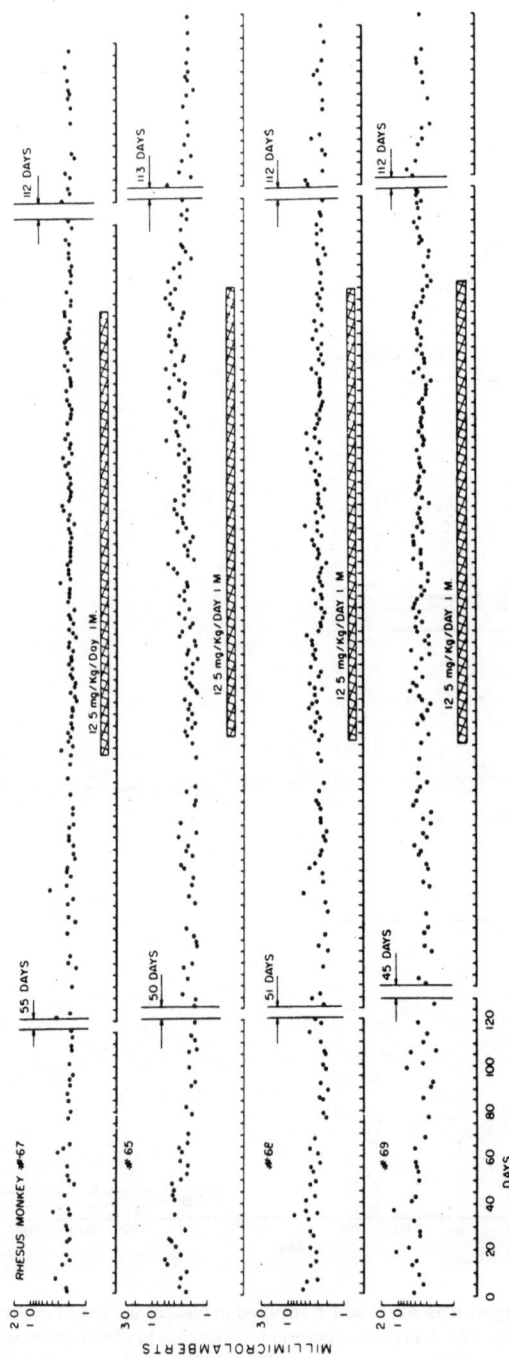

Figure 9. Effects of dosing with the agent studied in dogs (see Fig. 7 and 8) on visual thresholds in rhesus monkeys. The method of averaging data was the same as used in Fig. 7 as were the other details of the test situation. For simplicity, some control data were omitted from the figure.

rence of this unusual effect of this compound it perhaps does yield some information as to the functional significance of the tapidum lucidum. It would seem at least likely that this structure has an important, if not crucial, role to play in the dog visual system at low levels of illumination and that the unusual sensitivity of the dog eye to low levels of illumination is perhaps dependent on the presence of this structure.

For comparative purposes, four rhesus monkeys (the rhesus monkey does not possess a tapidum, but does have a cone-rich retina like man) were tested in a similar fashion to evaluate the effects of *trans* 11 - amino - 10,11-dihydro - 5 - (3 - dimethylaminopropyl) - 5,10 - epoxy - 5*H* - dibenzo[a,d]-cycloheptene dihydrochloride on sensitivity to light in this species. The data collected are shown in Fig. 9. The same averaging procedure was used as for the data shown in Fig. 7. The other details of the experiment were also similar. Even though drug was administered over an extended period, in excess of 5 months, no dramatic changes in sensitivity were seen. One monkey (#65) showed an as yet unexplained gradual increase in his absolute threshold as dosing progressed, which was significantly different during the last 60 days of testing compared to the preceding control period. This effect disappeared within 20 days after withdrawal of the drug. None of the other animals showed any effect as a result of drug administration. This period of drug administration was extreme compared to the data collected with the dog, i.e., in excess of 190 days. The data collected show a surprising degree of stability and reproducibility over a lengthy period, the total time of testing being almost 500 days. It would seem quite feasible in view of this stability to use this test system for the evaluation of other agents of

interest both subacutely and chronically.

I have outlined techniques used for the special study of compounds selected because of unusual toxicological manifestations on a sensory system. In the final analysis it would seem highly likely that there would be no more satisfactory way of detecting toxic effects on sensory systems than those analogous to the procedures used with man. The functional state of a sensory system in its entirety can be evaluated when behavioral responses are the dependent variable, which may not be true when local electrophysiological measures are used.

REFERENCES

1. CURRAN, C. S., AND R. H. THOMAS. Apparatus for luminance-threshold determination in animals. *J. Opt. Soc. Am.* 55: 727–728, 1965.
2. GOUREVITCH, G., M. H. HACK AND J. E. HAWKINS, JR. Auditory thresholds in the rat measured by an operant technique. *Science* 131: 1046–1047, 1960.
3. HANSON, H. M., J. J. WITOSLAWSKI AND E. H. CAMPBELL. Reversible disruption of a wavelength discrimination in pigeons following administration of pheniprazine. *Toxicol. Appl. Pharmacol.* 6: 690–695, 1964.
4. MYERS, E. N., AND J. M. BERNSTEIN. Salicylate ototoxicity, a clinical and experimental study. *Arch. Otolaryngol.* 82: 483–493, 1965.
5. STEBBINS, W. C. Studies of hearing and hearing loss in the monkey. In: *Animal Psychophysics: The Design and Conduct of Sensory Experiments*, New York: edited by W. C. Stebbins. Appleton-Century-Crofts, 1970, p. 41–66.
6. STEBBINS, W. C., S. GREEN AND F. L. MILLER. Auditory sensitivity in the monkey. *Science* 153: 1646–1647, 1966.
7. STEBBINS, W. C., J. M. MILLER. L.-G. JOHNSSON AND J. E. HAWKINS, JR. Behavioral measurement and histopathology of drug-induced hearing loss in subhuman primates. In: *Use of Nonhuman Primates in Drug Evaluation*, edited by H. Vagtborg. Austin: Univ. of Texas Press, 1968, p. 382–399.

Behavioral effects of mercury and methylmercury[1]

HUGH L. EVANS, VICTOR G. LATIES AND BERNARD WEISS

Department of Radiation Biology and Biophysics
School of Medicine and Dentistry
University of Rochester, Rochester, New York 14642

ABSTRACT

Intoxication by elemental mercury or by methylmercury is revealed primarily by changes in behavior and by neurological signs. Disorders of movement and posture have been most widely reported, both in animal experiments and in cases of human exposure. Specific sensory symptoms are also prominent in human methylmercury poisoning. Recent data indicate similar symptoms in monkeys during long-term exposure to methylmercury. Similar sensory impairment has not been described in experiments with sub-primates. Variations in the profile of behavioral and neurological effects are discussed in terms of differences in species and differences between acute and long-term exposure. The latter condition poses the most difficult questions for human health, yet has been less frequently studied. Procedures are suggested that may help to resolve these problems. In particular, tests of learned behavior hold great promise toward identifying specific symptoms and toward understanding how mercury compounds affect behavior.— EVANS, H. L., V. G. LATIES AND B. WEISS. Behavioral effects of mercury and methylmercury. *Federation Proc.* 34: 1858–1867, 1975.

Mercury and mercury compounds have increased as threats to environmental health in parallel with the rise of industrial civilization (22, 25). Epidemics of poisoning arising from food contamination by methylmercury demonstrate that the hazards associated with mercury are not confined to industrial workers (3, 15, 16, 31, 49).

Many reviews of the medical and toxicologic aspects of mercury poisoning have appeared recently (13, 14, 28, 36, 50). The central nervous system is most often identified as the primary site of damage. Even in persons exposed to very large quantities of mercury compounds, behavioral and neurological symptoms

[1] Supported in part by grants MH-11752 from the National Institute of Mental Health, NS-08048 from the National Institute of Neurological Disease and Stroke, GI-300978 from the RANN program of the National Science Foundation, and by a contract between the University of Rochester and the U.S. Atomic Energy Commission. Report No. UR-3490-564.

are virtually the only ones detectable (25). Early symptoms of intoxication include a variety of rather vague complaints, frequently of a psychological nature, that challenge the skills of the behavioral scientist. The precise measurement and analysis of behavior should prove useful in more clearly specifying symptoms and in following the course of intoxication. Identification of specific symptoms may also guide the search for mechanisms of action.

In this paper we review evidence from behavioral experiments, most of which have appeared in the last few years. Studies of neurological signs are included when they have been described in clear, behavioral terms. We begin with a brief examination of the behavioral effects of inorganic mercury, then proceed to methylmercury, which is the focus of the majority of publications. It is not surprising that nearly all behavioral experiments deal with either mercury vapor or methylmercury since mercury vapor and alkyl mercurials, such as methylmercury, most easily penetrate the brain.

INORGANIC MERCURY

Inhalation of mercury vapors

Workers exposed to mercury vapor in a thermometer factory presented numerous complaints including forgetfulness, headache, irritability, inability to concentrate, tremor, insomnia, unsteady gait and various gastrointestinal problems (51). This seemingly bewildering array of symptoms is typical of mercury intoxication (25). The workers often scored below the established norms on a series of standardized tests of intelligence, personality and perceptual-motor skills. The usefulness of these tests could have been more clearly established if they had also

been administered to co-workers not presenting symptoms and believed to be unexposed. Many of the test scores returned to normal when the workers were retested several months after exposure stopped, but there was no control procedure for possible improvement in test performance with repeated practice. The majority of the workers continued to exhibit poor short-term memory 20 months after exposure. The authors suggest that many of the psychological complaints may have been secondary to specific symptoms such as impaired memory and tremor (51).

Tremor seems to be the least ambiguous behavioral symptom of mercury vapor intoxication and is readily quantified. A survey of workers exposed to vapors from mercuric nitrate during the manufacture of felt hats illustrated that effects such as tremor are a joint function of dose and duration of exposure (37). Duration of exposure is important because mercury compounds are excreted slowly. Even a simple behavioral test could provide the orderly results shown in Fig. 1.

Tremor amplitude and its frequency spectrum are correlated with plasma mercury levels in individuals exposed to mercury vapor (54). This study used behavioral techniques and refined quantitative procedures to follow the course of recovery after exposure stopped. Together with previous observations, these data indicate that the probability of recovery is greater following inorganic mercury poisoning than after alkylmercury poisoning.

Exposure to mercury vapor produces a significant accumulation of mercury in the mammalian brain (8). Tests of conditioned behavior, previously employed to assess central nervous system drugs, can reveal effects of mercury vapor in animals prior to the appearance of tremor,

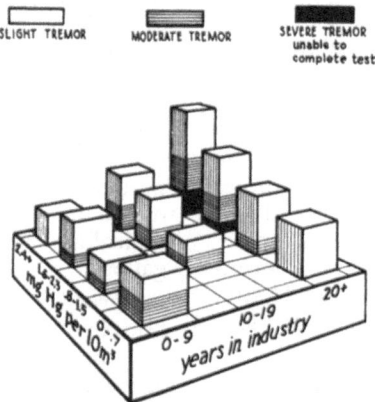

SLIGHT TREMOR   MODERATE TREMOR   SEVERE TREMOR
unable to
complete test

Figure 1. Incidence of tremor among workers in the felt-hat industry. The percentage of workers exhibiting tremor was a joint function of dose (concentration of mercury in the air at the work space) and duration of exposure (years of employment at the job). The height of the bars in the figure represents the percentage of workers showing tremor. For example, among those with more than 20 yr exposure, the incidence of tremor increased from 20% for workers exposed to less than 0.8 mg Hg/10 m^3 to 54% for workers exposed to more than 2.4 mg Hg/10 m^3. The test consisted of placing a small wooden block on one end of a flat wooden strip, raising the block with the strip, and then returning it to a table. This procedure could often elicit intention tremor when not otherwise apparent. (From Neal et al. (37).)

anorexia or other observable symptoms (2, 5). Rats exposed to 17 mg of Hg/m^3 for 2 hr, 5 days per week, showed a decline in pole-jump shock avoidance (5). An increase in spontaneous fighting among group-housed rats was also observed, but not quantified. All recovered following the 30-day exposure. Central nervous system histopathology was observed without evidence of lesions in other tissues.

The same regimen of mercury vapor exposure caused a decline in the response rates of pigeons performing two different tasks (2). In one, food was available once every 15 min for a peck on a green disk (fixed interval); in the other, food was available for every 60th peck on a red disk, regardless of the elapsed time (fixed ratio). The decline in response rate was largely the result of pausing that occurred with increasing frequency

under both conditions as exposure continued. The latency to onset of effects varied greatly among the eight pigeons studied, ranging from 3 to 30 weeks. Death occurred in the two cases in which exposure continued after responding had ceased. The other pigeons eventually recovered to preexposure baselines. In both of the above experiments, large doses changed behavioral performance several weeks prior to the appearance of overt signs of intoxication.

A major procedural problem, baseline drift, may have led to an inappropriate conclusion when a much lower concentration was assessed with the above technique (4). Two pigeons were exposed to 0.1 mg Hg/ m^3, which was the threshold limit value at that time. The authors concluded that 20 weeks of exposure was ineffective because there was no change from preexposure baselines.

However, the single control bird's index of fixed-interval curvature increased steadily during the course of the experiment (Fig. 2). With this as a comparison, the possibility cannot be excluded that mercury prevented a similar sharpening of the fixed-interval response patterns of the other birds. In view of Russian studies described by Friberg and Nordberg (20), which claim changes in the behavior of rats and cats at concentrations as low as 0.002 mg Hg/m³, the effects of low levels deserve more study.

Ingestion of mercuric chloride

Inorganic salts of mercury are less able to penetrate the brain than elemental mercury (8) and may exert their primary effect on the kidney (14). Nevertheless, a few publications have reported behavioral changes in animals exposed to mercuric chloride. Avoidance learning of rats was slightly retarded following a 2-week exposure to either 3.7 or 5.2 mg Hg/kg, administered daily as mercuric chloride[2] in the food (29). Control rats learned to avoid shock by running from a white compartment into a black compartment after a mean of five trials. Both mercury-treated groups required about seven trials to reach the avoidance criterion without displaying a change in body weight or general appearance.

In one of the few determinations of the joint effect of dose and duration of exposure, Weir and Hine (52) found that mercuric chloride produced an irreversible reduction in discriminative shock avoidance by goldfish. Twenty-four hours of exposure to water containing 0.01, 0.006 or 0.003 ppm Hg led to a de-

cline in the incidence of avoidance responses. A further decline occurred after 48-hr exposure to 0.006 and 0.003 ppm. The effective mercury concentrations were considerably below the lethal concentration. The unorthodox procedure of testing animals in groups complicates the interpretation of results. The usefulness of aquatic species for behavioral toxicology merits further investigation.

Mercuric chloride added to the drinking water led to a reduction in the frequency of mating behavior of both sexes of quail (48). Unfortunately, neither dose nor tissue concentrations were reported.

METHYLMERCURY: EFFECTS ON SUBPRIMATE ORGANISMS

Several properties of alkylmercurials render them the greatest poten-

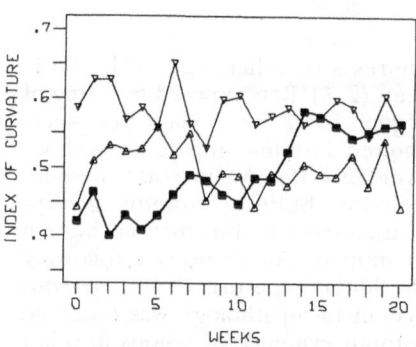

Figure 2. Indexes of curvature for fixed-interval responding of two pigeons exposed to mercury vapor ($\triangle$ —— $\triangle$) and (∇ —— ∇) and one control pigeon ($\blacksquare$ —— $\blacksquare$). The index, a measure of temporal discrimination, can vary from zero when responding is distributed evenly throughout the interval to 0.75 when responding is restricted to the end of the interval. Exposure was for 6 hr a day, 5 days a week for 20 wk. The progressive sharpening of the control pigeon's temporal discrimination raises the possibility that mercury vapor prevented the other two pigeons from undergoing a similar change. (Data from Beliles et al. (4).)

[2] To facilitate the comparison of doses reported by various authors, we have expressed all doses in terms of Hg rather than in terms of the whole compound.

tial hazard to man. The most important route of absorption is the digestive tract, where more than 90% of the dose is absorbed (14). Excretion is slow, thus allowing accumulation with repeated exposure to fairly small doses. Moreover, few overt neurological signs appear until advanced stages of toxicity, when therapeutic maneuvers are much less effective. In contrast to inorganic mercury poisoning, symptoms of alkylmercury poisoning are less likely to be reversible. And, as indicated by several reports to be reviewed, the "margin of safety" is relatively narrow between the minimally-effective dose and doses producing severe intoxication or death.

Neurological signs and unlearned behaviors

Neurological signs of methylmercury toxicity can be defined in rather specific behavioral terms. For example, Suzuki (46) used a rating scale to quantify results of neurological tests. He also determined the sequence in which symptoms appeared during a subacute course of exposure to methylmercury in food. Further refinement of the technique enabled Suzuki and Miyama (47) to determine symptoms associated with lower brain mercury concentrations. The earliest symptom shown by mice, inability to maintain the head in an erect position, appeared when the brain mercury concentration rose above 10 ppm (about $\frac{1}{3}$ the lethal concentration).

The above experiment also illustrates the importance of dosing procedure in determining critical brain concentrations associated with specific symptoms (Fig. 3). When mercury accumulated slowly, symptoms were associated with lower brain concentrations than when accumulation was rapid. A similar relationship has been observed with monkeys (9). The

long latency to onset of symptoms, coupled with rapid accumulation, probably produces super critical brain concentrations by the time clear symptoms are observed. This phenomenon contributed to the severity of the recent epidemic in Iraq (3). Another important aspect of dosing procedure is illustrated by Fig. 3: the intervals between appearance of successive symptoms increased as rate of methylmercury accumulation decreased.

The open field test is a frequently used procedure. In this test, an animal is placed in an open enclosure without being required to perform any specific task. An observer then counts the frequency of various classes of behavior emitted by the animal. Spyker et al. (45) uncovered subtle consequences of in utero exposure to methylmercury by observing behavior in the open field situation 30 days after birth. Swimming patterns also were affected. No abnormality was apparent from casual observations. The dose administered intraperitoneally (i.p.) to pregnant mice was approximately $\frac{1}{4}$ to $\frac{1}{3}$ of the LD_{50}. Spyker recently has reviewed progress in this field (43, 44).

Salvaterra et al. (41) gave adult mice a single i.p. injection of one of three doses. Doses bracketing the one used by Spyker et al. (45) reduced rearing and locomotion in an open field at 1 and 3 hours after injection but not at 72 hours. The highest dose killed 30% of the mice. Behavioral and neurochemical changes did not correlate with brain mercury concentration, which did not reach the peak of 5 ppm until 72 hours after injection. The rapid onset and reversal of symptoms argue against any significant effect of mercury in the central nervous system as the source of behavioral changes. Effects seen only within a day of exposure probably represent a different sort of intoxication than effects observed many days after a single exposure or after a prolonged low level exposure. Systemic injections of methylmercury are likely to produce local irritation (10, 18, 38) that complicates the interpretation of behavioral changes. Almost any behavioral test could reflect such nonspecific effects, and experimenters should endeavor to determine the specificity of observed changes in behavior. Similar factors of experimental design probably account for the brief appearance of symptoms reported by Diamond and Sleight (18).

A longer period between acute dosing and testing is preferable in assessing central nervous system symptoms. Post et al. (39), for example, found reduced locomotion and rearing in the open field in rats tested 22 days after oral administration of high doses. Unfortunately, the sequence of effects—onset, full development, possible reversibility, and so on—was not determined.

A change in the frequency of web construction occurred in spiders given food adulterated with methyl-

mercury for 2 weeks (32). The complexity of the web structure also appeared to change, but neither this nor bodily mercury concentrations were actually quantified.

Learned behaviors

The effects of chemical agents can be evaluated in terms of changes in previously-learned behavior or in terms of changes in the learning process itself. The study of learned behavior allows the experimenter to exert greater control over the test situation and to select from a greater variety of behaviors.

Rats required significantly more trials to learn a T-maze after oral administration of methylmercury in a dose[2] of 20 mg Hg/kg (39). Testing began 8 days after dosing. The same rats showed changes in open field activity (see above). Although this is a rather high dose, histology following behavioral testing revealed no brain lesions. The ability of rats to swim through a water maze was impaired following an indeterminate dose given at early stages of develop-

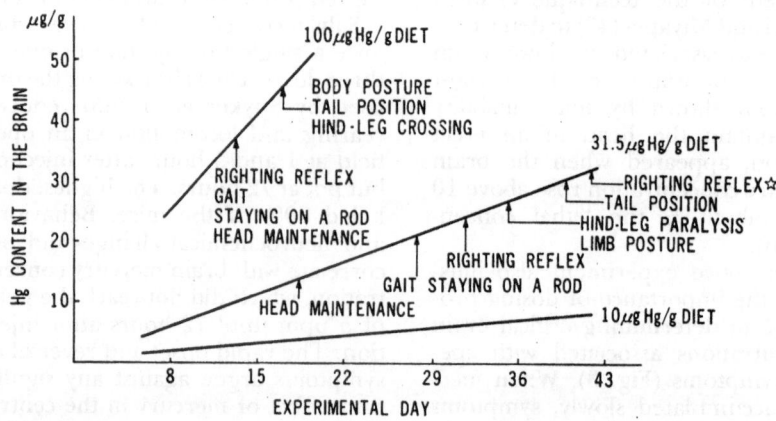

Figure 3. Brain mercury concentrations associated with the appearance of neurological symptoms in mice exposed to methylmercury in the diet. Exposure stopped after day 41 for the group exposed to the 10 μg/g concentration; exposure to the higher concentrations continued until death. (Reproduced with permission from Suzuki and Miyama (47).)

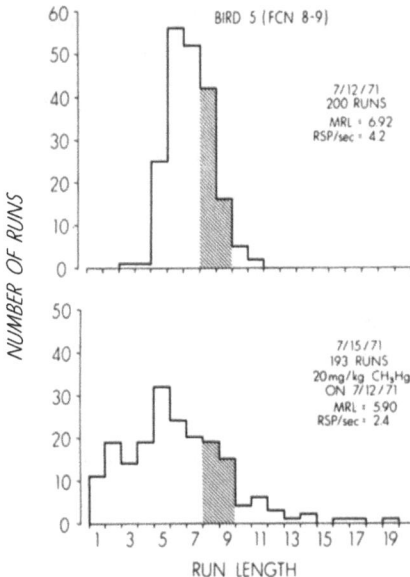

Figure 4. Disruption of a pigeon's fixed-con-secutive-number performance on the third day following an oral dose of methylmercury. Reinforcement was available for a response on the reinforcement key after either 8 or 9 responses had been made on the counting key (shown by the shaded bars). Runs shorter than 8 or longer than 9 were not reinforced. MRL = mean run length. RSP/sec = response rate during runs in responses per second.

ment (55). Chicks hatched from eggs containing 0.5 ppm methylmercury were somewhat slower than controls in learning to follow a detour course to obtain food (40). An injection of methylmercury (4 mg Hg/kg) following an opportunity to drink sweetened water is sufficient to establish in rats a conditioned aversion to the taste of the normally-preferred solution (10).

In our laboratory we have examined acute effects of methylmercury on the conditioned behavior of pigeons performing a type of counting task. The pigeon must peck a lighted disk a fixed number of times, then move over and peck a second

disk to activate the feeder (33, 35). There is no cue other than the bird's own behavior to signal the completion of the criterion number of pecks. If the criterion number of pecks has not been made when the bird pecks on the second key, reinforcement is omitted and a new series of pecks on the first key must be started.

We can determine how precisely a bird is performing the fixed-consecutive-number task by examining a histogram of the run lengths. The run length is the train of pecks emitted on the "counting" key before the bird switches over to the reinforcement key. Figure 4 shows the disruption of this counting behavior 72 hours after a single large dose of methylmercury. The normal performance of this bird is shown in the upper histogram. Immediately after this control test the pigeon received 20 mg Hg/kg. No effect was seen 24 hours later. Three days after the dose the bird made many runs that were far too short and far too long. Administration of the vehicle alone did not have this effect. Additional work with the fixed-consecutive-number schedule has been reported briefly elsewhere in this symposium (34). It appears that chronic administration of small doses of methylmercury also increases the variability of performance.

The beginnings of intoxication during subacute exposure can be revealed by subtle changes in learned behavior. We trained pigeons to obtain food by pecking one response key for a fixed number of times. This task, termed a fixed-ratio schedule, was also used to assess the effects of mercury vapor (2, 4). This behavior can be summarized in terms of a histogram of the precisely-measured times between key pecks. Each pigeon has a characteristic shape to its histogram, but by looking for gradual changes in the shape of each pigeon's histogram during the course of methylmercury exposure, one can detect the emergence of subtle behavioral changes preceding the appearance of overt symptoms (Fig. 5).

The control pigeon emitted mostly very short inter-response times (IRTs), on the order of 0.2 sec, and the form of the histograms remained unchanged during the period of placebo administration. The pigeon receiving methylmercury showed a decline in frequency of very short IRTs with a gradual increase in the number of IRTs in the vicinity of 0.5 sec. In the week before overt symptoms appeared, these changes were accompanied by a widening and

"blurring" of the histogram modes. This can be seen in the last two sessions in the upper right corner. Although the pigeon continued to peck, its behavior was no longer characterized by the orderly, rhythmic pattern typically observed. The decline in response rate caused by the occurrence of pauses in the stream of responding resembles effects of mercury vapor (2) and central nervous system drugs (53). In advanced stages of methylmercury intoxication, the pigeons show signs of motor impairment such as an inability to fly. Thus, the pigeon shares with cats (12) and rodents symptoms of methyl-

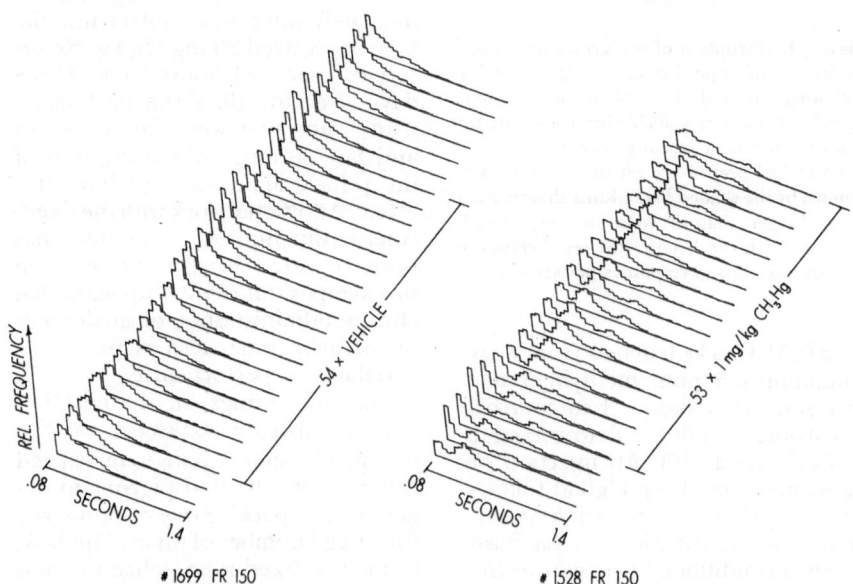

Figure 5. Changes in the pattern of fixed-ratio responding during long-term administration of methylmercury. Data consist of frequency distributions of interresponse times for a series of tests before, during and after 8 ml/kg vehicle administration (pigeon #1699 on the left) and a corresponding period of 4 ml/kg methylmercury administration (pigeon #1528 on the right). Doses were oral, 5 days a week, with the total number of doses indicated by the number within the brackets. The brackets mark tests conducted twice weekly during dosing. Pigeon #1528 displayed overt symptoms after the last test session (upper right) and thereafter ceased to perform.

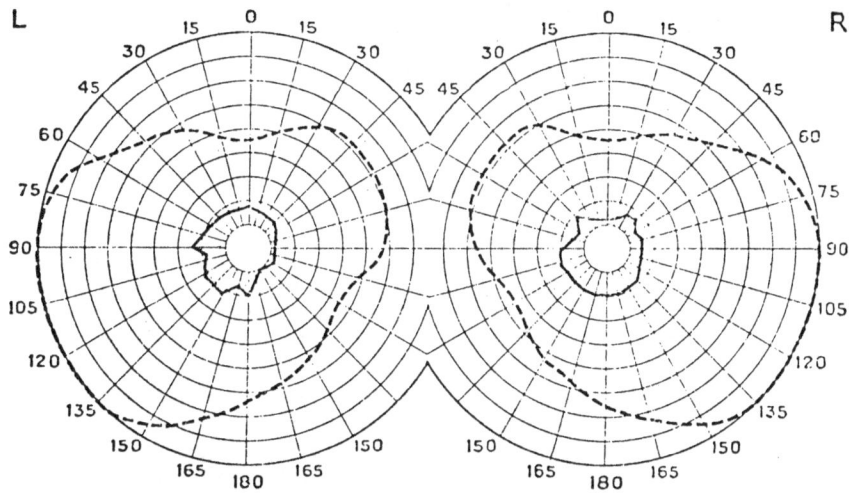

Case 7. Poisoning by Methyl Mercury Compounds. The visual fields (object, 2° white) 5 weeks after onset of first symptoms

Figure 6. Bilateral visual field constriction of a human after 4 mo occupational exposure to methylmercury (solid lines) compared to normal fields (dashed lines). This impairment was still present 25 yr later. (Reproduced with permission from Hunter (25).)

mercury intoxication that are dominated by postural and motor deficiencies.

METHYLMERCURY: EFFECTS ON PRIMATES

Sensory impairment is an important aspect of methylmercury poisoning in humans (3, 7, 49) not described by the animal experiments reviewed thus far. Hunter's pioneer observations suggested that the monkey may be more sensitive to methylmercury than the rat and that the pattern of neural damage in the monkey more closely resembles that found in humans (26, 27). The research reviewed below supports Hunter's observations and indicates that impairments of visual and cutaneous senses are prominent features of methylmercury intoxication in monkeys.

Visual effects

Severe loss of peripheral vision following occupational exposure to methylmercury is illustrated in Fig. 6. In our laboratory, macaques showed clear evidence of visual field constriction after chronic exposure to methylmercury. Even when hungry, they did not rapidly fix their eyes upon or attempt to obtain a piece of food until the food was placed in the direction of their gaze. Berlin et al. reported similar observations for squirrel monkeys (9).

Visual field constriction is probably the result of destruction of the visual cortex (9, 23, 26). In one human case, Hunter (26) observed the greatest cell loss in the region of the calcarine fissure that is thought to subserve peripheral vision. In animals, precise regional differences in amount of damage within the visual cortex have not been established. Since peripheral vision is represented by fewer cortical neurons, it may be more vulnerable to lesions than central vision (11). Given that the peripheral receptors, rods, are necessary for scotopic vision, and given that surgical ablation of visual cortex impairs discrimination of form more than simple discrimination of brightness (24, 30), a logical experiment would examine form discrimination under conditions of low illumination. Macaques were chosen as subjects for such a study because they closely resemble humans both in their visual capacities (17) and in methylmercury kinetics (9).

In order to avoid the problems associated with rapid accumulation, we established a specific mercury concentration in the blood by administering priming doses, then gave small weekly maintenance doses to sustain a concentration associated with visual symptoms in humans (3). Figure 7 provides an example. Prior to the start of the methylmercury administration, this monkey had been trained to perform two visual discrimination tasks. One was a form discrimination. On each trial, a two-dimensional form, either a circle, triangle or square, was projected on each of three rear-illuminated keys, with position randomized. The monkey could obtain a sip of fruit juice by pressing the key with the illuminated square. Pressing the keys illuminated with the circle or with the triangle, or a failure to respond within 10 sec, terminated the trial and also postponed the start of the next trial. A brightness discrimination task was interspersed among the form discrimination trials. For this, only one of the three keys was illuminated and the monkey simply was required to press the illuminated key. Throughout the experiment, tests were conducted with the stimuli at four different luminances. A preliminary experiment

had revealed that the proportion of correct discriminations declines as luminance is reduced.

A control monkey's discrimination performance showed no change over 34 weeks other than a slight improvement in form discrimination at the lowest luminances (Fig. 8). The results for the monkey that received methylmercury for 29 weeks are shown in Fig. 9. For about the first 10 weeks, performance remained normal; the monkey performed with less than 100% accuracy only at the lowest luminance levels. Starting at about the 12th week of exposure, we observed a gradual decline in the accuracy of form discrimination at the lowest luminance level. Note that the decline in brightness discrimination was not as severe as the decline in form discrimination.

The initial decline in performance was partly reversed during weeks 18 to 28, suggesting the action of some

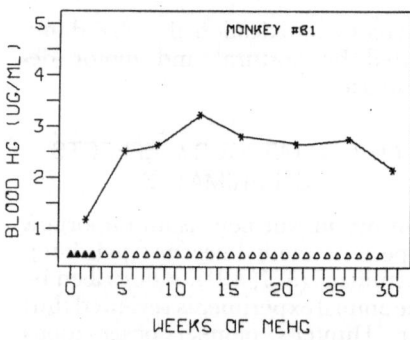

Figure 7. Blood mercury concentrations of a stumptailed macaque (*Macaca arctoides*) during long-term exposure to methylmercury. Solid triangles indicate administration of 1 mg Hg/kg priming doses at 5-day intervals. Open triangles indicate weekly maintenance doses of 0.5 mg Hg/kg. Methylmercury was added to Purina Monkey Chow and immediately fed to the monkey. Without the priming doses, more than one year of weekly 0.5 mg Hg/kg doses would have been required to attain a stable blood concentration. The monkey's visual discrimination performance is summarized in Figs. 9 and 10.

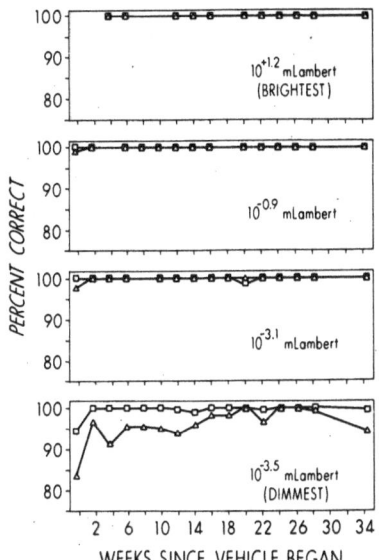

Figure 8. Visual discrimination of a control macaque (#70) administered sodium carbonate, the methylmercury vehicle, for 34 wk according to the procedure described in Fig. 7. Results from the brightness discrimination test are indicated by squares; the form discrimination by triangles. Numbers within each graph refer to the luminance of the stimuli. Tests were conducted in a dark chamber following a 10 min period of dark adaptation.

crimination performance also argues against motor impairment as a substantial cause of the deterioration in form discrimination; the monkey obviously was capable of reaching out and pressing the one illuminated key for weeks after form discrimination had disappeared. These findings illustrate the value of parametric determinations in behavioral toxicology. They also illustrate how such an

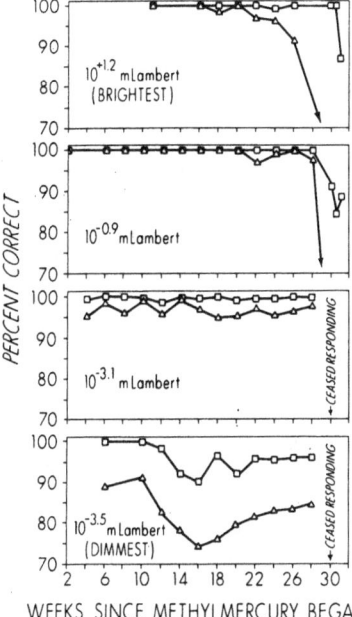

Figure 9. Effects of a 29-wk exposure to methylmercury on the visual discrimination of macaque #81. The dosing procedure and blood Hg concentrations are shown in Fig. 7. Testing of some luminances did not begin until blood concentrations had stabilized. Symbols are the same as in Fig. 8. The arrows in the two upper graphs signify a decline to 33% correct (chance). At the time of sacrifice at the 31st wk, the monkey had great difficulty in locating objects visually and finding its way around the home cage. Motor incoordination was minimal and seemed related to impairment of vision and touch (see Fig. 12). (Data from Evans et al. (19).) Histological findings are reported elsewhere (21).

compensatory mechanism. At about the 22nd week, the first errors occurred in form discrimination at the higher luminances. After the 28th week, the monkey's performance deteriorated markedly although blood mercury concentrations had remained steady since the 5th week (Fig. 7). Performance ceased under the two lowest luminance levels. At the two highest luminances responding continued, but form discrimination performance had declined to chance, 33% correct.

The monkey could not have been completely blind, because brightness discrimination remained well above chance with the high luminance levels. The quality of the brightness dis-

218   Hugh L. Evans et al.

approach can yield clues about appropriate clinical testing procedures.

The speed with which the monkey made its choices is shown in Fig. 10. This subject responded quickly to the brightest stimuli and took increasingly longer to respond to lower luminances. The mean response time remained remarkably steady during the period when the earliest visual changes shown in Fig. 9 were emerging. Only after the monkey could no longer see the dim stimuli, and was making many errors with the brighter stimuli, did response times increase consistently. The variability of the response times did not change consistently during the period of the experiment. These results have been replicated with other monkeys and will be reported in detail elsewhere.

Speed of responding can be affected by methylmercury in a task requiring considerably more manual dexterity and coordination. Berlin et al. (9) required squirrel monkeys to choose between two objects on the basis of color, form and size by using a chain to pull the appropriate object toward them. Four animals received an unspecified series of oral doses of methylmercury, resulting in different rates of accumulation. Slower accumulation delayed the appearance of symptoms and also caused symptoms to emerge more gradually. As mentioned above, this relationship also has been found in mice (47).

Changes in performance of the two monkeys with the slower accumulation consisted of a pronounced increase in the time required to complete a response and a slight decrease in the number of correct choices (Fig. 11). This was interpreted as an impairment of both visual and tactile senses and a disturbance of motor coordination.

A comparison of our findings with those of Berlin and co-workers (9)

suggests that the nature of the task probably influences the prominence with which specific effects are seen. Thus, Berlin et al. showed that methylmercury produced a marked increase in the time required to execute a complex response (Fig. 11) but, in our study, time to make the simpler response of pushing a key was not affected until vision was severely

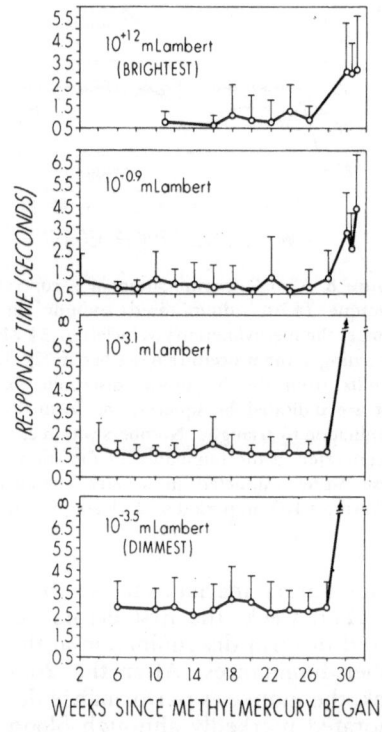

Figure 10. Mean response times of macaque #81 on the form discrimination task during methylmercury exposure. Response times on the brightness discrimination were similar. Response time is the interval between onset of the stimuli and the key press response. Note that the monkey responded more slowly with each decrease in luminance. The means and standard deviations, based on 100 to 300 trials per point, did not change consistently until the final week. Arrows in the two lower graphs indicate infinite response time when responding ceased.

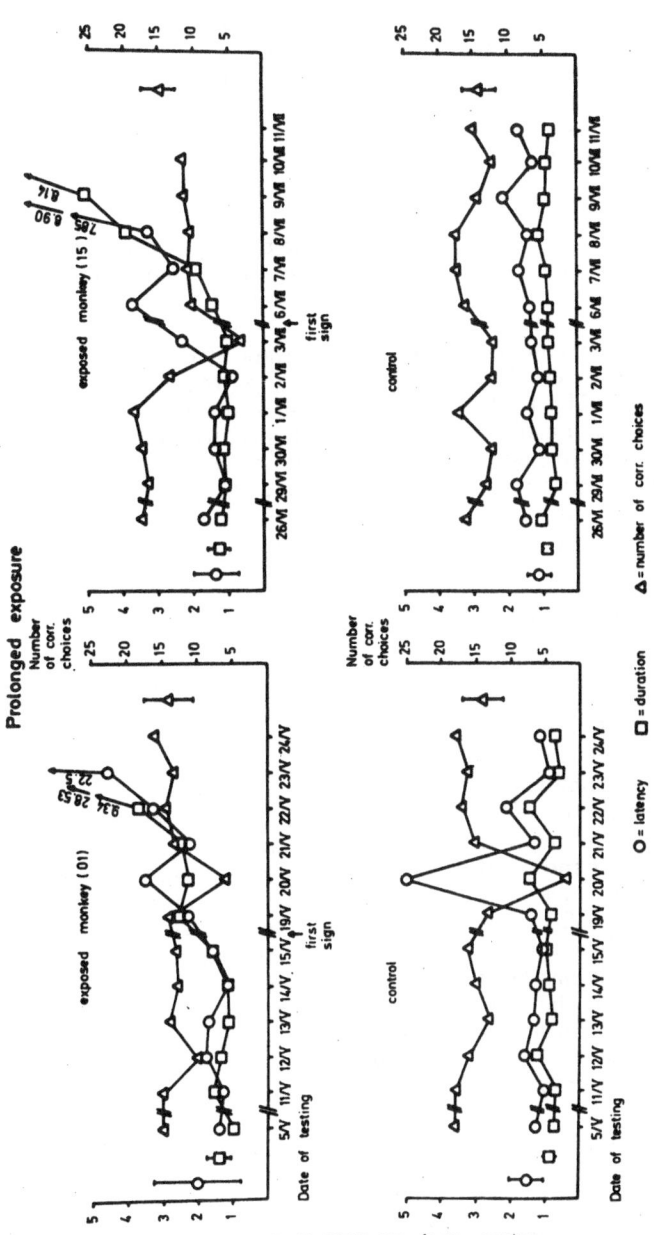

Figure 11. Two-choice visual discrimination performance of two squirrel monkeys receiving oral methylmercury and of two matched control monkeys. Blood mercury concentrations were 1.3 and 1.6 ppm at the time the first signs of intoxication appeared. Latency is the time from the start of the trial until the monkey initiated a response by picking up a chain. Duration is the time required to pull in the stimulus object with the chain. The mean and standard deviation of each measure before the onset of signs is shown next to the relevant axis. The marked increase in latency and duration in the final 6 days was attributed to loss of peripheral vision, which delayed the sighting of the test objects, and to a disturbance of coordination and, possibly, sensation in the fingers. During this period the number of correct choices remained near normal. Severe visual impairment prevented further testing after the last date shown. (Reproduced with permission of Berlin et al. (9).)

impaired (Fig. 10). In the same way, the ability to make visual discriminations was less vulnerable to methylmercury when the stimuli differed along three different dimensions and were probably brightly illuminated (Fig. 11) than when the stimuli were very dim and differed only in the dimension of form (Fig. 9).

Hypesthesia

Another sensory deficit in humans described by Hunter et al. (26) and observed in our monkeys is hypesthesia, an impaired sense of touch. This type of sensory impairment may underlie some of the effects described above and is illustrated in Fig. 12. In order to obtain a small marshmallow, the monkey had to reach through a hole in the cage wall. Normal monkeys first sight the marshmallow, then extend their entire arm through the hole in order to reach it. The monkey was not able to peer through the hole while reaching. Only rarely does the monkey fail to make contact with the marshmallow on the first reach, but on these occasions the normal monkey will scan the surface with its fingers, quickly locate the marshmallow and grasp it. A methylmercury-treated monkey will more often miss the marshmallow. The monkey will frequently close its hand as if grasping an object, then invariably return its empty, clenched fist to its mouth. An example is shown in the sequence in Fig. 12. This has been observed to happen even when the monkey had brushed against the marshmallow with its hand. Similar effects have been observed in squirrel monkeys (9, page 206).

Astereognosis, the inability to identify objects on the basis of touch, appears in humans poisoned by methylmercury (7) and by mercury vapor (51) and in monkeys. In the situation shown in Fig. 12, astereognosis is reflected by attempts to retrieve the round head of a screw, or

the experimenter's finger tip, as if it were a marshmallow. We have never seen such difficulties in normal monkeys. These symptoms do not seem to involve motor impairment since the monkey's grip remains strong and, once the marshmallow is located, the monkey has no difficulty in getting it into its mouth.

Motor effects

Although the human clinical data are not very clear, our evidence indicates that visual deficits precede motor effects in monkeys. One test, designed to measure static strength, required squirrel monkeys (weighing about 700 g) to maintain a criterion force of 200 g on a T-shaped bar attached to a strain gauge. Whenever the applied force fell below the 200 g minimum, a tone and light were presented, signaling that electric shock was to be delivered by tail electrodes. The warning signals and the shock, if it had occurred, were terminated when the force applied by the monkey once again exceeded the criterion. This task requires considerable exertion and should reflect muscular weakness and fatigue, symptoms often reported in the clinical literature.

A second procedure was designed to quantify changes in fine motor control. Humans have shown symptoms such as an inability to button clothes efficiently or an impairment of handwriting (25). Monkeys were trained to place a paw on a stationary platform attached to a strain gauge and to maintain a force between 5 and 15 g for 2 consecutive sec. A response that met these criteria produced delivery of a sucrose pellet. A similar force-tracking procedure had proved useful in quantifying finger tremor and in following the course of recovery of humans suffering from metallic mercury poisoning (54).

For both experiments, trained monkeys received a priming dose, orally, followed by a weekly dose calculated

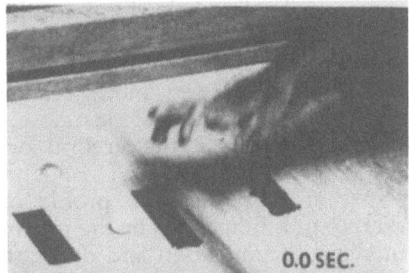

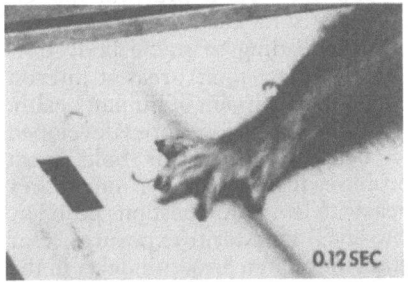

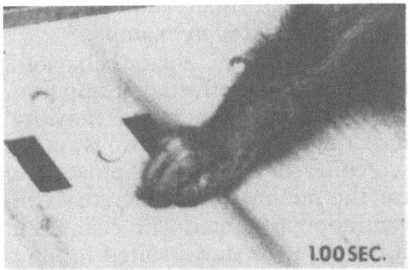

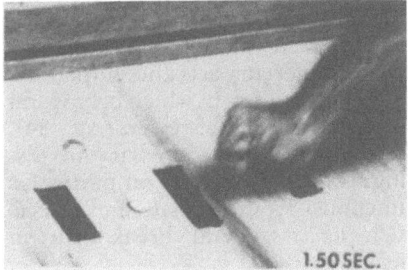

Figure 12. An example of hypesthesia in macaque #82. This type of symptom is first seen at about the same stage of methylmercury exposure as the early changes in visual discrimination (Fig. 9). The monkey is reaching for a miniature marshmallow placed on a wooden surface. Strips of black tape, 2 cm wide, indicate distances of 18, 25 and 32 cm from the cage. The monkey closes its fist and withdraws the clenched fist as if it has seized the marshmallow. Normal monkeys have never been observed to do this. The sequence was taken from a motion-picture film. The time sequence, with reference to the first frame, is indicated in the lower right corner of each frame. Methylmercury was administered by the same procedure shown in Fig. 7 except that 5 priming doses of 1.0 mg Hg/kg were given before the start of weekly maintenance doses of 0.5 mg Hg/kg. Blood mercury concentration ranged from 2.8 to 3.2 ppm after the maintenance doses began. The episode portrayed here occurred during the 21st week of methylmercury administration.

to maintain a relatively stable blood mercury concentration of either 0.5 or 1.0 ppm, levels below those previously associated with behavioral impairment in monkeys (9) and with paresthesia in humans (3). Monkeys have been maintained on this regimen for more than 18 months without showing any consistent changes in motor performance on either task. During this period a number of these monkeys developed signs of visual impairment including, in some cases, blindness (21).

CONCLUSIONS

The evaluation of mercury and its compounds is one of a few areas in toxicology and environmental health to which behavioral techniques have been applied. Progress so far has been modest compared to behavioral studies of drugs. Perhaps one reason is that given the clear need for dose-response determinations and given the slow turnover of most mercury compounds in animals, duration of exposure must be considered in addition to dose. Specifying the effective dose in terms of blood or brain mercury concentration is also essential. Without such data, one cannot correlate behavioral measures with chemical events because of large individual differences in excretion rates (1) and because of immense species differences in distribution. For example, brain/blood ratios of methylmercury appear to be about 0.1 for rats (6), 0.8 for mice (46) and pigeons (our unpublished observations), and 4.0 to 10.0 for squirrel monkeys (6) and macaques (our unpublished observations). Thus, differences in the administered dose necessary to affect different behaviors may be largely accounted for by the different species employed; critical brain mercury concentrations may be similar for most species (7).

The literature suggests species differences of another sort, in terms of the characteristic profile of symptoms. Experiments with rodents and cats suggest a profile of methylmercury symptoms dominated by disorders of movement and posture reflecting, perhaps, damage to cerebellum (12, 23, 46) and peripheral sensory nerves (42). Similar types of symptoms are seen in birds (48 and our unpublished observations). In

contrast, the few behavioral studies with primates mainly implicate sensory processes. Of these, changes in vision have been more readily quantified and seem to be most closely related to damage in the cerebral cortex (9, 21, 23, 26, 27).

Within a species, the profile of symptoms is influenced by the rate of mercury accumulation and duration of exposure. As a practical matter, the effects of chronic mercury exposures leading to stable body burdens generate their greatest interest from the standpoint of human health. It is here that procedures developed by behavioral scientists hold great promise for toxicology, since they deal with the main questions posed by low level, long-term exposures. Our observation of a 3-month delay in the appearance of visual symptoms in the monkey provides an example.

It is clear that systematic behavioral tests can reveal effects not apparent from gross observations. Behavioral tests can be applied in two ways. In screening for the early symptoms, and for the minimal brain mercury concentration associated with changes in function, procedures suited to long-term studies should be used. On the other hand, behavioral tests may be able to specify more precisely the way in which mercury acts and, hopefully, may suggest ways in which behavioral impairment can be treated (e.g., 34). In this case, it is necessary to distinguish specific effects on particular functions from nonspecific effects such as a general breakdown in health.

REFERENCES

1. AL-SHAHRISTANI, H., AND K. M. SHIHAB. Variation of biological half-life of methylmercury in man. *Arch. Environ. Health* 28: 342–344, 1974.
2. ARMSTRONG, R. D., L. J. LEACH, P. R. BELLUSCIO, E. A. MAYNARD, H. C. HODGE AND J. K. SCOTT. Behavioral changes in the pigeon following inhala-

tion of mercury vapor. *Am. Ind. Hyg. Assoc. J.*, 24: 366–375, 1963.

3. BAKIR, F., S. F. DAMLUJI, L. AMIN-ZAKI, M. MURTADHA, A. KHALIDI, N. Y. AL-RAWI, S. TIKRITI, H. I. DHAHIR, T. W. CLARKSON, J. C. SMITH AND R. A. DOHERTY. Methylmercury poisoning in Iraq. *Science* 181: 230–241, 1973.

4. BELILES, R. P., R. S. CLARK, P. R. BELLUSCIO, C. L. YUILE AND L. J. LEACH. Behavioral effects in pigeons exposed to mercury vapor at a concentration of 0.1 mg/M^3. *Am. Ind. Hyg. Assoc. J.* 28: 482–484, 1967.

5. BELILES, R. P., R. S. CLARK AND C. L. YUILE. The effects of exposure to mercury vapor on behavior of rats. *J. Toxicol. Appl. Pharmacol.* 12: 15–21, 1968.

6. BERGLUND, F., AND M. BERLIN. Risk of methylmercury cumulation in man and the relation between body burden of methylmercury and toxic effects. In: *Chemical Fallout*, edited by M. W. Miller and G. G. Berg. Springfield, Ill.: Thomas, 1969, p. 258–269.

7. BERGLUND, F., ET AL. Methylmercury in fish: Report of a Swedish Expert Group. *Nord. Hyg. Tidskr. Suppl.* 4, 1971.

8. BERLIN, M., J. FAZACKERLEY AND G. NORDBERG. The uptake of mercury in the brains of mammals exposed to mercury vapor and to mercuric salts. *Arch. Environ. Health* 18: 719–729, 1969.

9. BERLIN, M., G. NORDBERG AND J. HELLBERG. The uptake and distribution of methylmercury in the brain of *Saimiri sciureus* in relation to behavioral and morphological changes. In: *Mercury, Mercurials and Mercaptans*, edited by M. W. Miller and T. W. Clarkson. Springfield, Ill.: Thomas, 1973, p. 187–208.

10. BRAUN, J. J., AND D. R. SNYDER. Taste aversions and acute methylmercury poisoning in rats. *Bull. Psychon. Soc.* 1: 419–420, 1973.

11. BROWN, J. L. The structure of the visual system. In: *Vision and Visual Perception*, edited by C. H. Graham. New York: Wiley, 1965, p. 39–59.

12. CHARBONNEAU, S. M., I. C. MUNRO, E. A. NERA, R. F. WILLES, T. KUIPER-GOODMAN, F. IVERSON, C. A. MOODIE, D. R. STOLTZ, F. A. J. ARMSTRONG, J. F. UTHE AND H. C. GRICE. Subacute toxicity of methylmercury in the adult cat. *Toxicol. Appl. Pharmacol.* 27: 569–581, 1974.

13. CIACCIO, E. I. Mercury: therapeutic and toxic aspects. *Seminars Drug Treat.* 1: 177–194, 1971.

14. CLARKSON, T. W. Recent advances in the toxicology of mercury with emphasis on the alkylmercurials. *CRC Critical Reviews in Toxicology*. Cleveland, Ohio: Chemical Rubber Publ. Co., 1972, p. 203–234.

15. CURLEY, A., V. A. SEDLAK, E. F. GIRLING, R. E. HAWK, W. F. BARTHEL, P. E. PIERCE AND W. H. LIKOSKY. Organic mercury identified as the cause of poisoning in humans and hogs. *Science* 172: 65–67, 1971.

16. DERBAN, L. K. A. Outbreak of food poisoning due to alkyl-mercury fungicide. *Arch. Environ. Health* 28: 49–52, 1974.

17. DEVALOIS, R. L., H. C. MORGAN, M. C. POLSON, W. R. MEAD AND E. M. HULL. Psychophysical studies of monkey vision—I. Macaque luminosity and color vision tests. *Vision Res.* 14: 53–66, 1974.

18. DIAMOND, S. S., AND S. D. SLEIGHT. Acute and subchronic methylmercury toxicosis in the rat. *Toxicol. Appl. Pharmacol.* 23: 197–207, 1972.

19. EVANS, H. L., V. G. LATIES AND B. WEISS. Behavioral effects of methylmercury. *Proceedings of First Annual N.S.F. Trace Contaminants Conference*. Oak Ridge, Tenn.: U.S. Atomic Energy Comm., 1974, p. 534–540.

20. FRIBERG, L., AND G. NORDBERG. Inorganic mercury—a toxicological and epidemiological appraisal. In: *Mercury, Mercurials and Mercaptans*, edited by M. W. Miller and T. W. Clarkson. Springfield, Ill.: Thomas, 1973, p. 5–22.

21. GARMAN, R. H., B. WEISS AND H. L. EVANS. Alkylmercurial encephalopathy in the monkey. *Acta Neuropath.* In press.

22. GOLDWATER, L. J. *Mercury: a History of Quicksilver*. Baltimore: York Press, 1972.

23. GRANT, C. A. Pathology of experimental methylmercury intoxication: Some problems of exposure and response. In: *Mercury, Mercurials and Mercaptans*, edited by M. W. Miller and T. W. Clarkson. Springfield, Ill.: Thomas, 1973, p. 294–310.

24. HUMPHREY, N. K., AND L. WEISKRANTZ. Vision in monkeys after removal of the striate cortex. *Nature* 215: 595–597, 1967.

25. HUNTER, D. *The Diseases of Occupations*. Boston: Little, Brown, 1969, 4th ed.

26. HUNTER, D., R. R. BOMFORD AND D. S. RUSSELL. Poisoning by methylmer-

cury compounds. *Q. J. Med.* 9: 193–213, 1940.

27. HUNTER, D., AND D. S. RUSSELL. Focal cerebral and cerebellar atrophy in a human subject due to organic mercury compounds. *J. Neurol. Neurosurg. Psychiatry* 17: 235–241, 1954.

28. JOSELOW, M., D. LOURIA AND A. A. BROWDER. Mercurialism: environmental and occupational aspects. *Ann. Intern. Med.* 76: 119–130, 1972.

29. KLEIN, S. B., AND E. J. ATKINSON. Mercuric chloride influence on active-avoidance acquisition in rats. *Bull. Psychon. Soc.* 1: 437–438, 1973.

30. KLÜVER, H. An analysis of the effects of the removal of the occipital lobes in monkeys. *J. Psychol.* 2: 49–61, 1936.

31. KOJIMA, K., AND M. FUJITA. Summary of recent studies in Japan on methylmercury poisoning. *Toxicology* 1: 43–62, 1973.

32. LAHUE, R. A sensitive bio-behavioral assay for methylmercury. *Bull. Environ. Contam. Toxicol.* 10: 166, 1973.

33. LATIES, V. G. The modification of drug effects on behavior by external discriminative stimuli. *J. Pharmacol. Exp. Ther.* 183: 1–13, 1972.

34. LATIES, V. G. The role of discriminative stimuli in modulating drug action. *Federation Proc.* 34: 1880–1888, 1975.

35. MECHNER, F. Probability relations within response sequences under ratio reinforcement. *J. Exp. Anal. Behav.* 1: 109–122, 1958.

36. MILLER, M. W., AND T. W. CLARKSON, editors. *Mercury, Mercurials and Mercaptans*. Springfield, Ill.: Thomas, 1973.

37. NEAL, P. A., ET AL. Mercurialism and its control in the felt-hat industry. *U.S. Public Health Bulletin* No. 263, 1941.

38. NORSETH, T. Biotransformation of methylmercuric salts in the rat with chronic administration of methylmercuric cysteine. *Acta Pharmacol. Toxicol.* 31: 138–148, 1972.

39. POST, E. M., M. G. YANG, J. A. KING AND V. L. SANGER. Behavioral changes of young rats force-fed methyl mercury chloride. *Proc. Soc. Exp. Biol. Med.* 143: 1113–1116, 1973.

40. ROSENTHAL, E., AND S. B. SPARBER. Methylmercury dicyandiamide: retardation of detour learning in chicks hatched from injected eggs. *Life Sci.* 11: part I, 883–892, 1972.

41. SALVATERRA, P., B. LOWN, J. MORGANTI AND E. J. MASSARO. Alterations in neurochemical and behavioral parameters in the mouse induced by low doses

of methyl mercury. *Acta Pharmacol. Toxicol.* 33: 177–190, 1973.

42. SOMJEN, G. G., S. P. HERMAN AND R. KLEIN. Electrophysiology of methylmercury poisoning. *J. Pharmacol. Exp. Ther.* 186: 579–592, 1973.

43. SPYKER, J. M. Behavioral teratology and toxicology. In: *Behavioral Toxicology*, edited by B. Weiss and V. G. Laties. New York: Plenum, 1975, p. 311–344.

44. SPYKER, J. M. Assessing the impact of low level chemicals on development: Behavioral and latent effects. *Federation Proc.* 34: 1835–1844, 1975.

45. SPYKER, J. M., S. B. SPARBER AND A. M. GOLDBERG. Subtle consequences of methylmercury exposure: behavioral deviations in offspring of treated mothers. *Science* 177: 621–623, 1972.

46. SUZUKI, T. Neurological symptoms from concentration of mercury in the brain. In: *Chemical Fallout*, edited by M. W. Miller and G. G. Berg. Springfield, Ill.: Thomas, 1969, p. 245–256.

47. SUZUKI, T., AND T. MIYAMA. Neurological symptoms and mercury concentration in the brain of mice fed with methylmercury salt. *Ind. Health* 9: 51–58, 1971.

48. THAXTON, J. P., AND C. R. PARKHURST. Abnormal mating behavior and reproductive dysfunction caused by mercury in Japanese quail. *Proc. Soc. Exp. Biol. Med.* 144: 252–255, 1973.

49. TOKUOMI, H., T. OKAJIMA, J. KANAI, M. TSUNODA, Y. ICHIYASU, H. MISUMI, K. SHIMOMURA AND M. TAKABA. Minamata disease. *World Neurol.* 2: 536–545, 1961.

50. VOSTAL, J. J., AND T. W. CLARKSON. Mercury as an environmental hazard. *J. Occup. Med.* 15: 649–656, 1973.

51. VROOM, F. Q., AND M. GREER. Mercury vapour intoxication. *Brain* 95: 305–318, 1972.

52. WEIR, P. A., AND C. H. HINE. Effects of various metals on behavior of conditioned goldfish. *Arch. Environ. Health* 20: 45–51, 1970.

53. WEISS, B., AND C. T. GOTT. A microanalysis of drug effects on fixed-ratio performance in pigeons. *J. Pharmacol. Exp. Ther.* 180: 189–202, 1972.

54. WOOD, R. W., A. B. WEISS AND B. WEISS. Hand tremor induced by industrial exposure to inorganic mercury. *Arch. Environ. Health* 26: 249–252, 1973.

55. ZENICK, H. Behavioral and biochemical consequences of methylmercury chloride toxicity. *Pharmacol. Biochem. Behav.* 2: 709–713, 1974.

Schedule-controlled behaviors as determinants of drug response

W. H. MORSE

Department of Psychiatry, Harvard Medical School
Boston, Massachusetts 02115

ABSTRACT

Schedule-controlled performances began to be used in assessing the behavioral effects of drugs because of practical advantages over other techniques for studying behavior. Schedule-controlled behavior is, however, of fundamental importance in behavioral pharmacology. It has been found repeatedly that the effects of many drugs depend critically upon the patterns of responding engendered by different schedule contingencies. These dependencies of the effects of drugs on schedule-controlled behavior occur because ongoing behavior is itself an important determinant of drug action.—MORSE, W. H. Schedule-controlled behaviors as determinants of drug response. *Federation Proc.* 34: 1868–1869, 1975.

THe topic of schedule-controlled behavior is an integral part of this symposium on behavioral pharmacology. Much material on the use of schedules of reinforcement in behavioral pharmacology has already been presented in the earlier sessions. The present session deals mainly with the use of schedule contingencies as analytical tools in investigating specific topics of specialistic interest to those actively working in this field. As a field develops and inevit- ably becomes more technical, it loses some of the excitement of the initial search for scientifically useful vari- ables. Because schedule-controlled behavior has been important in establishing the structure of facts in behavioral pharmacology, I would like briefly to recount this history.

The earliest work in behavioral pharmacology, long before the use of schedules to study drugs, indicated that behavioral effects of drugs were dependent on the environmental situation—that so-called nondrug factors had big effects. A good example is the work of Chance on the lethal toxicity of amphetamines. The LD_{50} of amphetamine in mice could be changed by a factor of 10 (from over 100 mg/kg to 10 mg/kg) by changing the ambient temperature, ambient noise, the size of the cage, and the number of animals per cage (1, 2). Yet such findings do not mean that the lethality of amphetamines is capriciously variable. Under constant conditions the results were reproducible, but they depended very much on the prevailing environmental conditions. The situation here is not altogether different from other areas of pharmacology. Even in studying the effects of drugs in an isolated organ bath, the temperature or the type of recording lever can determine the result obtained. For behavioral pharmacology, however, the study of the nature of these nondrug environmental determinants of behavior is essential.

There were always a few pharmacologists interested in the effects of central nervous system drugs on overt behavior, but the discovery of the major tranquilizer, chlorpromazine, created an intense interest in the behavioral effects of drugs and a demand for ways to study them. Fortunately, some excellent ways of studying behavior had already been developed by the experimental psychologist, B.F. Skinner. These scheduling techniques were promising because they were objective, quantitative, provided reproducible patterns of behavior, could be studied over long periods of time, and were sensitive to the effects of various interventions. Understandably, these techniques were promoted by scientific entrepreneurs for their practical value. Many laboratories in the mid 1950's began

using these schedule techniques to study drugs. At that time, no one suspected that schedule-controlled behavior would be of fundamental significance for drug action.

In retrospect, the results from experiments on schedule-controlled behavior indicated, from the very beginning, an important relation—that ongoing behavior was a powerful determinant of the behavioral effects of drugs—but it was several years before this point began to be appreciated. I noted that the effects of drugs had been shown to depend on the environmental conditions and that results were likely to be different unless the conditions were exactly the same. Yet, from laboratory to laboratory, the results with comparable schedules were the same. It was not necessary to control the many features of the environment. Simply specifying a particular schedule condition (fixed interval or fixed ratio) was usually sufficient to make the effects of drugs reproducible in different laboratories, using different types of equipment, maintenance events, and the like. Why was this? Clearly there was something about schedules that controlled important aspects of behavior with respect to the actions of drugs. Knowing that drugs affect rates and patterns of ongoing responding, it is now easy to see why schedules were so powerful. Schedules control rates and patterns of responding, and therefore have a fundamental as opposed to a practical significance for drug action.

Another early clue was provided by the finding that fixed-interval schedules are particularly sensitive to the behavioral effects of some drugs. These schedules engender an increasing number of responses from the beginning to the end of each interval, so in this one schedule a range of rates can be obtained. Gradually, the notion of behavior itself as a determinant of drug action developed to the point of making quantitative predictions about the effects of drugs on the basis of ongoing rates of responding—and, as we all will hear today, has now further developed to the point of being able to specify the exceptional conditions of rate dependencies.

Contingencies of reinforcement are powerful means of controlling behavior and can be used to produce all sorts of different behaviors. This aspect of reinforcement, that the consequences of behavior can be used to shape new and unusual response patterns, has made and continues to make behavioral pharmacology an exciting field. One can specify what has been done when behavior is changed with schedules. Part of the initial appeal and continued promise for using schedule contingencies in behavioral pharmacology was their power as analytical tools in isolating and controlling environmental factors

precisely. Much of the present symposium deals with such analysis. In the earliest days of behavioral pharmacology, it was possible to talk colloquially about drug effects on behavior. Invariably, the important features of environmental situations turned out to be different from the common conceptions of ordinary language and concepts based more on procedural arrangements have evolved. These schedule concepts may be less easily understood by those not directly involved, but actually they are simpler in terms of predictive generality and in providing valid accounts of the behavioral effects of drugs.

REFERENCES

1. CHANCE, M. R. A. Aggregation as a factor influencing the toxicity of sympathomimetic amines in mice. *J. Pharmacol. Exp. Ther.* 87: 214–219, 1946.
2. CHANCE, M. R. A. Factors influencing the toxicity of sympathomimetic amines to solitary mice. *J. Pharmacol. Exp. Ther.* 89: 289–296, 1947.

Determinants of drug effects on punished responding[1]

D. E. McMILLAN

Department of Pharmacology, School of Medicine,
University of North Carolina, Chapel Hill, North Carolina 27514

ABSTRACT

The effects of drugs on punished responding depend on interactions among a large number of experimental variables. Among these variables are the drug history of the animal, the dose of the drug administered, the type of stimulus used to punish responding, the intensity and duration of the punishing stimulus, the schedule of presentation of the punishing stimulus, the control rate and pattern of punished responding, the schedule of positive reinforcement maintaining the punished responding, the species of animal, the deprivation state of the animal, the behavioral history of the animal, and the nature of the required response. Although it is not known how all of these variables interact to determine the effect of drugs on punished responding, there is evidence that many of these variables are important as determinants of drug effects. The task facing behavioral pharmacologists studying drug effects on punished responding is to determine under what conditions drugs produce their characteristic effects on punished responding. —McMILLAN, D. E. Determinants of drug effects on punished responding. *Federation Proc.* 34: 1870–1879, 1975.

Punishment is a process defined by "a reduction of the future probability of a specific response as a result of the immediate delivery of a stimulus for that response" (4). A stimulus that reduces the future probability of occurrence of the responses that produce it usually is referred to as a punishing stimulus. In experiments on the effects of drugs on punished responding, responding usually is established and maintained by a schedule of food presentation before the punishing stimulus is introduced. The rate of

[1] Supported in part by Grant No. 1-ROI-MH 19440 from the National Institute of Mental Health and by a grant from Hoffmann-La Roche Inc.

Abbreviations: VI, variable interval; FI, fixed interval; FR, fixed ratio; DRL, differential reinforcement of low rate; CRF, continuous reinforcement.

responding is subsequently suppressed by scheduling the responding maintained by food presentation to also produce the punishing stimulus. Typically, the effects of drugs on punished responding are studied when stable rates and patterns of punished responding are observed.

Much of the early work on the effects of drugs on punished responding was done by Geller and his associates (14–19). In the Geller procedure, responding is reinforced with food under a variable-interval schedule (VI schedule). Periodically, a tone is introduced, during which each response produces both food and electric shock. Responding under the VI schedule occurs at a steady and moderate rate, but the rate of punished responding during the tone is much lower. Geller and his colleagues (17–19) reported that barbiturates and drugs used clinically as minor tranquilizers increased rates of responding during tone, or in other words, increased punished responding.

In more recent studies on the effects of drugs on punished responding, the same schedule of positive reinforcement has been used to maintain both punished and unpunished responding (28, 35, 43). From these experiments, as well as those of Geller and his colleagues, a fairly clear picture of the effects of drugs on punished responding seems to emerge. Drugs used clinically as minor tranquilizers, including meprobamate (18, 19, 35) and several benzodiazepines (17, 35, 43), increase rates of punished responding, as do the barbiturates. However, nonbarbiturate sedative hypnotics apparently do not increase punished responding (37). Usually, major tranquilizers such as the phenothiazines, CNS stimulants such as the amphetamines, and strong analgesics such as morphine, do not

increase rates of punished responding (28).

Unfortunately, there are many exceptions to these generalizations about the effects of drugs on punished behavior. For example, under some conditions amphetamines (13, 35, 38, 39), morphine (31, 35), and chlorpromazine (9, 36) all can increase rates of responding suppressed by punishment, and when punishment does not markedly suppress responding (17, 36) minor tranquilizers and barbiturates have little tendency to increase rates of punished responding. These exceptions, as well as the effects of other drugs on punished responding, are summarized in Table 1.

If the experiments that have produced conflicting data on the effects of drugs on punished responding are examined closely, it becomes apparent that the experimental parameters vary widely. If behavior is a function of its consequences, and if the consequences of behavior vary from one experiment to another, it should not be too surprising to find that drug effects on punished behavior vary from one experiment to another.

The task of sorting out the variables that contribute to the interaction between drugs and punished behavior is not an easy task, because there are so many variables that might be important. The purpose of the present review is to summarize the manner in which some of these variables contribute to drug-behavior interactions to determine the effects of drugs on punished responding.

NATURE OF
THE PUNISHING STIMULUS

In practically all of the experiments on the effects of drugs on punished responding, electric shock has been used as the punishing stimulus. When

TABLE 1. Summary of drug effects
on punished responding

Drugs that "increase" punished responding
Barbiturates (18, 19, 23, 35, 36, 42)[a]
Meprobamate (7, 18, 19, 23) ,
Benzodiazepines (7, 15, 17, 23, 35, 42, 43)
Caffeine (39)
Trimethadione (14)
Hedonal, emylcamate and urethane (14)
Tryptamine antagonist (21)

Drugs that "do not increase" punished
responding

Most phenothiazines (7, 17, 40, 41)
Nicotine (39)
Scopolamine and benactyzine (23, 41)
Imipramine (35, 43)
α-Methyltryptamine (21)
Diphenylhydantoin (20)
Propranolol, ethchlorvynol and chloral
hydrate (37)
Hallucinogens (6, 35)

Drugs that have been reported both to increase and
not to increase punished responding

Amphetamine (13, 24, 35, 39)
Morphine (16, 28, 31, 35)
Reserpine and chlorpromazine (6, 7, 17, 22,
27, 40)
Ethanol (6, 24, 27)

[a] References.

punishing stimuli other than electric
shock have been studied, the effects
of drugs on punished responding
usually have been similar to the effects
of these same drugs on shock-
punished responding.

One approach to studying the effects of drugs on responding punished by stimuli other than electric shock is that of Margules and Stein (32), who added quinine sulfate to sweetened condensed milk presented to rats that were not deprived of food and water. Normally the rats drank 20 to 25 ml of milk, but with quinine added to the milk they drank only about half as much. A 10 mg/kg dose of oxazepam, a benzodiazepine minor tranquilizer, increased the consumption of quinine-adulterated milk to almost normal amounts. This restoration of "quinine-punished" licking by oxazepam is similar to the effects of oxazepam on electric-shock punished responding.

Figure 1 shows the effects of pentobarbital on quinine-punished licking by rats. The two rats were conditioned to lick a water spout to obtain food under a fixed interval (FI) 90-sec schedule of food-pellet presentation. After responding had stabilized, 500 mg/liter of quinine hydrochloride was added to the drinking water. Addition of the quinine reduced the rate of licking to about half of the control rate of responding; however, a 10 mg/kg dose of pentobarbital produced large increases in the rate of licking, just as pentobarbital increases rates of responding punished by electric shock (18, 35, 36).

Response-produced time-out has also been used as a punished response

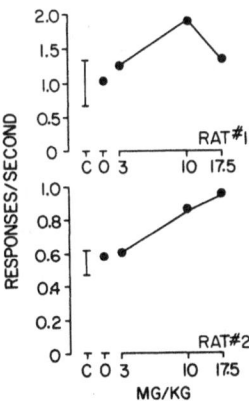

Figure 1. Effects of pentobarbital on "quinine-punished" responding of two rats. *Abscissa*: mg/kg dose, log scale. *Ordinate*: rate of licking in responses (licks)/second. The brackets at C show the range of values during four control sessions. The points at 0 show the effect of control injections. Each point is the mean rate of licking from an entire session in one rat.

in drug experiments (34). In these experiments, squirrel monkeys were conditioned to respond for food under a VI 1-min schedule. In the presence of a white light, each response produced a 40- or a 60-sec time-out during which food was not available. Responding during the pre-time-out stimulus occurred at a much lower rate than did responding under the VI schedule, suggesting that response-produced time-out functioned as a punisher. Pentobarbital produced small increases in the rate of responding during the pretime-out stimulus.

These experiments using time-out and quinine taste as punishing stimuli suggest that increases in punished responding produced by drugs can occur regardless of the type of punishing stimulus used to suppress responding. This suggestion is consistent with the experiments on positive reinforcement that have shown drug effects to be independent of the type of positive reinforcer (29).

INTENSITY AND DURATION OF THE PUNISHING STIMULUS

It has been suggested that minor tranquilizers and barbiturates have little tendency to increase the rate of punished responding when the punishment intensity is low and the amount of response suppression is small (17). On the other hand, Cook and Catania (7) have suggested that benzodiazepines cannot increase rates of punished responding when the shock intensity is very severe.

Figure 2 shows the effects of pentobarbital and chlordiazepoxide on punished responding under a procedure that permits the determination of drug effects on responding suppressed by increasing intensities of electric shock. The schedule is a mult FI 5, FI 5, FI 5, FI 5 with pro-

gressive increases in the punishment intensity during each FI component. The session begins with a green key light, during which responses were reinforced with food under a FI 5-min schedule. In the second FI 5-min component, a blue key light was on and each response produced a 1.5 mA shock to the pigeon through electrodes implanted around the pubis bone. In the third FI 5 component the key light was orange and the shock intensity was 3.0 mA, and in the fourth component the key light was red and the shock intensity was 4.5

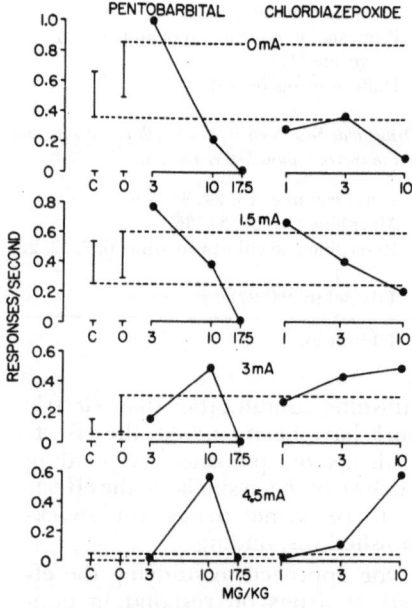

Figure 2. Effects of pentobarbital and chlordiazepoxide on responding punished with increasing shock intensity during successive FIs in a mult FI FI FI FI schedule. *Abscissa*: mg/kg dose, log scale. *Ordinate*: rate of key pecking in responses/second. The brackets at C show the range of values during six control sessions. The brackets at 0 show the range of three control injections. Each point is a mean rate of single observations in four birds during an entire session.

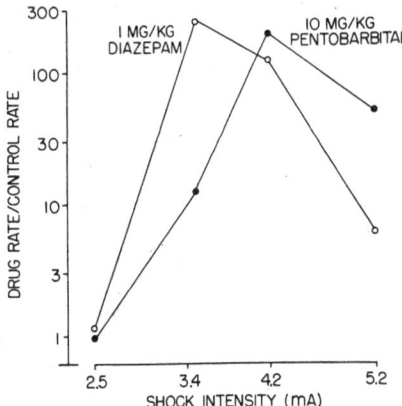

Figure 3. Interaction of shock intensity with diazepam and pentobarbital on responding maintained by a FI schedule. *Abscissa:* increasing (shock) intensity in milliamperes. *Ordinate:* ratio of drug rate to control rate, log scale. Each point is a mean of single observations in each of three birds for an entire session for one dose of each drug.

mA. The sequence was repeated twice within a session. Figure 2 shows that as the shock intensity increased, the nondrug rate of responding decreased. However, as the intensity of the shock increased, both the relative and the absolute increase in the rate of punished responding after pentobarbital and chlordiazepoxide increased.

McMillan (36) studied the effects of a number of drugs on responding punished with different intensities of electric shock. When the shock intensity was low (2.5 mA) and the responding maintained by a mult fixed ratio (FR) 30 FI 5 schedule of food presentation was only slightly suppressed, pentobarbital and diazepam showed little tendency to increase the rate of punished responding. When the shock intensity was increased and responding was almost completely suppressed, pentobarbital

and diazepam produced large increases in the rate of punished responding. A further increase in the shock intensity did not suppress the near zero rates of responding further, but did attenuate the effects of pentobarbital and diazepam in increasing punished responding. It was suggested that the relationship between shock intensity and the drug effect for pentobarbital and diazepam was that of an inverted U, with increasing punishment intensity first leading to larger increases in punished responding, but with higher shock intensities attenuating these increases. Figure 3 shows a reanalysis of some of these data, along with data at an additional shock intensity not reported previously. The doses of pentobarbital and diazepam shown are those that caused the largest increases in the rate punished responding maintained by the FI schedule. Figure 3 shows that as the intensity of the punisher increases, the drug effect first increases and then decreases.

Figure 4 shows that responding in pigeons maintained by a differential reinforcement of low rate (DRL) 15-sec schedule of food presentation decreased when responses were punished with 5 mA shocks of 40-msec duration. Increasing the duration of the shock to 285-msec without changing the shock intensity decreased the rate even further. Chlordiazepoxide did not increase the rate of responding under the DRL schedule when responding was not punished, or when responding was punished with 40-msec shocks, but the lower rates of responding punished by 285-msec shocks were increased by 3 mg/kg of chlordiazepoxide. Although the data are very limited, the relationship of the effect of chlordiazepoxide to the duration of the punishing stimulus appears to be much the same as that described for diazepam and pentobarbital for shock intensity. That is, when the shock duration is short and the amount of suppression by punishment is small, chlordiazepoxide has little tendency to increase punished responding. As the shock duration increases and the rate of punished responding decreases further, the rate increasing effects of chlordiazepoxide on punished responding become larger. Whether or not at very long shock durations the effects of chlordiazepoxide on punished responding are attenuated, as occurred with high shock intensity, has not yet been determined.

Although the present section has concentrated on the relationship between drugs that increase punished responding and the shock intensity and shock duration, these parameters can also interact with drugs that usually do not increase punished responding. For example, Foree et al. (13) have suggested that amphetamines can increase punished responding only when the shock intensity is low and the degree of suppression produced by punishment is very small, which is a relationship quite different than that between pentobarbital and shock intensity.

SCHEDULE OF PRESENTATION OF THE PUNISHING STIMULUS

It is not necessary that every response be punished in order to suppress responding, since intermittent punishment can also decrease rates of responding. Over a wide range of shock intensities, the amount of suppression produced by response-contingent shock is inversely related to the punishment frequency (1). Unfortunately, there appear to be few experiments where the schedule of the punishing stimulus was systematically varied during the study of drugs. Therefore, most comparisons that can be made are across studies by dif-

ferent investigators. Unfortunately, in these experiments other parameters besides the schedule of shock presentation have been simultaneously varied.

It appears that drugs which increase punished responding do so regardless of whether the schedule of punishment presentation is continuous or intermittent. For example, chlordiazepoxide increases rates of punished responding when the schedule of punishment presentation is a continuous-punishment schedule (17, 35), a fixed-ratio punishment schedule (43), or a variable-ratio punishment schedule (23).

In one of the few studies where drug effects were studied using dif-

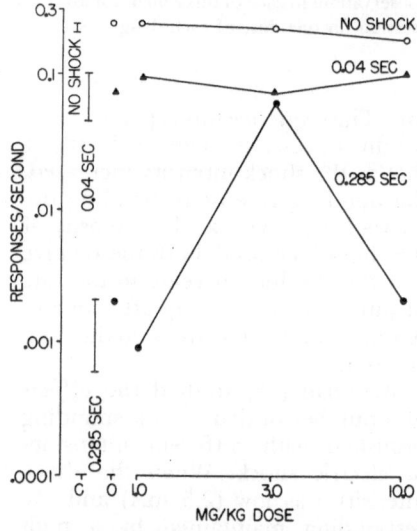

Figure 4. Effects of chlordiazepoxide on responding by pigeons under a DRL 15-sec schedule and punished with shocks of different durations. *Abscissa*: mg/kg dose, log scale. *Ordinate*: rate of key pecking in response/second, log scale. The brackets at C show the range of values during four sessions. The points at 0 show the effects of control injections. Each point is a mean of single observations in three birds for an entire session.

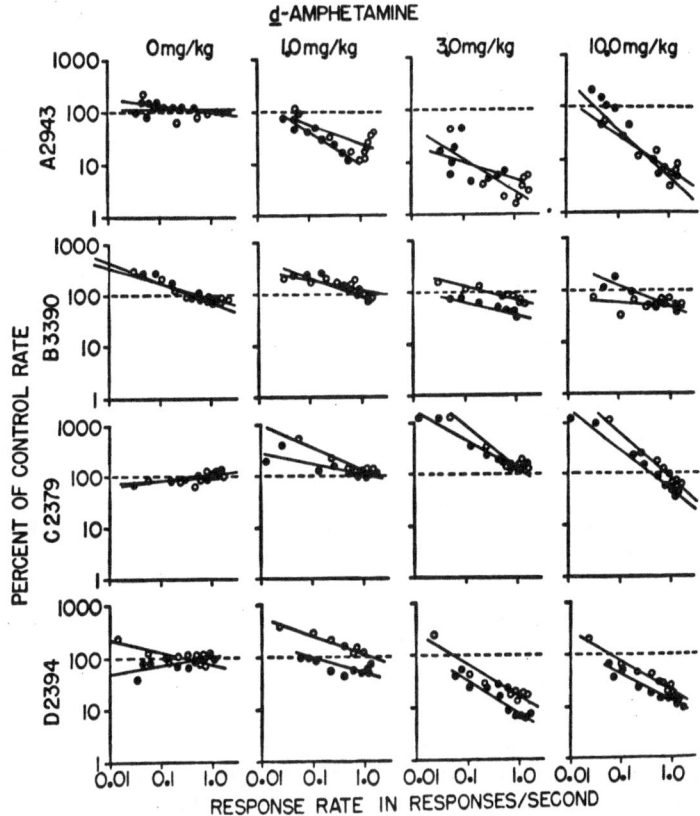

Figure 5. Log-log plots for the rate-dependent effect of d-amphetamine on punished responding of four birds (A2943, B3390, C2379 and D2394). *Abscissa*: mean rate of responding (responses per second) during 30-sec segments of the FI 5-min averaged over 10 noninjection control sessions, log scale. *Ordinate*: per cent of the mean rate of responding during noninjection control sessions, log scale. The filled circles are for the FR 1 shock component and the open circles are for the FR 30 shock component. The broken horizontal line in 100% represents no drug effect. Regression lines were fitted to the points by the least-squares method. The data at each dose are the mean of two determinations/bird. (Figure from ref 13 by permission of the Society for the Experimental Analysis of Behavior, Inc.)

ferent schedules of punishment presentation, Foree et al. (13) trained pigeons under a mult FI 5 FI 5-min schedule of food presentation and punished every response during one component and every 30th response during another component. Overall rates of responding were lower in the component where every response was punished than in the component where responses were punished under the FR 30 schedule of punishment presentation. Analysis of the rate-dependent effects of d-amphetamine on the punished responding suggested that a given rate of responding under FR-30 punished responding was increased more or decreased less than a matched rate of FR 1 punished responding (Fig. 5).

On the basis of these experiments, as well as their experiments on the effects of d-amphetamine on responding punished at different shock intensities, Foree et al. suggested that d-amphetamine increases punished responding only when the rate of responding is low and the intensity of the punishing stimulus is not too high. Increasing the degree of suppression by increasing the shock intensity (13, 39) or the shock frequency (13) will attenuate the rate-increasing effects and potentiate the rate-decreasing effects of amphetamines. Faiure of amphetamines to increase punished responding when the shock duration is long and suppression is great (23) may also reflect a relationship between punishment severity and d-amphetamine (13).

At the beginning of this review punishment was defined as a process characterized by a reduction of the future probability of a specific response as a result of the immediate delivery of a stimulus for that response. However, response-independent electric shock also produces suppression of responding (2, 11). Suppression of responding by electric shock that is not dependent upon responding is often called conditioned suppression. Kelleher and Morse (29) have indicated that the effects of drugs in the conditioned suppression procedure are not as reliable as the effects of drugs in the punishment procedure, and they suggest that it is because the relationship between responding and the presentation of electric shock is adventitious in the conditioned suppression procedure.

Generally, in the conditioned suppression procedure responding is maintained by a schedule of positive reinforcement. Superimposed on the pattern of responding maintained by this schedule is a stimulus that terminates with unavoidable electric shock. After repeated exposure to the stimulus that terminates with electric shock, responding is greatly suppressed during the stimulus. The procedure may be contrasted with most punishment procedures where more than one shock is received as punished responding occurs and the amount of shock received usually varies directly with the rate of responding. In an effort to make the conditioned suppression and the punishment procedures more analogous, a yoked punishment-conditioned suppression procedure was used to study the role of the response-dependent shock and response-independent shock in determining drug effects. Under this procedure, responding by pigeons was maintained under a FI 5-min schedule of food presentation. Each response by one pigeon (response-dependent shock) produced electric shock to that pigeon, as well as to a second pigeon (response-independent shock) whose responding was also maintained by a FI 5 schedule. To make sure that the FIs remained in phase for both pigeons, a limited hold and a time-out occurred at the end of the FI. After 5 min had elapsed the bird had 1 min in which to respond (limited hold). If the bird responded and received food before 1 min elapsed, the remaining portion of the minute was spent in darkness with no programmed contingencies (time-out). If no responses occurred during the 1-min limited hold, there was a 0.2-sec time-out at the beginning of the FI.

Response-dependent shock suppressed responding slightly more than did response-independent shock (Fig. 6). The pigeons were carefully matched with the birds with higher rates of punished responding yoked to birds with high rates receiving

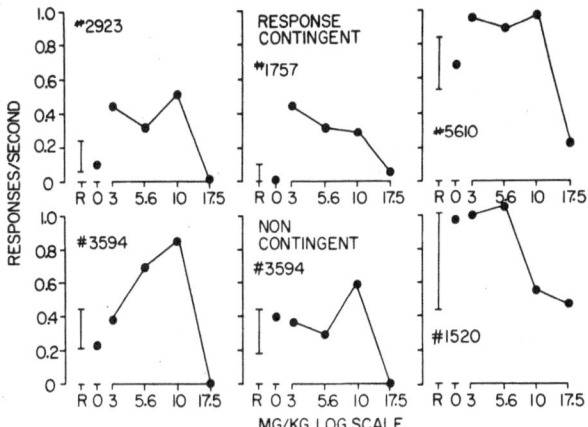

Figure 6. Effects of pentobarbital on punished responding and responding suppressed by shock that was not response contingent. *Abscissa*: mg/kg dose, log scale. *Ordinate*: rate of key pecking in responses/second. The brackets at R are a range of four control sessions and the points at 0 show the effects of control injection. Each point is a single observation in a bird for an entire session. The birds in the lower frames were yoked to the ones directly above them in the upper frames.

noncontingent shocks. Figure 6 shows that pentobarbital increased rates of responding in both groups of birds. When 10 mg/kg of pentobarbital increased punished responding, the resulting increase in shock rate to birds receiving nonresponse contingent shock did not prevent pentobarbital from increasing the rate of responding in these birds. Thus, pentobarbital seems as effective in increasing responding suppressed by noncontingent electric shock as it is in increasing responding suppressed by response-contingent shock. These data are consistent with previous reports that the barbiturates increase both punished responding (18, 19, 35, 36, 40) and responding suppressed by noncontingent electric shock (30).

RATE-DEPENDENT EFFECTS OF DRUGS ON PUNISHED RESPONDING

As Kelleher and Morse (28, 29) have indicated, many of the drugs that in-

crease low rates of punished responding have a tendency to increase low rates of responding maintained by conditions other than punishment. It is quite possible that the increase in punished responding produced by some drugs are merely a reflection of a general tendency of these drugs to increase low rates of responding, rather than any specific interaction between these drugs and the behavioral processes controlling punished responding.

One of the first attempts to evaluate specific interactions between drugs and punished responding versus rate-dependent drug effects was that of Cook and Catania (7), who used a multiple variable-interval 2-min, variable-interval 6-min schedule of food presentation and punished responses during the variable interval 2-min component. Overall rates of responding were approximately equal during the punishment component and the component without punishment. Chlordiazepoxide increased punished responding more than matched rates of unpunished responding, suggesting that chlordiazepoxide's effects on punished responding are influenced by factors other than the control rate of responding.

More recently, Wüttke and Kelleher (43) studied the effects of diazepam, chlordiazepoxide and nitrazepam on responding under fixed-interval schedules where the responses of some birds were punished under a FR 30 schedule of shock presentation, while responses of other birds were not punished. These benzodiazepines produced increases in rates of responding during the FI that were inversely proportional to control rates of responding. Where the control rates of punished and unpunished responding were nearly equal, the effects of the

benzodiazepines on punished and unpunished responding were approximately the same. These data suggested that the benzodiazepines increase low rates of responding, regardless of how these rates are established, a conclusion opposite to that of Cook and Catania (7).

McMillan (35, 36) also used fixed-interval schedules to generate inhomogeneous rates of punished and unpunished responding to study rate-dependent drug effects. In the first series of these experiments, pigeons responded for food under mult FI 5 FI 5-min schedules of food presentation (35). During one component of the schedule, each response was punished with electric shock, but responses were not punished during the other component. Figure 7 shows that chlordiazepoxide, diazepam and pentobarbital all increased overall mean rates of responding during the

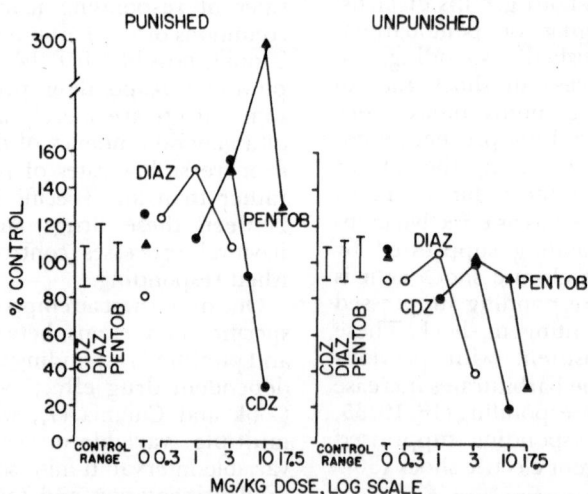

Figure 7. Effects of pentobarbital, diazepam and chlordiazepoxide on punished and unpunished responding. *Abscissa*: mg/kg dose, log scale. *Ordinate*: rate of responding as a percentage of the mean rate of responding on noninjection control days during an entire session for each component. The brackets represent the range of rates of responding on the days immediately preceding injections. Each range is based on at least five observations. Each point is the mean of a single injection in each of four birds.

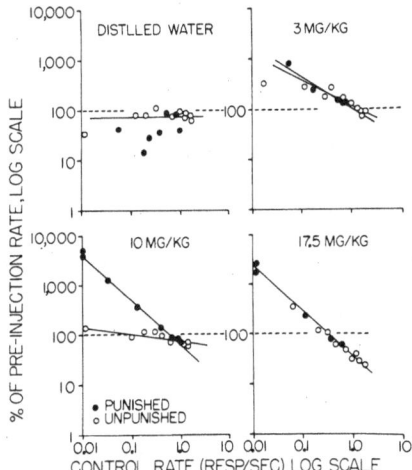

Figure 8. Log-log plots for the rate-dependent effects of pentobarbital on punished and unpunished responding in a single bird. *Abscissa*: control rate of responding during the session immediately before drug, log scale. *Ordinate*: rate after drug as a percentage of the control rate, log scale. Open circles are for unpunished responding and filled circles are for punished responding. Each point is the mean rate of responding during one of the 10 successive 30-sec segments of the fixed interval, averaged over the session for a single bird. Points were not plotted for control rates less than 0.01 responses/sec.

these drugs on both punished and unpunished responding depended on the control rate of responding. Low rates of both punished and unpunished responding were increased, while higher rates were increased less, or decreased. Yet, for both of these drugs, there were differences in the rate-dependency functions for punished and unpunished responding. This was especially true at the dose levels (10 mg/kg for pentobarbital and 1 mg/kg for diazepam) that produced the maximum increases in punished responding. At high control rates, similar rates of punished and unpunished responding were affected simi-

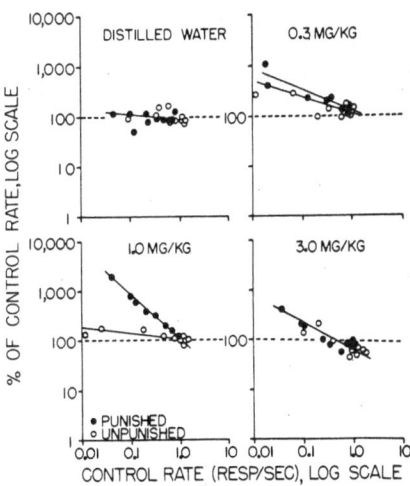

Figure 9. Log-log plots for the rate-dependent effects of diazepam on punished and unpunished responding in a single bird. *Abscissa*: control rate of responding during the session immediately before drug, log scale. *Ordinate*: rate after drug as a percentage of the control rate, log scale. Open circles are for unpunished responding and filled circles are for punished responding. Each point is the mean rate of responding during one of the 10 successive 30-sec segments of the fixed interval, averaged over the session for a single bird. Points were not plotted for control rates less than 0.01 respones/sec.

punishment component, but not during the component where responses were not punished. Because the rates of punished responding in these experiments were only about half of the rates of unpunished responding, the increases in punished responding might have occurred because of the lower control rates of punished responding. Therefore, an analysis of the rate-dependent effects of these drugs was done for both punished and unpunished responding.

Figures 8 and 9 show the changes in local rates of responding for a single bird after pentobarbital and diazepam, respectively. The effects of larly, but when the control rates were low, rates of punished responding were increased more than similar rates of unpublished responding. Although it must be emphasized that the effects of these drugs are rate dependent with the drug effects being inversely related to the control rate of responding for both punished and unpunished responding, the larger increases in low rates of punished responding after intermediate doses of pentobarbital and diazepam suggest that there is a specific interaction between some drugs and the behavioral processes underlying punishment.

The failure to find significant increases in the higher rates of punished responding at the end of fixed intervals is in agreement with the evidence reviewed previously that showed that rates of responding only slightly suppressed by punishment do not show much increase after barbiturates and minor tranquilizers. Figures 8 and 9 show that with higher

rates of responding the drug rate is usually close to 100% of the control rate, except at very high doses.

Figure 10 shows the effects of *d*-amphetamine and chlorpromazine on overall rates of punished and unpunished responding. In these experiments, *d*-amphetamine produced small increases in the rate of punished responding, but chlorpromazine did not. *d*-Amphetamine produced large increases in the rate of unpunished responding, but again chlorpromazine did not. A rate-dependency analysis of the local rates of responding within the fixed interval shows that the effects of both *d*-amphetamine and chlorpromazine are inversely related to the control rates of both punished and unpunished responding (Figs. 11 and 12). In contrast to the rate-dependent effects of diazepam and pentobarbital, similar rates of unpunished responding were increased more by *d*-amphetamine and chlorpromazine than were similar rates of punished responding. Pun-

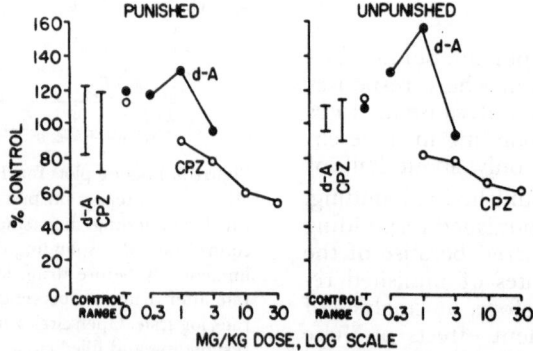

Figure 10. Effects of *d*-amphetamine (d-A) and chlorpromazine (CPZ) on punished and unpunished responding. *Abscissa*: mg/kg dose, log scale. *Ordinate*: rate of responding as a percentage of the mean rate of responding on noninjection control days during an entire session for

each component. The brackets represent the range of rates of responding on the days immediately preceding injections. Each range is based on at least five observations. Each point is the mean of a single injection in each of four birds.

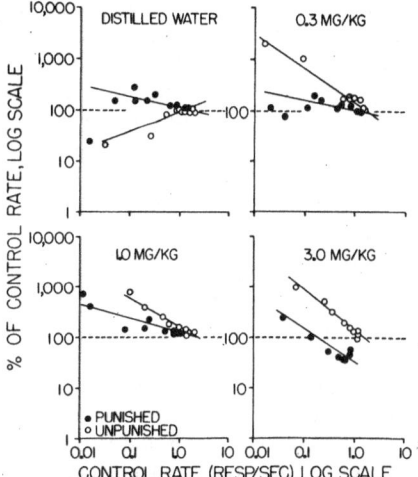

Figure 11. Log-log plots for the rate-dependent effects of d-amphetamine on punished and unpunished responding. *Abscissa:* control rate of responding during the session immediately before drug, log scale. *Ordinate:* rate after drug as a percentage of the control rate, log scale. Open circles are for unpunished responding and filled circles are for punished responding. Each point is the mean rate of responding during one of the 10 successive 30-sec segments of the fixed interval, averaged over the session for a single bird. Points were not plotted for control rates less than 0.01 responses/sec.

ishment clearly modifies the usual rate-dependent effect of these drugs.

Figure 10 showed that d-amphetamine increased overall rates of both punished and unpunished responding, while chlorpromazine did not. A comparison of Figs. 11 and 12 shows that the rate-dependent effects of these two drugs were very similar. How can the differences between the two drugs in their effects on overall rates of responding be explained in view of their similar rate-dependent effects? Figure 11 shows tht d-amphetamine was capable of increasing much higher control rates of responding than was chlorpromazine (Fig.

12). Control rates of unpunished responding of about 1 response/sec were increased by d-amphetamine, but only much lower control rates of responding were increased by chlorpromazine.

SCHEDULE OF POSITIVE REINFORCEMENT MAINTAINING PUNISHED RESPONDING

Although there have been some drug experiments where "unlearned" responses have been punished (41), in most drug experiments investigators have maintained responding under a schedule of positive reinforcement

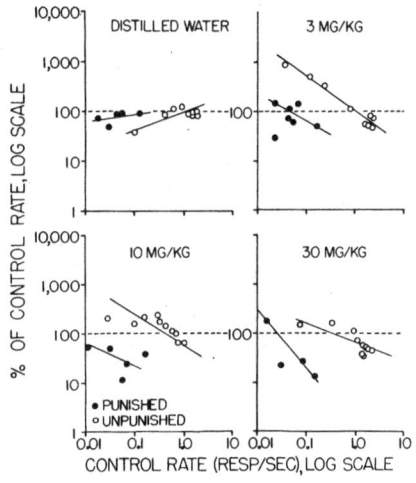

Figure 12. Log-log plots for the effects of chlorpromazine on punished and unpunished responding. *Abscissa:* control rate of responding during the session immediately before drug, log scale. *Ordinate:* rate after drug as a percentage of the control rate, log scale. Open circles are for unpunished responding and filled circles are for punished responding. Each point is the mean rate of responding during one of the 10 successive 30-sec segments of the fixed interval, averaged over the session for a single bird. Points were not plotted for control rates less than 0.01 responses/sec.

and subsequently punished these responses. Since the effects of drugs on responding under positive reinforcement schedules depend largely on the particular schedule used and the pattern of responding it generates, it might be expected that the schedule of positive reinforcement maintaining the punished responding would have some role in determining the effect of a drug on punished responding.

Drugs used clinically as minor tranquilizers (benzodiazepines and propanediols), as well as the barbiturates, have been shown to increase rates of punished responding when responding is maintained by continuous reinforcement (CRF) schedules (17–20, 32), by VI schedules (23, 40), by FI schedules (35, 43), by FR schedules (28, 36), and by DRL schedules (Fig. 4).

Although these experiments suggest that the drugs that increase punished responding do so regardless of the schedule of positive reinforcement maintaining the punished responding, there is some evidence that the schedule of positive reinforcement can be an important determinant of the effects of drugs on punished responding. Foree et al. (13) maintained responding under a mult Fr 30 FI 5 schedule of food presentation, where each response during both schedule components was punished with electric shock. The effects of d-amphetamine were studied in pigeons under this procedure. Figure 13 shows a rate-dependency analysis of the effects of d-amphetamine on responding using this punishment paradigm. As has been observed in previous experiments, low rates of punished responding which occurred near the beginning of the FI were increased by d-amphetamine, while the higher rates near the end of the FI were increased less, or more often were decreased. In Fig. 13, a regression line has been drawn to fit the FI rate-dependency data. Note that the filled triangles in Fig. 13, which represent punished responding maintained by the FR component of the schedule, fit the FI regression line under control conditions. Yet after d-amphetamine, the rate of punished responding maintained by the FR component does not fall on the regression line derived from the FI data for birds 6678 and 3547. Since the only difference was the schedule of positive reinforcement maintaining the punished responding, this schedule difference must account for the failure of the FR data to fit the FI regression line in two of the three birds.

In very few instances have investigators studied more than one schedule of positive reinforcement maintaining punished responding. In no instance has a study been dedicated to the investigation of drug effects, as the parameters of one or more schedules of positive reinforcement maintaining punished responding are systematically varied. The need for such experiments is obvious.

THE NATURE OF THE RESPONSE

In most punishment experiments with monkeys and rats, the animals have been conditioned to lever press, while pigeons have been conditioned to peck a response key. Although pigeons have been trained to step on a treadle to obtain positive reinforcement and rats and monkeys have been trained to push a key for positive reinforcement, there have been no experiments on punished responding where the type of response required has been varied systematically as drug effects were studied. Benzodiazepines, meprobamate, and various

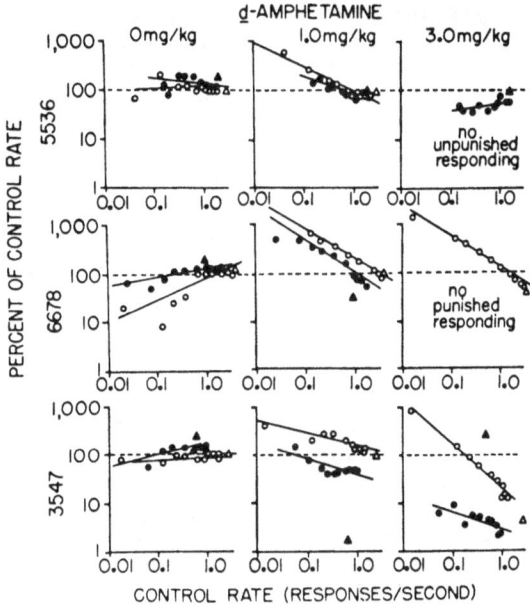

Figure 13. Log-log plots for the rate-dependent effect of *d*-amphetamine on punished (2.5 mA) and unpunished (0 mA) responding of three birds (5536, 6678, and 3547). *Abscissa*: mean rate of responding (responses per second) during 30-sec segments of the FI 5-min, averaged over non-injection control sessions, log scale. *Ordinate*: per cent of the mean rate of responding during noninjection control sessions, log scale. The filled symbols are for punished responding and the open symbols are for unpunished responding. The circles represent the ten 30-sec segments of the FI component and the triangles are for the FR components. The broken horizontal line at 100% represents no drug effect. Regression lines were fitted to the points by the least-squares method. (Figure from ref 13 by permission of the Society for the Experimental Analysis of Behavior, Inc.)

barbiturates can increase punished responding when the required response is licking a drinking tube (32, 41; also Fig. 1). In these experiments responding was not explicitly conditioned, yet increases in punished responding occurred.

On the basis of these experiments it seems that drug effects on punished responding do not depend on the nature of the required response, nor do the effects depend on whether or not the responses are explicitly conditioned. However, it must be remembered that there have been no attempts

to systematically investigate these parameters and the role they might play in determining drug effects on punished responding.

Several authors (8, 10, 12, 26) have referred to changes in fundamental dimensions of the operant during punishment. Included among the changes observed were "abortive responding," changes in response location and response topography, and changes in response duration. The effects of drugs on these changes in the fundamental dimensions of the operant have not been studied.

ORGANISMIC VARIABLES

As discussed previously, barbiturates and minor tranquilizers increase punished responding. These increases have been demonstrated in pigeons, rats, cats, and monkeys (14, 15, 17–19, 23, 27–29, 35, 36, 40, 41). So far no important differences in the effects of drugs on punished responding have been demonstrated to be species dependent.

In most punishment experiments all of the experimental animals are studied at a fixed level of food or water deprivation. It is well known that the rate of punished responding is sensitive to slight changes in the level of food deprivation (3, 5). It seems especially unfortunate that there have been no experiments where the food deprivation has been varied during a study of the effects of drugs on punished responding.

One way in which organismic variables can be studied is to manipulate the drug history of the experimental subjects. Holtzman and Villarreal (25) studied the effects of several drugs during morphine abstinence in rhesus monkeys that had previously been made physically dependent on morphine. They found that morphine produced dose-related decreases in rates of both punished and unpunished responding maintained by VI and FR schedules of food presentation in monkeys not previously exposed to drugs. Morphine was then administered chronically and tolerance occurred to the rate-decreasing effects of morphine so that after repeated morphine administration, morphine no longer decreased rates of either punished or unpunished responding. After 8 weeks of morphine administration the drug was no longer given. During morphine abstinence, punished responding emitted at high baseline rates was increased,

while punished responding administered at low baseline rates was decreased. Holtzman and Villarreal studied the effects of a number of drugs on punished responding during morphine abstinence. Morphine, which initially decreased punished responding and had no effect on the punished responding of the drug-tolerant monkey, increased the rate of punished responding in the monkey undergoing morphine abstinence. Meperidine and levorphanol had similar effects. d-Amphetamine which usually decreases punished responding, also decreased punished responding in the abstinent animal; however, diazepam which usually increases punished responding further decreased punished responding in the morphine abstinent animal. Thus, the effects of drugs in these experiments depended on the drug history of the animal, with different drug effects depending on whether the animal was drug naive, morphine tolerant, or undergoing morphine abstinence.

The role of drug tolerance in determining the effects of drugs on punished responding also has been studied by Margules and Stein (33), who found that a 20 mg/kg dose of oxazepam initially suppressed the punished responding in rats, but produced large increases in the rate of punished responding by rats that had received oxazepam for a 10-day period.

OTHER PARAMETERS

There are a variety of other factors that might influence the effects of drugs on punished responding. For example, Wüttke and Kelleher (43) found that the effects of benzodiazepines on punished responding could be explained entirely in terms of rate-dependency, while McMillan

(35) found evidence that some of these same drugs increased low rates of punished responding more than matched rates of unpunished responding. In both experiments pigeons were reinforced with food under an FI 5-min schedule. However, there were differences in a number of parameters in these two experiments. For example, Wüttke and Kelleher studied simple FI schedules with different birds to compare drug effects on punished and unpunished responding, while McMillan studied the effects of benzodiazepines on responding maintained by a multiple schedule with punishment and non-punishment components. The use of the multiple schedule raises the possibility that behavioral contrast and induction might contribute to the effects of drugs on punished responding, while the use of simple schedules raises the possibility that individual differences may contribute to the results.

Although it is almost too obvious to discuss, the dose of a drug can be an important determinant of whether or not a drug increases punished responding. Some drugs do not increase punished responding at any dose level, and even drugs that do increase punished responding do so only at certain dose levels. Very low doses of drugs that can increase punished responding may not be large enough to produce significant effects. As the dose is increased, increases in punished responding occur but further increases in the dose begin to produce ataxia, anesthesia and toxicity, thereby reducing the rate of punished responding. Therefore, the dose-effect curve describing the effects of drugs that increase punished responding usually takes the form of an inverted U (23, 35, 36, 43).

There are many other factors that might also influence the effects of drugs on punished responding. For example, pairing response-produced electric shock with positive reinforcement might increase the rate of punished responding. Whether or not punished responding interacts with drugs in the usual manner when punishment is a discriminative stimulus correlated with positive reinforcement is not known.

SUMMARY AND CONCLUSIONS

The effects of drugs on punished responding depend on a variety of factors. Among these factors are the type and the dose of the drug, the animal's drug history, the nature of the punishing stimulus, the intensity and duration of the punishing stimulus, the schedule of presentation of the punishing stimulus, the control rate and pattern of punished responding, the schedule of positive reinforcement maintaining the punished responding, the species, and the type of punished response. Some of these factors, for example the type of punishing stimulus, have not yet been shown to be important determinants of drug effects on punished responding. Other factors, such as the animal's drug history and the punishment intensity, are very important determinants of the effects of drugs on punished responding.

Because of the large number of factors that potentially might influence the effects of drugs on punished responding, as well as the smaller number of factors that have already been demonstrated to contribute to drug effects on punished responding, any simple characterization of a drug as being a drug that "increases" or "decreases" punished responding seems inappropriate. It has been obvious for some time that the effects of drugs on positively reinforced responding are determined by many variables. In a like manner the effects of drugs on punished responding also depend on a variety of experimental parameters. Much more important than the question of whether or not a drug increases or decreases punished responding is the question of determining the conditions under which drugs produce increases and decreases in punished responding. It is toward this latter question that subsequent research in the behavioral pharmacology of punished responding should be directed.

The author wishes to thank Dr. J. David Leander for helpful suggestions in revising the manuscript.

REFERENCES

1. APPEL, J. B. Fixed-interval punishment. *J. Exp. Anal. Behav.* 11: 803, 1968.
2. AZRIN, N. H. Some effects of two intermittent schedules of immediate and nonimmediate punishment. *J. Psychol.* 42: 3, 1956.
3. AZRIN, N. H. Effects of punishment intensity during variable-interval reinforcement. *J. Exp. Anal. Behav.* 3: 123, 1960.
4. AZRIN, N. H., AND W. C. HOLZ. Punishment. In: *Operant Behavior: Areas of Research and Application*, edited by W. K. Honig. New York: Appleton-Century-Crofts, 1966, p. 380.
5. AZRIN, N. H., W. C. HOLZ AND D. F. HAKE. Fixed-ratio punishment. *J. Exp. Anal. Behav.* 6: 141, 1963.
6. BARRY, H. B., III., S. A. WAGNER AND N. E. MILLER. Effects of several drugs on performance in an approach-avoidance conflict. *Psychol. Rep.* 12: 215, 1963.
7. COOK, L., AND A. C. CATANIA. Effects of drugs on avoidance and escape behavior. *Federation Proc.* 23: 818, 1964.
8. DINSMOOR, J. A. A discrimination based on punishment. *Q. J. Exp. Psychol.* 4: 27, 1952.
9. DINSMOOR, J. A., AND D. A. LYON. The selective action of chlorpromazine on behavior suppressed by punishment. *Psychopharmacologia* 2: 456, 1961.
10. DUNHAM, P. J., A. MARINER AND H. ADAMS. Enhancement of off-key pecking by on-key punishment. *J. Exp. Anal. Behav.* 12: 789, 1969.
11. ESTES, W. K., AND B. F. SKINNER. Some quantitative properities of anxiety. *J. Exp. Psychol.* 29: 390, 1941.
12. FERRARO, D. P., AND K. M. HAYES. Variability of response duration during punishment. *Psychol. Rep.* 21: 121, 1967.
13. FOREE, D. D., F. H. MORETZ AND D. E. MCMILLAN. Drugs and punished responding II: *d*-Amphetamine-induced increases in punished responding. *J. Exp. Anal. Behav.* 20: 291, 1973.
14. GELLER, I. Use of approach avoidance behavior (conflict) for evaluating depressant drugs. In: *The First Hahnemann Symposium on Psychosomatic Medicine*, edited by J. H. Nodine and J. H. Moyer. Philadelphia: Lea and Febiger, 1962, p. 267.
15. GELLER, I. Relative potencies of benzodiazepines as measured by their effects on conflict behavior. *Arch. Int. Pharmacodyn. Ther.* 149: 243, 1964.
16. GELLER, I., E. BACHMAN AND J. SEIFTER. Effects of reserpine and morphine on behavior suppressed by punishment. *Life Sci.* 4: 226, 1963.
17. GELLER, I., J. T. KULAK, JR. AND J. SEIFTER. The effects of chlordiazepoxide and chlorpromazine on a punishment discrimination. *Psychopharmacologia* 3: 374, 1962.
18. GELLER, I., AND J. SEIFTER. The effects of meprobamate, barbiturates, *d*-amphetamine and promazine on experimentally induced conflict in the rat. *Psychopharmacologia* 1: 482, 1960.
19. GELLER, I., AND J. SEIFTER. The effects of mono-urethanes, di-urethanes and barbiturates on a punishment discrimination. *J. Pharmacol. Exp. Ther.* 136: 284, 1962.
20. GOLDBERG, M. E., AND V. B. CIOFALO. Effect of diphenylhydantoin sodium and chlordiazepoxide alone and in combination on punishment behavior. *Psychopharmacologia* 14: 233, 1969.
21. GRAEFF, F. G., AND R. I. SCHOENFELD. Tryptaminergic mechanisms in punished and non-punished behavior. *J. Pharmacol. Exp. Ther.* 173: 277, 1970.
22. GROSSMAN, S. P. Effects of chlorpromazine and perphenazine in an approach-avoidance conflict. *J. Comp. Physiol. Psychol.* 54: 517, 1961.
23. HANSON, H. M., J. J. WITOSLAWSKI AND E. H. CAMPBELL. Drug effects in squirrel monkeys trained on a multiple schedule with a punishment contingency. *J. Exp. Anal. Behav.* 10: 565, 1967.
24. HENDRY, D., AND C. VAN TOLLER. Fixed-ratio punishment with continuous

reinforcement. *J. Exp. Anal. Behav.* 7: 293, 1964.

25. HOLTZMAN, S. G., AND J. E. VILLARREAL. Operant behavior in the morphine-dependent rhesus monkey. *J. Pharmacol. Exp. Ther.* 184: 528, 1973.

26. HUNT, H. F., AND J. V. BRADY. Some effects of punishment and intercurrent "anxiety" on a simple operant. *J. Comp. Physiol. Psychol.* 48: 305, 1955.

27. JACOBSEN, E. The effects of psychotropic drugs under psychic stress. In: *Psychotropic drugs*, edited by S. Garattini and V. Ghetti. Amsterdam: Elsevier, 1957, p. 119.

28. KELLEHER, R. T., AND W. H. MORSE. Escape behavior and punished behavior. *Federation Proc.* 23: 808, 1964.

29. KELLEHER, R. T., AND W. H. MORSE. Determinants of the specificity of the behavioral effects of drugs. *Ergeb. Physiol. Biol. Chem. Exp. Pharmakol.* 60: 1, 1968.

30. LAUNER, H. Conditioned suppression in rats and the effect of pharmacological agents thereon. *Psychopharmacologia* 4: 311, 1963.

31. LEAF, R. C., AND S. A. MULLER. Effects of shock intensity, deprivation and morphine in a simple approach-avoidance conflict situation. *Psychol. Rep.* 17: 819, 1965.

32. MARGULES, D. L., AND L. STEIN. Neuroleptics vs. tranquilizers: Evidence from animal behavior studies of mode and site of action. *Excerpta Med. Int. Congr. Ser.* 129: 108, 1966.

33. MARGULES, D. L., AND L. STEIN. Increase of "antianxiety" activity and tolerance of behavioral depression during chronic administration of oxazepam. *Psychopharmacologia* 13: 74, 1968.

34. MCMILLAN, D. E. A comparison of the punishing effects of response-produced shock and response-produced time out. *J. Exp. Anal. Behav.* 10: 439, 1967.

35. MCMILLAN, D. E. Drugs and punished responding I: Rate dependent effects under multiple schedules. *J. Exp. Anal. Behav.* 19: 133, 1973.

36. MCMILLAN, D. E. Drugs and punished responding III: Punishment intensity as a determinant of drug effect. *Psychopharmacologia* 30: 61. 1973.

37. MCMILLAN, D. E. Drugs and punished responding IV: Effects of propranolol, ethchlorvynol, and chloral hydrate. *Res. Commun. Chem. Pathol. Pharmacol.* 6: 167, 1973.

38. MICZEK, K. A. Effects of scopolamine, amphetamine and chlordiazepoxide on punishment. *Psychopharmacologia* 28: 373, 1973.

39. MORRISON, C. F. The effects of nicotine on punished behavior. *Psychopharmacologia* 14: 221, 1969.

40. MORSE, W. H. Effects of amobarbital and chlorpromazine on punished behavior in the pigeon. *Psychopharmacologia* 6: 286, 1964.

41. NAESS, J., AND E. W. RASMUSSEN. Approach-withdrawal responses and other specific behavior reactions as screening tests for tranquillizers. *Acta Pharmacol.* 15: 99, 1958.

42. STEIN, L. The chemistry of reward and punishment. In: *Psychopharmacology: A review of progress, 1957–1967*, edited by D. Efron. Washington; U.S. Public Health Service, 1968, p. 105.

43. WÜTTKE, W., AND R. T. KELLEHER. Effects of some benzodiazepines on punished and unpunished behavior in the pigeon. *J. Pharmacol. Exp. Ther.* 172: 397, 1970.

The role of discriminative stimuli in modulating drug action[1]

VICTOR G. LATIES

*Departments of Radiation Biology and Biophysics, Pharmacology
and Toxicology, and Psychology, University of Rochester
Rochester New York 14642*

ABSTRACT

Behavior reinforced in the presence of a stimulus comes under the control of the stimulus.
A drug can then modify that control and, therefore, modify the behavior itself.
Studies over the past 2 decades have shown that the nature of the controlling (or
discriminative) stimulus can govern the degree to which drugs change performance. These
experiments usually have compared behavior on various schedules of reinforcement with
and without added discriminative stimuli. For instance, pigeons that had been trained on a
fixed-interval schedule showed great changes in response distribution after amphetamine
and scopolamine. The same birds, when performing on a fixed-interval schedule to
which time-correlated discriminative stimuli had been added, showed smaller changes in
response distribution. Other pigeons were trained to make a minimum number of
consecutive responses on one key before a peck on a second key would be reinforced;
d-amphetamine and scopolamine led to pronounced increases in premature switching.
Adding a discriminative stimulus when the response requirement was fulfilled increased
the likelihood that a switch would occur only after the appropriate number of pecks
had been emitted. It also attenuated the effects of the drugs. The presence of discriminative stimuli did not make as large a difference in performance in either of these experiments when chlorpromazine and promazine were studied. In general, work with other
schedules of reinforcement supports the conclusion that behavior under strong external
stimulus control is less apt to be readily affected by many drugs. Addition of the discriminative stimulus can also "improve" the behavior of pigeons that have been given enough

[1]Partially supported by Grant MH-11752
from the National Institute of Mental Health,
by Grant GI-300978 from the RANN program
of the National Science Foundation, and by a
contract with the U. S. Atomic Energy Commission at the University of Rochester Atomic
Energy Project, and has been assigned Report
No. UR-3490-545.

Abbreviations: FCN, fixed consecutive number; DRL, differential reinforcement of low
rate; FR1, fixed ratio one.

methylmercury to increase greatly the variability of their performance.—LATIES,
V. G. The role of discriminative stimuli in modulating drug action. *Federation
Proc.* 34: 1880–1888, 1975.

If a particular stimulus is present
during a particular schedule of reinforcement, the stimulus may come to
control the response pattern peculiar
to the schedule (10, 12, 35). Behavioral pharmacologists have been
actively concerned with the ways
drugs modify this control for 2
decades, having turned to such
problems first in the mid-1950's
(2, 4–6, 30, 32, 34).[2]

Probably the most influential early
papers were by Dews (4, 5) and
appeared as the first two in the
important series that he published
in the *Journal of Pharmacology and
Experimental Therapeutics.* In the first
(4) he pointed to the possible role of
discriminative control in accounting
for the difference in how fixed-
ratio and fixed-interval performances
were affected by a drug. But he did
not manipulate stimulus control
directly (23). In the second paper
(5) he explicitly examined the effects of several drugs on two instances of stimulus control. He did this
by training pigeons to respond on a
key only when the key was red.
When the key was blue, responses
were not reinforced and the bird
eventually stopped pecking the key.
He found that this simple discrimination was quite resistant to effects of
methamphetamine (Fig. 1), although

response rate (not shown in the
figure) was reduced markedly. He
also trained other birds to perform a
more complex conditional discrimination. Here they had to learn to
respond only in the presence of
various combinations of key color and
presence or absence of another light
in the chamber. Even with the more
complex discrimination, the birds
learned to do very well under control
conditions. The discriminative ratio
was close to 1, which means that they
made very few pecks when such
pecks were inappropriate. However,
the drug produced a somewhat
greater effect on this complex discrimination than it had on the
simpler discrimination. The dose-

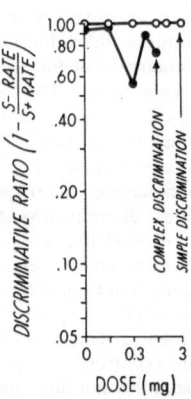

Figure 1. Effects of methamphetamine on
simple and complex discrimination performances. The curves represent mean data from
two pairs of pigeons. Total doses are given;
the birds weighed between 400 and 500 gms.
(The figure has been constructed from data
given in Dews (5), Tables 3 and 4.)

---

[2] However, the general problem of environmental influence on drug action can be traced
at least to the work of Pavlov's student,
Zavadskii (40), who contrasted the effects
of several drugs on examples of both internal
and external inhibition, using the conditioned
salivary response.

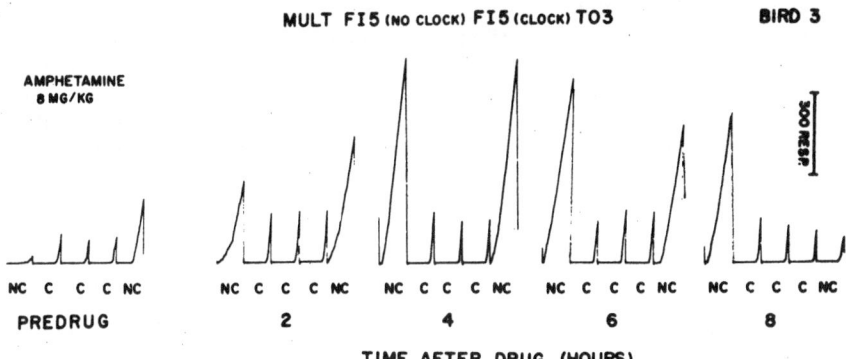

Figure 2. Cumulative records of performance on a multiple schedule consisting of two types of fixed-interval (FI) schedules. In each sequence of five intervals, the middle three are FI 5-min (clock) and the end two are FI 5-min (no clock). The pen reset to the baseline at reinforcement. A 3-min time-out occurred after each reinforcement. During this time the pigeons usually did not respond and the recording paper did not move. (These records are taken from Fig. 6 in Laties and Weiss (22).)

response function was somewhat ragged, but the drug clearly affected the complex discrimination more than the simple. It was this type of finding that led other workers to explore other ways in which variations in stimulus control would lead to variations in drug effect.[3]

Our own experience with this problem started with some work on the fixed-interval schedule. Weiss and I contrasted the effects of several drugs on two variants of the schedule (22). In one, a regular fixed-interval schedule, the first response made at least 5 min after the beginning of the interval was reinforced. The response key remained red from the start of the interval until the reinforced response. In the other, external discriminative stimuli were provided the birds during the 5-min fixed interval. These stimuli, a series of symbols (o, ×, △, +, □) projected onto the response key, changed every

minute and soon came to control responding. These discrete "clock" stimuli (33) controlled a near zero rate during the first 4 min of the interval. With the presentation of the stimulus always associated with reinforcement, a high rate of responding would usually start.

To the left in Fig. 2 is displayed the type of performance that usually occurred when the two fixed-interval schedules were alternated irregularly as parts of a multiple schedule. The three "clock" intervals show almost complete suppression of responding in the presence of the first four added stimuli projected on the key. The first and last intervals are examples of performance under the usual fixed-interval schedule, with no "clock" stimuli present. Performance varied greatly, with birds occasionally starting work very early in an interval, occasionally very late, occasionally showing gradually increasing responding but more often a relatively abrupt transition to a high rate.

The effect of a large dose of amphetamine on this performance is

---

[3] See (10) for an incisive discussion of the concept of stimulus control and drug action. The two papers by Dews are also reprinted there.

also shown in Fig. 2. Note that even at the time of the maximal effect, the performance under control of the discriminative stimuli projected onto the key remained intact whereas the regular fixed-interval performances at the ends of this block of five intervals had been changed considerably.

If this effect is quantified by calculating the Index of Curvature (13) for each interval and then averaging these indexes for a period of time after the drug has been given, one gets the results shown in Fig. 3. 0.80 would be the maximum Index of Curvature possible for our conditions. This would mean that the pigeon had confined its responses to the last minute of the 5-min interval. An Index of Curvature of zero would mean that the subject had responded equally often in the five-fifths of the interval. With amphetamine, pentobarbital, and scopolamine the performance with the added discriminative stimuli was not affected

very much in comparison to what happened with the regular fixed-interval performance. Two phenothiazines, chlorpromazine and promazine, were also used. With them the presence of the added clock stimuli usually did not make as much of a difference; the curvature of the intervals was degraded regardless of whether or not the clock stimuli had been added. Cumulative records showing the action of promazine are displayed in Fig. 4.

These are interesting effects but their interpretation is not completely unambiguous, partly because the two fixed-interval schedules produced very different response rates, and the underlying response rate itself is known to be quite important in determining the effects of drugs (7, 8, 19). The influence of rate can be minimized by choosing a reinforcement schedule and a method of adding external stimuli that will not themselves produce large changes in baseline rate (21, 38). One such

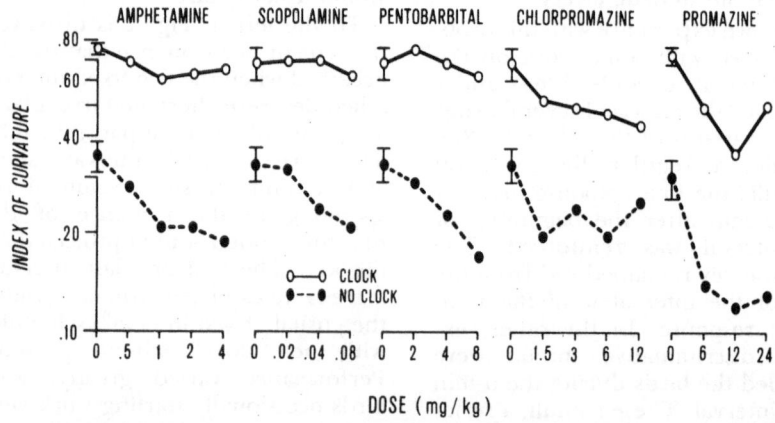

Figure 3. Effects of drugs on the Index of Curvature during the first 2-hr period after treatment. The points represent means of 3 pigeons. Control values derive from the saline control data collected closest in time to the data collected on each drug. The vertical lines indicate ± SE. These values were estimated from the range of the three saline control means (one for each bird) according to the techniques described by Wilcoxin and Wilcox (39). (From Laties and Weiss (22). Copyright 1966 by the Williams & Wilkins Co., Baltimore. Reprinted by permission of publisher.)

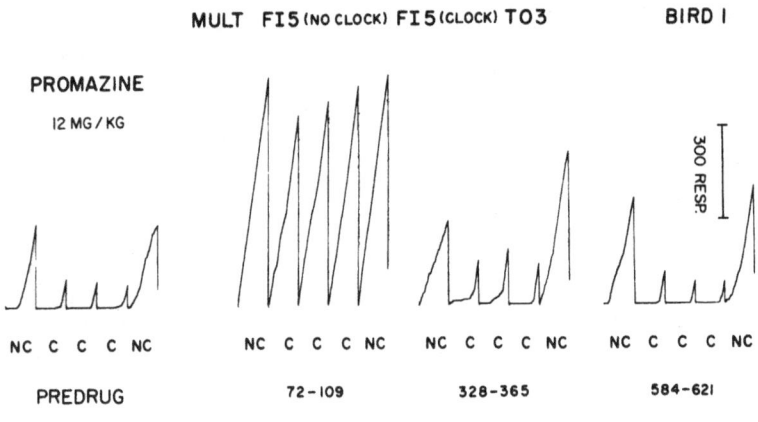

Figure 4. Cumulative records of performance on a multiple schedule consisting of two types of fixed-interval (FI) schedules. For further details see the legend of Fig. 2. (Abstracted from Figure 7 in Laties and Weiss (22).)

schedule, fixed consecutive number (FCN), requires the subject to make a specified minimum number of consecutive responses on one key before a response on a second key will be reinforced (1, 28, 29). If the subject switches too early in its series of responses (or run), it is not rewarded and must start its run on the first key all over again. In the case of the version used here, eight or more pecks on the left key were required (FCN 8) before the switch to the right key would be reinforced (21).

In normal pigeons the performance one finds is shown in Fig. 5 (upper panel). The predominant run length tends to be close to the minimum required value with fairly wide dispersion about this value. If one changes the conditions slightly by adding a discriminative stimulus, so that the key being pecked turns red with the eighth response (FCN 8-S^D), one gets a markedly different distribution of run lengths (Fig. 6, upper panel). Now the bird will tend to make the minimum number

of responses necessary to set up reinforcement for a peck on the second key. However, the addition of this signal does not change response rate. The four birds used in

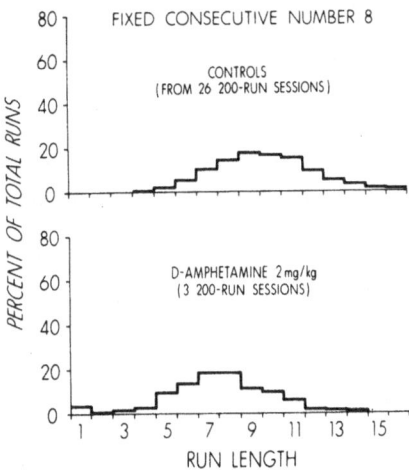

Figure 5. Effects of d-amphetamine on performance of a pigeon (#2965) on the fixed-consecutive-number schedule of reinforcement. Run lengths of eight or more, followed by a response on a second key, were reinforced with access to mixed grain.

Laties (21; see Table 1) had an over-all rate of 1.78 responses per sec without and 1.82 responses per sec with the added discriminative stim-ulus. Their response rates during the runs themselves, as measured from the first peck on the first key to the peck on the second key that occurred at the end of each run, averaged 3.02 and 3.00 responses per second, respectively. Thus pat-tern of responding, so important in determining drug effects (9, 10, 19, 31), remained quite similar.

Amphetamine appears to interact with degree of stimulus control in the same way on this schedule as it did on the fixed-interval schedule with and without the added clock. Representative results are shown in the lower panels of Figs. 5 and 6. With the regular FCN 8 schedule, d-amphetamine moves the run length distribution to the left, the mean run length now being a little more than one shorter than it was before. The analogous data for the FCN 8-S^D schedule show the drug effect to have disappeared.

The same general point can be made for a wider range of doses by examining conditional probability functions (Fig. 7). Here the ordinate values give the probability that the pigeon will switch and peck the reinforcement key after having made the particular number of responses given on the abscissa. For instance, under control conditions with the added discriminative stimulus (Fig. 7, left panel, the shaded portion), it was extremely unlikely that this bird would switch after five, six or seven responses. But the probability rose abruptly to about 0.70 after the eighth response. In other words, given a key that turns red with the eighth peck, the bird quite sensibly usually switched after making 8 pecks. On the other hand, the shaded por-

tion of the right panel shows that the control conditional probability func-tion was much more gently sloped for the FCN 8 schedule. The various doses of d-amphetamine again are seen to have made an earlier switch much more probable for the FCN 8 schedule while not having much effect if reinforcement availability had been signaled.

Just about the same thing is seen for scopolamine (Fig. 8). Again, there were large changes in the probability of a switch early in a run only when no added cue was controlling the switch. The same type of effect on the same two schedules had been shown previously for scopolamine by Wagman and Maxey (38) in the rat.

Chlorpromazine gives a somewhat different picture, with large doses producing an increase in the prob-ability of a switch even when the

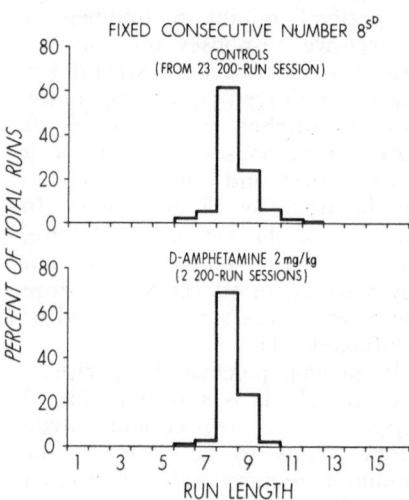

Figure 6. Effects of d-amphetamine on per-formance of a pigeon (#2965) on the fixed-consecutive-number schedule of rein-forcement so modified that a discriminative stimulus (S^D) was presented when the bird made its eighth response. The upper panel of this figure should be compared to that in Fig. 5 to see the effect of adding the external signal.

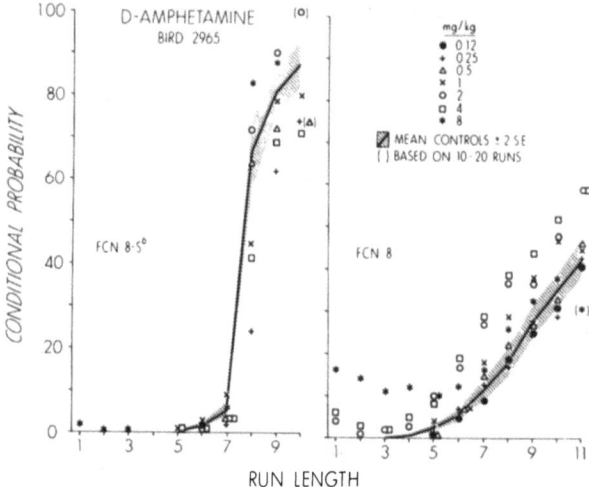

Figure 7. Effects of *d*-amphetamine on the conditional probability functions of one pigeon. The ordinate gives the probability that the bird will stop after the number of consecutive responses on the abscissa and peck the reinforcement key. The FCN8-S[D] control curve is based on 23 experiments, each 200 runs long; the FCN8 control curve on 26 such experiments. Points have not been plotted at zero probability nor when they would have had to be based on fewer than 10 runs. (From Laties (21). Copyright 1972 by the Williams & Wilkins Co., Baltimore. Reprinted by permission of publisher.)

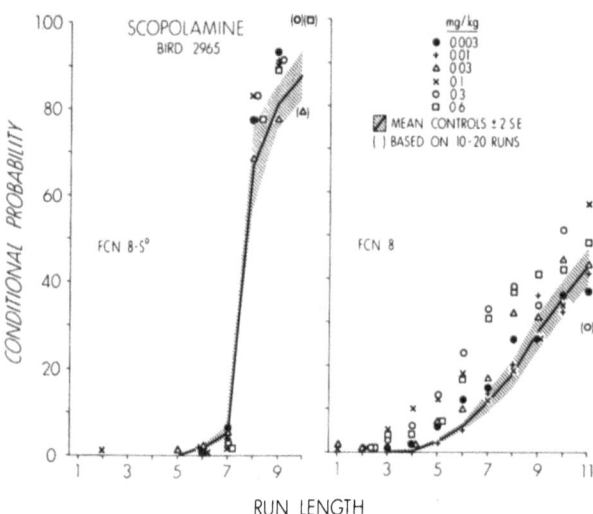

Figure 8. Effects of scopolamine on the conditional probability functions of one pigeon. (For further details, see the legend to Fig. 7.) (From Laties (21). Copyright 1972 by the Williams & Wilkins Co., Baltimore. Reprinted by permission of publisher.)

added discriminative stimulus is present (Fig. 9). This result resembles what we found in the fixed-interval schedule work described previously.

Perhaps a better feel for how a seemingly slight change in stimulus control can modify a drug effect can be obtained from a few cumulative records. Figure 10, right, shows the effects of scopolamine on the FCN 8 schedule. The top record is from a saline control session during which the pigeon was allowed to make 40 runs, most of which on this occasion were long enough to be reinforced. The middle record shows a session with methylscopolamine, which was given as a control for the peripheral effects of scopolamine. It had little effect (see 38). The bottom record shows the effects of scopolamine. Besides decreasing rate, the main effect was to increase the number of runs too short to be reinforced. The actions of the same drugs on the FCN 8-S^D schedule

produces a low response rate, one which is easily increased by the amphetamines. Figure 11 shows the increase that they found (left upper panel). This change in response rate are displayed to the left of the figure. Scopolamine's effect on run length has disappeared.

The same general diminution of drug effects when a performance is under strong stimulus control has been shown in several other situations (e.g., 3, 14–16, 18, 20, 26, 27, 37). Only two more demonstrations will be examined here. Carey and Kritkausky (3) trained rats to work on a schedule of reinforcement on which responses on a lever produced a drop of water only if they followed pauses without responding of at least 22 sec (DRL 22 sec). This schedule typically was accompanied by a decrease in the reinforcement rate (left lower panel). Carey and Kritkausky also trained their rats to work on the same general schedule but arranged for the light in the experimental box to go on

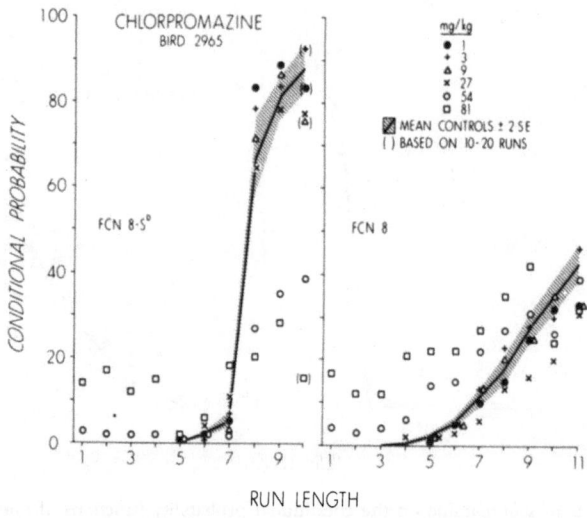

Figure 9. Effects of chlorpromazine on the conditional probability functions of one pigeon. (For further details, see the legend to Fig. 7.) (From Laties (21). Copyright 1972 by the Williams & Wilkins Co., Baltimore. Reprinted by permission of publisher.)

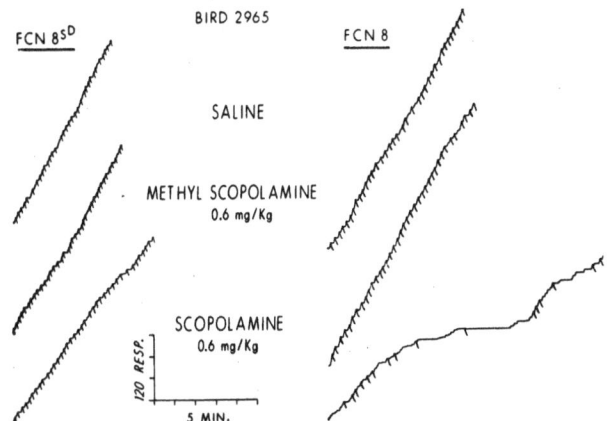

Figure 10. Cumulative records showing the effects of scopolamine and methylscopolamine on performance on the FCN8-S^D and FCN8 schedules of reinforcement. The oblique marks on the records indicate reinforcements; a maximum of 40 were possible in any of the sessions shown here. These records start 10 min following intramuscular injections. (See Laties (21) for further details).

when 22 sec had elapsed without a response and a response on the lever would indeed be reinforced. Under this schedule, with the light serving as a discriminative stimulus, the rats responded even more slowly and showed no effects from the same dose of the drug (Fig. 11, right panels).

A very recent study by Thompson and Corr (37) shows a similar attenuation of a drug effect with pigeons working on a variable interval 1-min schedule. Responding on this basic schedule was increased some-what by very small doses of *d*-amphetamine and decreased by larger ones, as shown in Fig. 12, upper panels. Thompson and Corr's subjects worked on this schedule for 5-min periods that were alternated with 5-min periods on a slightly different schedule, one on which the availability of reinforcement was signaled by a stimulus change. The lower panels show what happened in the signaled condition; no rate-en-

hancing effect can be seen. As in the differential reinforcement of low rate (DRL) schedule work described above, the response rate was much lower when the subjects had a discriminative stimulus associated with reinforcement availability.

How can all this be put together? It looks like placing an animal's performance under strong external stimulus control diminishes the extent to which that performance will be changed by drugs. This effect may sometimes be mediated by changes in the patterns of behavior that are associated with the additional controlling stimuli. We know that the constituents of the behavior pattern— the rate, the distribution of responses in time, even the intensive and topo-graphical characteristics of the responses—all can modify drug effects. Framing an explanation of how drug action is modulated by explicit discriminative stimuli may require us first to determine just what types of behavioral changes are produced by the addition of particular environmental stimuli at particular times. But that course may itself be insufficient. For instance, in several of the examples discussed, the added discriminative stimuli served to induce profound overall rate decreases by marking the times when responding would be reinforced (3, 22, 37). Low (but not zero) rates are usually more sensitive to drugs and are likely to show drug-induced increases (19). Yet these low rates did not show such increases, the overall performances being more resistant to drugs when the added discriminative stimuli were present (but see Thomas (36) for a conspicuous exception). We also reported that performances equal in rate with and without external discriminative stimuli also showed the same kind of differential sensitivity (21, 38). These two facts lead us to conclude that rate dependencies may be less important when the behavior is controlled by powerful discriminative stimuli. And conversely, baseline rate becomes more important when control by such discriminative stimuli is weak.

Perhaps we should be looking more closely at momentary rate. Thompson and Corr (37) have suggested, for instance, that their variable-interval schedule with an added signal can be considered as a multiple schedule consisting of extinction and fixed-ratio 1 components. The fixed-interval schedule with the added "clock" stimuli could be considered to consist of four 1-min extinction periods followed by a 1-min fixed-interval schedule. The FCN 8-S^D schedule could be thought of as being a chained schedule consisting of a fixed-ratio 8 component that is followed by (and maintained in strength by) a stimulus in the presence of which a single response on a second key is reinforced by food (FR1).

In each case the added discriminative stimulus is associated with a higher reinforcement density. Response rate in the presence of the stimulus would therefore be expected to be higher (17); rate would, however, have to be inferred from latencies for the FR1 condition. In this type of analysis rates would then have to be considered for each component separately in examining drug effects. This approach would not challenge evidence from response-based schedules as directly as it would that from time-based schedules. But it is obviously the direction in which one should

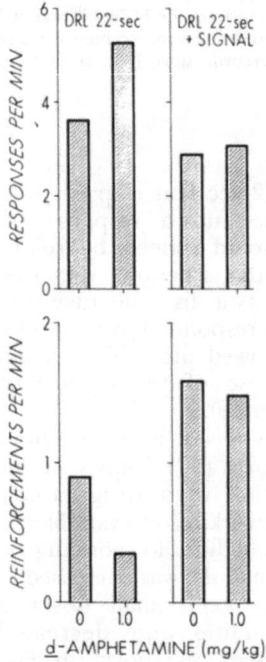

Figure 11. Effects of d-amphetamine on response rate (top panels) and reinforcement rate (bottom panels) of rats on DRL 22-sec schedules of reinforcement. The right panels summarize data collected when a change in external stimulation was correlated with the end of each 22-sec pause without responding. The left panels summarize data collected when no such change occurred. (Constructed from data in Carey and Kritkausky (3).)

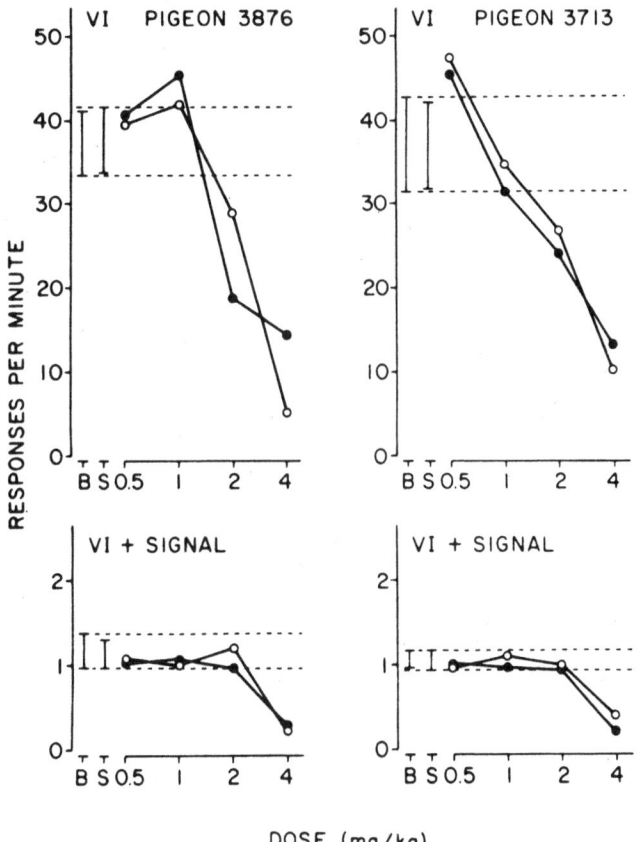

Figure 12. Effects of *d*-amphetamine on response rate of pigeons working on variable-interval (VI) schedules of reinforcement. In the case of the data summarized in the two lower panels, a stimulus change occurred whenever reinforcement became available. The dose-response function has been determined twice for each bird. The brackets and the dashed horizontal lines indicate the ranges for the baseline (B) and saline (S) sessions. (After Thompson and Corr (37).)

move in order to tease apart the interactions between stimulus control and drug variables.

It is still too early to make many strong generalizations about how specific drugs differentially affect behavior under different degrees of stimulus control. It does look as if scopolamine and amphetamine show the most pronounced interaction with level of stimulus control: when stim-ulus control is strong, the drug effects are minimal; when control is weak, the drug effects are greater. Chlorpromazine and promazine, on the other hand, are relatively in-sensitive to changes along this dimen-sion, with great changes easily produced by these drugs even in the face of strong stimulus control. In fact, the attenuation of stimulus control that has been suggested as the main mode of action of the drugs (7) may only be important in the presence of strong discriminative control against which the drugs can act. However, there are some experimental findings that do not fit. One example, drawn from work already discussed, will have to suffice to make the point. Dews (5) examined scopolamine's actions on the same two different levels of discriminative complexity that were described above. He did not find any differential effects, even with doses that produced extremely low response rates in both cases. But we have seen that such differences could be demonstrated using several other methods of varying discriminative control (21, 22, 38).

Let us turn briefly to the way the addition of a discriminative stimulus can modify a performance that has already been degraded. Recently, we have been involved in some work with methylmercury, work initially somewhat unrelated to problems of stimulus control (11). The

schedule of reinforcement was fixed consecutive number again. This time the pigeon had to peck at least eight but no more than nine times on one colored disc before a switch to a second disc would produce food. A switch before too few or after too many pecks merely reset the response counter and the bird had to start responding all over again. The type of performance produced is shown in Fig. 13. Run length distributions are shown for three sessions before methylmercury treatment was begun. In these histograms the reinforced runs, i.e., those either 8 or 9 responses long, are darkened in. Right after the first three histograms, a 105-day course of methylmercury was given that kept the blood level at about 15–20 ppm. Another 100 days elapsed before the second series of histograms shown here. The dominant effect of the poison was to increase the amount of variability and induce a switch after shorter runs. This may be a permanent effect, since more than 95%

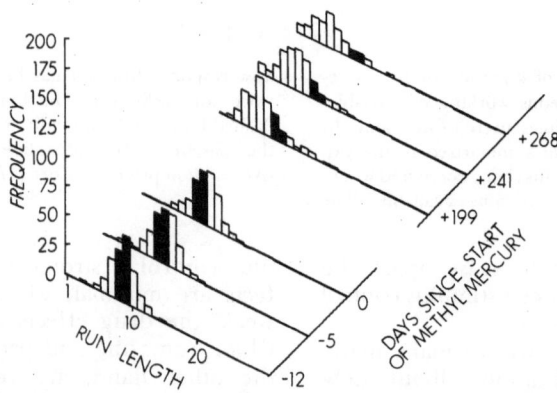

Figure 13. Effects of methylmercury on the performance of a pigeon working on a fixed-consecutive-number (FCN) schedule in which only runs of eight and nine responses served to make possible reinforcement for a single response on a second key. (No stimulus change was correlated with these, or any other, run lengths.) The dark bars indicate reinforced runs. Four 200-run sessions occurred each week. Dosage regimen: 2.5 mg Hg/kg by mouth daily Monday through Friday for 2 wk and 2.2 mg Hg/kg Monday and Friday thereafter. Further dosage details are in Evans et al. (11).

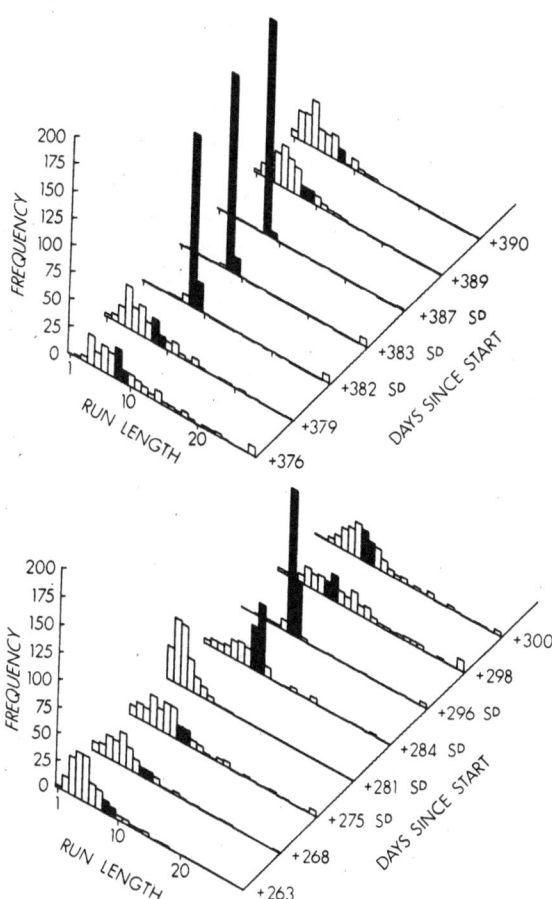

Figure 14. Changes induced in run length distributions by the addition of a discriminative stimulus (S^D). This stimulus was present after the eighth and before the tenth response on the FCN schedule of reinforcement during sessions labelled "S^D." During other sessions, no stimulus was correlated with any response runs. In all cases, only run lengths of eight or nine were reinforced; these are indicated by the dark bars.

of the methylmercury would have disappeared from the pigeon's body in this time, the half-life being about 20 days.

It occurred to us in the midst of this work that it might be interesting to do some therapy, as it were, with this pigeon once its behavior had stabilized. A cue was added with the eighth peck at the key (the key now turned red) and removed at the tenth peck. In other words, the presence of a red key after run lengths of 8 and 9 signaled to the pigeon that a switch to the second key would be reinforced with grain. Rather than discriminating the amount of behavior emitted, the bird now had only to discriminate the presence or absence of a distinct external stimulus. The first two run length distributions in Figure 14 (lower panel) show again what the behavior looked like before the addition of the stimulus. The next four distributions are selected from 17 sessions with the discriminative stimulus, each 200 runs long, that occurred next. (These are actually the 1st, 7th, 11th, and 17th sessions.) Note that by the 11th session control by the stimulus over switching was becoming obvious and the bird was being rewarded for more than half its runs. By the 17th session, it almost always made the reinforced number of responses. The cue was then removed with results shown in the last two run length distributions.

The upper panel of Fig. 14 shows what happened when the cue was again added a few months later. The first two run length distributions show that the effects of the exposure to methylmercury were still present, provided no discriminative stimulus was there. (Recall that no methylmercury had been given since day 105.) As soon as the stimulus was added, however, performance was markedly changed, with most runs now being of the appropriate length. Again, the removal of the added stimulus brought the bird back to its past performance, showing that we do not have a "cure" for the damage done but only a way of alleviating the symptom.

We have here, in a sense, a behavioral prosthesis for the pigeon, to steal a term from O. R. Lindsley (24, 25). Lindsley has pointed out that one very important way to deal with the deficits shown by the retarded, the physically handicapped, the old, and the like, is to focus on their interaction with the environment, and to make the changes necessary to insure the appearance of appropriate behaviors. Here, a pigeon devastated by a chronic regimen on a poison has to some extent been made whole again in the presence of a slight but influential change in the environment.

In the other experiments described above, the supplementary external stimuli helped attenuate the effects of several drugs—to protect the organism, so to speak. Now the same type of environmental change has been shown to reverse, temporarily, the ravages of methylmercury.

In looking at the drug–behavior interaction, we should keep our eye on the importance of behavioral variables. I know of no drug that would have produced such a great improvement in this pigeon's behavior. It should be sobering to behavioral pharmacologists to realize that our area has yet to uncover many chemical agents that can readily produce selective behavioral changes as safely and as quickly as some relatively small environmental manipulations.

REFERENCES

1. BERRYMAN, R., W. WAGMAN AND F. S. KELLER. Chlorpromazine and the discrimination of response-produced cues. In: *Drugs and Behavior,* edited by L. Uhr and J. G. Miller. New York: Wiley 1960, p. 243–249.
2. BLOUGH, D. S. Technique for studying the effects of drugs on discrimination in the pigeon. *Ann. N.Y. Acad. Sci.* 65: 334–344, 1956.
3. CAREY, R. J., AND R. P. KRITKAUSKY. Absence of a response-rate-dependent effect of *d*-amphetamine on a DRL schedule when reinforcement is signaled. *Psychonomic Sci.* 26: 285–286, 1972.
4. DEWS, P. B. Studies on behavior. I. Differential sensitivity to pentobarbital of pecking performance in pigeons depending on the schedule of reward. *J. Pharmacol. Exp. Ther.* 113: 393–401, 1955.
5. DEWS, P. B. Studies on behavior. II. The effects of pentobarbital, methamphetamine and scopolamine on performances in pigeons involving discriminations. *J. Pharmacol. Exp. Ther.* 115: 380–389, 1955.

6. DEWS, P. B. Modification by drugs of performance on simple schedules of positive reinforcement. *Ann. N.Y. Acad. Sci.* 65: 268–281, 1956.

7. DEWS, P. B. Analysis of effects of psychopharmacological agents in behavioral terms. *Federation Proc.* 17: 1024–1030, 1958.

8. DEWS, P. B. Studies on behavior. IV. Stimulant actions of methamphetamine. *J. Pharmacol. Exp. Ther.* 122: 137–147, 1958.

9. DEWS, P. B. Pharmacology of positive reinforcement and discrimination. In: *Pharmacology of Conditioning, Learning and Retention,* edited by M. Ya. Mikhel'son and V. G. Longo. New York: Macmillan, 1965, p. 91–96.

10. DEWS, P. B. Drug–behavior interactions. In: *Behavioral Analysis of Drug Action,* edited by J. A. Harvey. Glenview, Ill.: Scott, Foresman, 1971, p. 10–43.

11. EVANS, H. L., V. G. LATIES AND B. WEISS. Behavioral effects of mercury and methylmercury. *Federation Proc.* 34: 1858–1867, 1975.

12. FERSTER, C. B., AND B. F. SKINNER, *Schedules of Reinforcement.* New York: Appleton-Century-Crofts, 1957.

13. FRY, W., R. T. KELLEHER AND L. COOK. A mathematical index of performance on fixed-interval schedules of reinforcement. *J. Exp. Anal. Behav.* 3: 193–199, 1960.

14. HEARST, E. Drug effects on stimulus generalization gradients in the monkey. *Psychopharmacologia* 6: 57–70, 1964.

15. HEISE, G. A., N. LAUGHLIN AND C. KELLER. A behavioral and pharmacological analysis of reinforcement withdrawal. *Psychopharmacologia* 16: 345–368, 1970.

16. HEISE, G. A., AND N. L. LILIE. Effects of scopolamine, atropine, and *d*-amphetamine on internal and external control of responding on non-reinforced trials. *Psychopharmacologia* 18: 38–49, 1970.

17. HERRNSTEIN, R. J. On the law of effect. *J. Exp. Anal. Behav.* 13: 243–266, 1970.

18. HOLLOWAY, F. A., AND R. A. WANSLEY. Factors governing the vulnerability of DRL operant performance to the effects of ethanol. *Psychopharmacologia* 28: 351–362, 1973.

19. KELLEHER, R. T., AND W. H. MORSE. Determinants of the specificity of behavioral effects of drugs. *Ergeb. Physiol. Biol. Chem. Exp. Pharmakol.* 60: 1–56, 1968.

20. KSIR, C. J., JR. Scopolamine effects on two-trial delayed-response performance in the rat. *Psychopharmacologia* 34: 127–134, 1974.

21. LATIES, V. G. The modification of drug effects on behavior by external discriminative stimuli. *J. Pharmacol. Exp. Ther.* 183: 1–13, 1972.

22. LATIES, V. G., AND B. WEISS. Influence of drugs on behavior controlled by internal and external stimuli. *J. Pharmacol. Exp. Ther.* 152: 388–396, 1966.

23. LATIES, V. G., AND B. WEISS. Behavioral mechanisms of drug action. In: *Drugs and the Brain,* edited by Perry Black. Baltimore: The Johns Hopkins Press, 1969, p. 115–133.

24. LINDSLEY, O. R. Direct measurement and prosthesis of retarded behavior. *J. Educ.* 147: 62–81, 1964.

25. LINDSLEY, O. R. Geriatric behavioral prosthetics. In: *New thoughts on old age,* edited by R. Kastenbaum. New York: Springer-Verlag, 1964, p. 41–60.

26. MCKEARNEY, J. W. Rate-dependent effects of drugs: Modification by discriminative stimuli of the effects of amobarbital on schedule-controlled behavior. *J. Exp. Anal. Behav.* 14: 167–175, 1970.

27. MCMILLAN, D. E. The effects of ethyl alcohol on temporally spaced responding in humans. *J. Pharmacol. Exp. Ther.* 171: 159–165, 1970.

28. MECHNER, F. Probability relations within response sequences under ratio reinforcement. *J. Exp. Anal. Behav.* 1: 109–122, 1958.

29. MECHNER, F., AND M. LATRANYI. Behavioral effects of caffeine, methamphetamine, and methylphenidate in the rat. *J. Exp. Anal. Behav.* 6: 331–342, 1963.

30. MORSE, W. H., AND R. J. HERRNSTEIN. Effects of drugs on characteristics of behavior maintained by complex schedules of intermittent positive reinforcement. *Ann. N.Y. Acad. Sci.* 65: 303–317, 1956.

31. MORSE, W. H., AND R. T. KELLEHER. Schedules as fundamental determinants of behavior. In: *The Theory of Reinforcement Schedules,* edited by W. N. Schoenfeld. New York: Appleton-Century-Crofts, 1970, p. 139–185.

32. RUTLEDGE, L. T., AND R. W. DOTY. Differential action of chlorpromazine on reflexes conditioned to central and peripheral stimulation. *Am. J. Physiol.* 191: 189–192, 1957.

33. SEGAL, E. F. Exteroceptive control of fixed-interval responding. *J. Exp. Anal. Behav.* 5: 49–57, 1962.

34. SIDMAN, M. Drug–behavior interaction. *Ann. N.Y. Acad. Sci.* 65: 282–302, 1956.

35. SKINNER, B. F. *The Behavior of Organisms.* New York: Appleton-Century-Crofts, 1938.

36. THOMAS, J. R. Differential effects of two phenothiazines on chain and tandem performance. *J. Pharmacol. Exp. Ther.* 152: 354–361, 1966.

37. THOMPSON, D. M., AND P. B. CORR. Behavior parameters of drug action: signalled and response-independent reinforcement. *J. Exp. Anal. Behav.* 21: 151–158, 1974.

38. WAGMAN, W. D., AND G. C. MAXEY. The effects of scopolamine hydrobromide and methyl scopolamine hydrobromide upon the discrimination of interoceptive and exteroceptive stimuli. *Psychopharmacologia* 15: 280–288, 1969.

39. WILCOXIN, F., AND R. A. WILCOX. *Some Rapid Approximate Statistical Procedures.* Pearl River, N. Y.: Lederle Laboratories, 1964.

40. ZAVADSKII, I. V. Experience with the application of the conditioned reflexes method to pharmacology. Toward the problem of the effects of certain drugs (alcohol, morphine, cocaine, and caffeine) on the function of the higher regions of the central nervous system. *Tr. Obshchestva Russkikh Vrachei—St. Petersburg* 75: (9), 269–287, 1908.

Reinforcement schedules and extrapolations to humans from animals in behavioral pharmacology

LEONARD COOK AND JERRY SEPINWALL

Department of Pharmacology, Research Division
Hoffmann-La Roche Inc., Nutley, New Jersey 07110

ABSTRACT

Behavior controlled by various schedules of reinforcement is useful for characterizing drugs as well as for analyzing the mechanisms of action of their effects on behavior. Conditioned avoidance techniques have been useful for studying neuroleptics and for predicting their clinical antipsychotic activity; the possible involvement of dopaminergic mechanisms in the effects of neuroleptics on avoidance behavior is discussed. Tricyclic antidepressant agents have been studied in assays involving interactions with other agents, such as cocaine, amphetamine and tetrabenazine. One type of operant behavior, Sidman avoidance, has been used as a particularly sensitive assay for such drug interactions. Another schedule, in which "observing" responses in pigeons are measured, seems to provide a method for studying antidepressants without involving drug interaction phenomena. For tricyclic compounds, facilitation of observing responses and weak potency of conditioned avoidance inhibition constitute a pharmacological profile that seems to have some predictive value for clinical imipramine-like antidepressant activity. "Conflict" (punishment) schedules have been useful for predicting antianxiety activity in man. Although the degree of anticonflict effect observed is consistent with Dews' rate dependency hypothesis, this principle does not fully account for the observed drug effects. In the conflict model, the actions of benzodiazepines differ in drug-naive versus drug-experienced animals. Experiments with parachlorophenylalanine have not yet provided clear support for the postulated role of serotonin in related phenomena.— COOK, L., AND J. SEPINWALL. Reinforcement schedules and extrapolations to humans from animals in behavioral pharmacology. *Federation Proc.* 34: 1889–1897, 1975.

Abbreviations: VI, variable interval; FR, fixed ratio; CER, conditioned emotional response; PCPA, parachlorophenylalanine.

Psychopharmacological agents modify behavior. Their clinical use is to affect behavior and emotionality. Although analyses and probes into the physiological and biochemical brain mechanisms that underlie normal or abnormal behavior are valuable and necessary, it is equally important to be able to experimentally control, develop and quantify various aspects of behavior. Behavioral studies certainly provide relevant measures of the animal's total response to the net action of psychopharmacological agents. The behavioral pharmacology literature contains examples of many techniques that have been employed to study behavior, ranging from simple, quickly acquired tasks to complex and time consuming methods. Usually, the procedures subsumed under the heading of "schedules of reinforcement" belong to the group of more complex techniques. An investigator using these methods must frequently invest a certain amount of time to train his animal subjects. This is balanced, however, by the advantages possessed by schedule procedures: they uniquely can provide very stable baselines for drug studies; animals can serve as their own controls; and multiple schedules that combine several different measures and patterns of behavior allow for drug studies in which one can illustrate selectivity and specificity of drug action on the various ongoing patterns. Schedules of reinforcement permit a continuous measure of behavior, usually lasting over several hours, thus providing for a continuous measure of onset and duration of drug effect.

Kelleher and Morse (30) have pointed out,

> An outstanding feature of behavioral pharmacology is that the behavioral effects of drugs are largely determined by the environmental circumstances. Such dependence is not qualitatively unique; the effects of a drug on any biological system depend upon the state of the system. For example, the effect of epinephrine on blood pressure depends in part on the height of the blood pressure when the drug is administered. The effects of past circumstances are, however, much greater in behavioral pharmacology. The profound effects of environmental factors have not been fully recognized because physiological and pharmacological experiments on intact animals usually attempt to hold environmental determinants constant and thereby obscure their importance. For behavioral pharmacology, however, the study of the nature of the environmental determinants of behavior is essential.

It is within this context that schedules of reinforcement have been most useful in advancing our knowledge concerning the effects of drugs on behavior.

In this paper we shall discuss some psychotherapeutically important drugs in terms of the schedules of reinforcement that perhaps best measure their behavioral effects. It should be stated at the outset, however, that certain problems exist. As tempting as it is to develop animal models of human behavior, especially psychopathological behavior, limitations are obvious. The most important barrier is the lack of definition of the clinical syndrome in operational terms that can be applied to animals studies. Therefore, the relevance of animal behavioral techniques in drug studies rests on empirical correlations with established clinical effects of drugs. In the ensuing discussion, we shall deal with three classes of drugs: neuroleptics, antidepressants, and antianxiety agents.

NEUROLEPTICS

The most important therapeutic property of members of this class is antipsychotic activity. In animals, neuroleptics exert several different effects, among which are production of catalepsy and ptosis, and antagonism of effects induced by amphetamine or apomorphine (27); however, our main focus will be on behavioral situations that involve schedules of reinforcement. Conditioned avoidance procedures are the operant conditioning methods employed most commonly to study neuroleptics. There are two types of conditioned avoidance procedures generally used. In a discrete avoidance task a stimulus precedes a shock in discrete trials, and a subject can avoid the shock by responding during the conditioned stimulus (11). The other procedure, frequently called Sidman avoidance (46), has no external stimuli to warn the subject of shocks that are scheduled for presentation at fixed intervals, e.g., every 20 sec. Responses at intervals of less than 20 sec will avoid shock delivery. These avoidance schedules have been used extensively, and both their advantages and limitations have been elucidated.

Operant conditioning procedures have provided information that allows correlations to be made between pharmacologic effects in animals and in human clinical situations, and we shall be considering such correlations. We can note first, however, that a direct translation from animal to human results can be made within operant conditioning designs. Cook and Catania (6) reported that tests conducted with several drugs in a human Sidman avoidance situation yielded qualitatively similar results to those obtained in animal avoidance procedures. Chlorpromazine, a prototypical neuroleptic, had the same characteristic effect in all species: a decrease in avoidance responding, while escape behavior was maintained. This selective effect was not produced by even high doses of

chlordiazepoxide, meprobamate or phenobarbital.

The predictive value of animal data, as measured by conditioned avoidance blocking potency, to clinical potency for treatment of severe mental and emotional disorders has been recognized for several years (6, 8, 27, 39, 40). It is well known that in the clinic most neuroleptics produce other neurological effects, such as extrapyramidal symptoms (EPS), in addition to their antipsychotic actions. In considering extrapolations to humans from animals, it is therefore pertinent to consider the extent to which behavioral animal models are uniquely predictive of antipsychotic efficacy. Might conditioned avoidance techniques be related instead more closely to the extrapyramidal symptom effects? To consider this question, reference must be made to one current hypothesis about the biochemical mechanism of action of neuroleptics. Several authors have proposed that neuroleptics block brain dopamine receptors, particularly in the striatum (5, 12, 41). Within this context, Van Rossum (51) has indicated that the hypothesis of dopamine receptor blockade by the neuroleptics means that their parkinsonian side effects cannot be disconnected from their neuroleptic action.

A number of studies involving conditioned avoidance bear directly on this issue. Apparently the ability of such neuroleptics as perphenazine, chlorpromazine, trifluoperazine, haloperidol, thioridazine and chlorprothixene to inhibit the conditioned avoidance response is antagonized by agents having antiparkinsonian effects (15, 24, 38, 43). Included among these have been several anticholinergic agents, (e.g., atropine) as well as agents such as amantadine, apomorphine and l-dopa, which apparently act via dopamine receptors. One interpretation of these studies suggests that avoidance procedures may reflect the extrapyramidal symptoms liability of neuroleptics.

Nevertheless, it would be difficult to conclude that conditioned avoidance techniques are not relevant as predictors of antipsychotic efficacy. This point can be elaborated by a consideration of two clinically effective antipsychotic agents, thioridazine and clozapine. Thioridazine is equipotent clinically to chlorpromazine but has a low extrapyramidal symptom incidence (21). Although in the rat it is only one-fourth as potent a conditioned avoidance blocker as chlorpromazine (39), in the squirrel monkey it is equipotent (3). The squirrel monkey data thus correlate well with thioridazine's antipsychotic potency. On some measures which relate to dopamine, however, thioridazine differs from chlorpromazine and these differences may be related to a separation between the antipsychotic and extrapyramidal system effects of neuroleptics. In the monkey, as well as in some other species, thioridazine was relatively inactive compared to chlorpromazine in altering brain dopamine metabolism (35). In rats, thioridazine was equipotent to chlorpromazine in reducing spontaneous motor activity yet was relatively inactive in a "turning" model that has been used to identify striatal dopamine receptor blockers (13).

Clozapine, a comparatively new agent, is reported to be a clinically effective antipsychotic agent with little or no extrapyramidal symptom liability (16, 22, 25). It is an active conditioned avoidance blocker, being one-half to one-fifth as potent as chlorpromazine, and this was an important criterion for originally classifying and predicting its clinical activity (48, 50). Clozapine, however, has no cataleptic properties and no antiapomorphine activity in rats (1, 48, 50). Therefore, potential antipsychotic properties are identified perhaps more reliably in conditioned avoidance tests, as illustrated by the examples of thioridazine and clozapine, than in other animal models for neuroleptics, e.g., catalepsy. It would be of great interest to determine whether anticholinergic and other antiparkinsonian agents affect the conditioned avoidance blocking properties of thioridazine and clozapine. (It has been suggested that clozapine and thioridazine have low extrapyramidal symptom liability because they possess in vivo or in vitro antimuscarinic properties (47, 48). Clozapine was inactive, however, in an in vivo assay that was believed to measure antimuscarinic effects specifically in the nigrostriatal system (1).) The effects of clozapine in other types of operant procedures should be studied to determine similarities and differences with other antipsychotic agents.

ANTIDEPRESSANTS

The evaluation of antidepressant pharmacological properties has been fraught with difficulties, both clinically and preclinically. One technique pharmacologists use involves drug interactions, e.g., antagonism of reserpine induced ptosis. Operant conditioning methods have also been used to measure drug interactions to identify antidepressant activity. For example, Sidman avoidance in rats or monkeys is especially sensitive as a measure of drug interactions (44). In these studies, tetrabenazine, *d*-amphetamine or cocaine were used at doses that were inactive in affecting avoidance response rates. Doses of the tricyclic antidepressants being evaluated were also subthreshold and had no effect, alone, on avoidance rates. Under these conditions, antidepressants appeared to potentiate the inherent stimulant properties of cocaine and d-amphetamine, and presumably also interacted with the norepinephrine relased by low doses of tetrabenazine to cause stimulation of avoidance behavior. In such drug interaction studies, one advantage provided by the Sidman avoidance schedule of reinforcement is sensitivity, both to dose-response and time-response relationships. The antidepressants were active at unusually low doses relative to other assays, and the continuous 5-hr behavioral baseline clearly showed the onset and duration of these sensitive drug interactions.

Although it has been difficult to show qualitative differences between tricyclic antidepressants and tricyclic neuroleptics on the basis of single drug studies, certain operant conditioning experiments with pigeons have provided interesting data in this regard. Cook and Kelleher (9) showed in "observing-response" studies that pigeons would respond to imipramine in a characteristic manner. These studies involved two response keys; pecking responses on one were reinforced with food intermittently and most unpredictably (29). Responses on another (observing) key provided information to the pigeon as to whether a period of food payoff was in effect. Under control conditions, pigeons typically worked predominantly on the food key and only occasionally switched over to the observing key to determine whether a food payoff period was in effect. It was found that imipramine, desipramine, amitriptyline, and nortriptyline all increased these observing responses (9, and unpublished data). (It is tempting to speculate that these drugs caused the pigeons' overall behavior to become more proficient and relevant to the environmental contingencies.)

The observing response was not changed by other types of substances, such as amphetamine, chlordiazepoxide and meprobamate, nor by many neuroleptics, e.g., trifluoperazine and haloperidol. Two neuroleptics, chlorpromazine and promazine, did affect the observing response. However, the principle we wish to present is that there appears to be a high correlation between clinical imipramine-like antidepressant activity and the ability of a compound to increase observing response rates in pigeons, while being weakly potent or ineffective in blocking conditioned avoidance responses. Table 1 provides specific data relevant to this suggestion. Imipramine increased the observing responses at low doses, yet was poorly effective as an avoidance inhibitor. Whereas chlorpromazine also was effective in increasing the observing response, it was a potent inhibitor of avoidance. This correlation between potent increases of observing responses in pigeons and clinically effective antidepressant properties is intriguing.

ANTIANXIETY AGENTS

Operant conditioning methods have been most useful for studying the behavioral effects of antianxiety agents. The most commonly employed technique is often referred to as a "conflict" model. The basic concept consists of training an animal to work for a desirable goal object, such as food, and then introducing punishment, such as electric footshock, when the animal responds to obtain food. By pairing appetitive and aversive reinforcement, the experimenter creates a situation for the animal in which there is an inferred "conflict" between approach and avoidance tendencies. A commonly

TABLE 1. Behavioral effects of chlorpromazine and imipramine[a]

	Rats– decrease of conditioned avoidance (pole-climb), ED_{50}[b]	Pigeons– increase of responses (observing behavior), MED[c]
Chlorpromazine	11	1.7
Imipramine	>90, 180 (inactive)	2.9

Values represent oral dosage in milligrams per kilogram body weight: [a] Methodology refs: (9, 11, 29).   [b] Mean effective dose.   [c] Minimum effective dose.

employed method of this type was initially described by Geller and Seifter (20). In some of the research to be discussed below, a modification developed by Davidson and Cook (14) was used. This consisted of a multiple schedule of reinforcement in which the alternating components were a variable interval 30 sec (VI 30″) schedule reinforced by food in the presence of a white houselight, and a fixed ratio 10 response (FR10) schedule reinforced simultaneously with food and footshock in the presence of a red houselight (45).

The development of suppression, after shock was introduced during the FR components, can be seen in Fig. 1. As the shock intensity was gradually increased the FR response rates, which were initially higher than the VI rates, decreased to approximately 10% of their original values. At the same time, there was an increase in unpunished VI rates, i.e., a contrast effect. The suppressed FR rates were operationally designated as "conflict" behavior. Psychotropic agents were studied for the ability

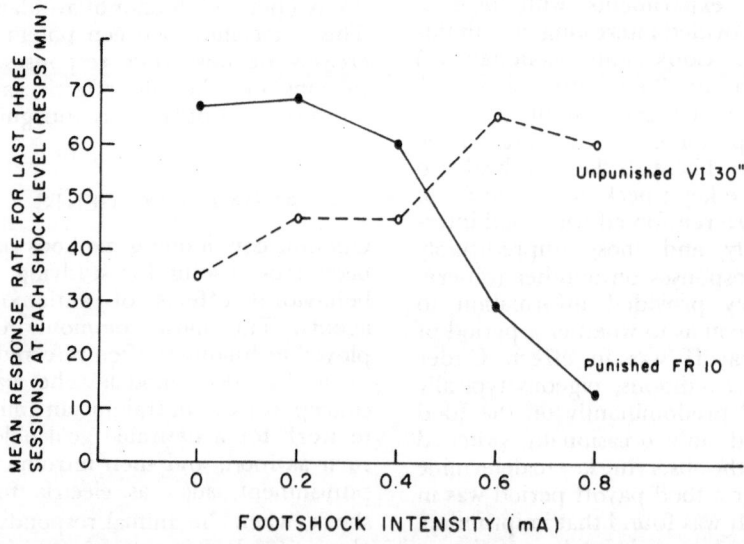

Figure 1. Development of suppression in multiple VI 30″ FR 10 conflict schedule. The figure presents representative data for one rat. The intensity of the footshock punishment was gradually adjusted to produce the desired degree of FR suppression. The number of sessions during which the shock was maintained at given intensities differed both within and between animals. Each point represents the mean response rate for the last three sessions at each shock level. As the FR response rate decreased, a contrast effect was observed consisting of an increased VI response rate.

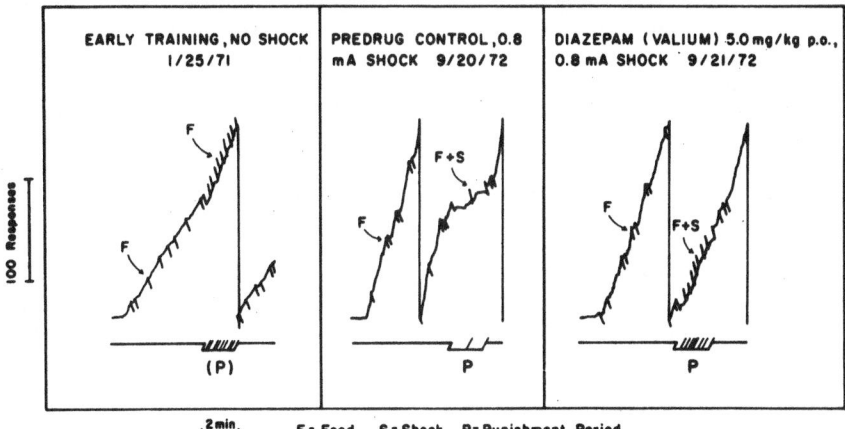

Figure 2. Rat multiple VI 30″ FR 10 conflict schedule. The figure illustrates typical training sequence and diazepam effect for one rat; cumulative records of lever pressing responses are shown. *Left panel*: during variable interval (VI) 30″ condition responses occasionally produced food (F); during the FR 10 periods (P) every tenth response resulted in food delivery. *Center panel*: responding is suppressed during P stimulus condition when every tenth response now results in food plus footshocks (F + S). *Right panel*: typical effect of diazepam in restoring responding for food under punishment conditions without any change in unpunished responding. (From *Emotions—Their Parameters and Measurement*, edited by L. Levi, Raven Press, New York, copyright 1974.)

to attenuate the conflict by increasing responding during the punishment periods. Figure 2 shows the characteristic effect exerted in this procedure by diazepam, a compound with clinical antianxiety activity. Diazepam administration produced increased responding during the punishment periods so that FR response rates recovered toward the level at which they had formerly been during this rat's preshock history.

Cook and Davidson (7) described the kinds of compounds that were active or inactive in this model. Increases in responding during the punishment periods were produced most markedly by benzodiazepines, carbamic acid esters, and some barbiturates. Increases in responding were obviously not due to any general stimulant property since amphetamine, for example, actually produced a greater amount of suppression during FR segments. Inactive compounds included neuroleptics, such as chlorpromazine and haloperidol, various antidepressants, an antihistaminic agent, and morphine.

Figure 3 illustrates the dose-response profile of chlordiazepoxide in this procedure. Punished FR responding was increased in a dose related manner while unpunished VI behavior was only slightly changed over much of the effective anticonflict dose range. At the upper end of the range, VI responding was decreased, an effect that possibly may be related to the onset of sedative actions. Although there was less of an increase in FR responding at high doses, these response rates were still considerably higher than control levels. Thus, the effects of agents active in this procedure can be characterized on the basis of three parameters: the minimum effective

Leonard Cook and Jerry Sepinwall

anticonflict dose; the range of doses across which anticonflict activity is maintained; and the dose at which VI responding is first decreased. Occasionally, compounds that are not typically characterized as antianxiety agents, for example, trifluoperazine which is a phenothiazine neuroleptic, may produce a moderate increase in FR responding at one or two dose levels (14). The smaller magnitude of their peak effects and the size of their effective dose ranges, however, appear to distinguish them from members of standard antianxiety classes of agents. Anticonflict potency in the multiple VI/FR conflict procedure was reported by Cook and Davidson (7) to correlate highly with clinical potency in cases of psychoneurotic disorders. This operant conditioning model, therefore, not only appeared to be selective for so-called minor tranquilizers but also had predictive validity with respect to the clinical antianxiety application of these compounds.

At this point, it is appropriate to consider a theoretical postulate that has been advanced to describe the effects of drugs on schedule-controlled behavior. It has often been suggested from an operational approach that the actions of psychotropic agents are "rate-dependent," i.e., the magnitude and quality of the pharmacological effects are dependent upon the predrug pattern of responding (17). One principal postulate of this hypothesis is that low predrug response rates will be increased by many psychotropic agents whereas high response rates will be increased less or will be decreased. Some investigators have attempted to apply a rate-dependency analysis to the effects of anxiolytics in the conflict paradigm (54). According to this interpretation, the ability of antianxiety agents to increase responding that has been suppressed by

punishment is merely one instance of a general tendency these agents have to increase low response rates, irrespective of how those low rates originated. It is clear that anxiolytics have produced response rate increases in many situations, including some that involved appetitive reinforcement only (23, 33).

For several reasons, however, response-contingent punishment conditions remain unique with respect to measuring what many investigators

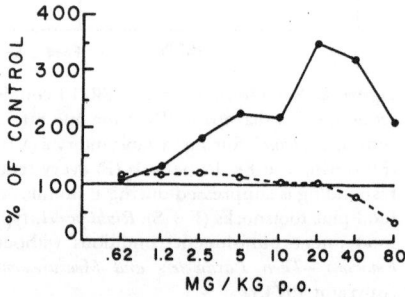

Figure 3. Dose response curve for chlordiazepoxide on punished fixed ratio (• —— •) and unpunished variable interval (o – – – o) behavior. Doses of chlordiazepoxide hydrochloride, in water solution, were administered orally 40 to 50 min prior to testing. Effect on treatment days is expressed as a percentage of the mean baseline value for the 3 control days preceding each treatment. The solid horizontal line at 100% represents the control level. Fifteen rats served as subjects, with at least 10 rats tested at every dose level except for 80 mg/kg where only 7 rats were used. The minimum effective anticonflict dose was in the range of 1.25–2.5 mg/kg and significant attenuation of punished behavior continued up through 40 mg/kg. Across the same dose range unpunished responding was either increased approximately 20%, unchanged, or decreased 20%. At 80 mg/kg, unpunished responding was decreased to 44% of control whereas the effects on punished responding were mixed: three rats continued to show large response increases but the other four had decreased rates. (From: *Emotions—Their Parameters and Measurement*, edited by L. Levi, Raven Press, New York, copyright 1974.

interpret as preclinical predictors of antianxiety properties. First, under certain conditions amphetamine, chlorpromazine and imipramine can increase low rates of responding maintained by food reinforcement, yet these agents do not typically increase low rates that have resulted from response-contingent punishment (30). Some recent experiments that have attempted to explore this issue more thoroughly have shown that amphetamine can sometimes increase these low rates (18; McKearney and Barrett, ms in preparation). The conditions under which this can occur, however, seem to be limited. This limitation is consistent with recent studies that have shown that amphetamine cannot increase certain food-maintained low response rates that have been suppressed by schedule manipulations in experiments *not* involving punishment (32, 49).

Second, in some studies a multiple or concurrent schedule was used in which each animal was exposed to both unpunished and punished periods and conditions were arranged so that matched low rates of unpunished and punished responding occurred. In those cases where amphetamine and chlorpromazine produced response increases, these two agents increased unpunished rates more than punished rates (36). In contrast, antianxiety agents raised the punished rate to a greater extent (6, 36). In Fig. 4, the effects of two doses of chlordiazepoxide on both punished and unpunished responding as a function of baseline response rates in the multiple VI30″ FR10 conflict schedule are displayed. It can be seen that while there were rate-dependent effects at a maximally effective dose (20 mg/kg orally; see Fig. 3), clear attenuation of the punishment-induced response suppression was also observed without any

rate-dependency features at a dose (2.5 mg/kg) that was low, but effective, in this procedure. Thus, even though the degree to which anxiolytics increase punished responding is inversely related to baseline rates (7, 36), there appears to be a specific effect that these agents exert on punished behavior that cannot entirely be explained in terms of rate-dependence.

The use of multiple schedules of reinforcement has led to observations that are relevant to the clinical application of benzodiazepine antianxiety agents and that begin to suggest something about underlying neurochemical mechanisms. Margules and Stein (34) reported that a given dose of a benzodiazepine produced different effects after its first administration as compared to effects with repeated treatment. The first administration of oxazepam to rats that had never before been treated with any psychotropic agent produced the following: there was a marked decrease in unpunished responding and only a moderate increase in responding during the punishment periods. Over the next three or four consecutive daily administrations the decrements in unpunished responding gradually ceased while the anticonflict effect increased to an asymptotic level. It took approximately four treatments, therefore, for the characteristic features of that given dose's profile to be attained. Similar results have been reported with flurazepam hydrochloride (30-100 mg/kg orally), even when the repeated treatments were given at 2- or 3-day intervals rather than daily (4). The decrease in unpunished responding was considered to be correlated with the sedative properties of benzodiazepines (4, 34). (We have previously suggested (10) that this interpretation warrants further study.) The anticonflict or response disinhibitory effects were interpreted as being masked by the initial sedative actions and as subsequently becoming apparent when tolerance to the sedative effects occurred (4, 34). Furthermore, these results were considered to be analogous to similar findings observed when such agents were used to treat patients with psychoneuroses.

We have extended these findings by showing that the same phenomenon occurred when chlordiazepoxide (10 mg/kg p.o.) was administered at 7-day intervals to conflict-trained drug-naive rats (10). For example, one group of rats was trained on the multiple VI30″ FR10

conflict schedule until control levels were stable, after which they were given one treatment per week for 6-wk, approximately 40 min prior to testing sessions. The first two treatments consisted of water injections (2 ml/kg orally) and the last four of chlordiazepoxide (10 mg/kg orally). The decrease in unpunished responding disappeared and the maximum anticonflict effect appeared only after the third chlordiazepoxide administration. Another group of similarly trained rats was exposed to a series of six consecutive chlordiazepoxide (10 mg/kg orally) treatments at weekly intervals. The first two treatments, however, were given ap-

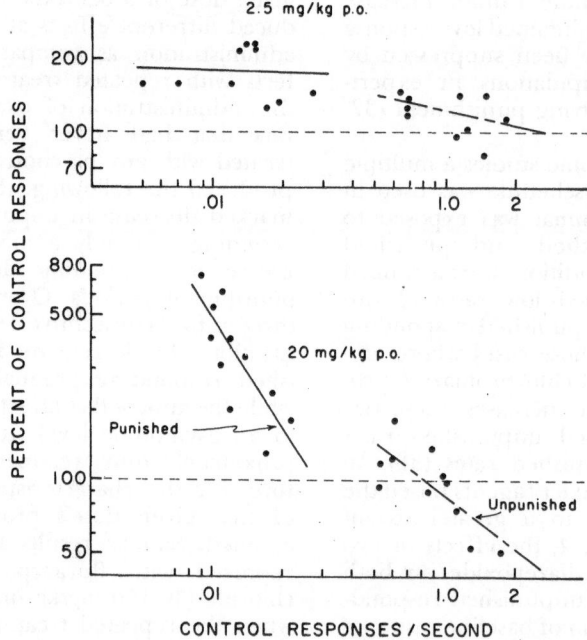

Figure 4. Effect of chloridazepoxide as a function of baseline response rate on multiple VI 30″ FR 10 conflict schedule. *Ordinate*: same as in Fig. 3, but note logarithmic scale. *Abscissa*: baseline response rate calculated as mean of three control sessions prior to drug session. Curves were drawn according to regression analysis calculations. Top portion: data for 10 rats tested at 2.5 mg/kg dose where a significant increase in punished responding occurred without a steep rate-dependent slope (each rat contributed a pair of points: one punished FR value and one unpunished VI value). Bottom portion: data for 11 rats tested at 20 mg/kg dose; a steep rate-dependent slope was observed.

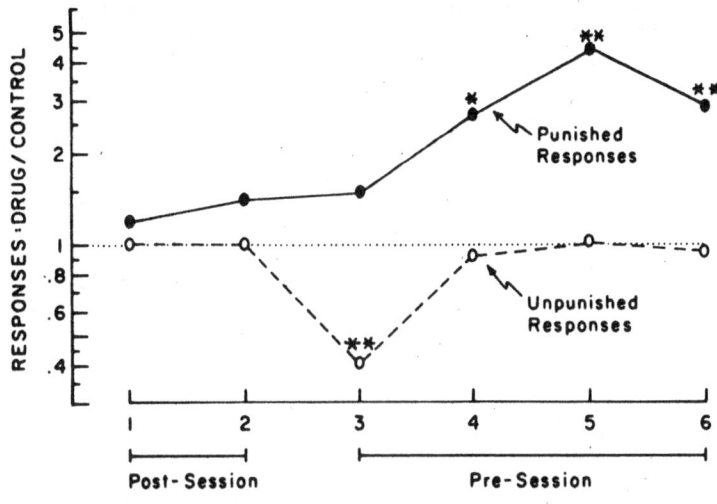

Figure 5. Effect of chlordiazepoxide in conflict-punishment test in rats with history of two prior extrasituational chlordiazepoxide treatments. Once a week for 6 wk, five pharmacologically naive rats received chlordiazepoxide (10 mg/kg by mouth). Ordinate details similar to Fig. 3 but given as ratio of drug/control, not as percentage. The first two treatments were given 15 min *after* testing had been completed, whereas the last four were given 40–50 min *prior* to testing.

The first presession treatment (#3) was still associated with a significant decrease in unpunished VI responding. No real increase in punished FR responding was seen during this session, and it was not until the third presession treatment (treatment #5) that a peak in FR responding was seen. (* $P < 0.02$; ** $P < 0.005$) (From: *Emotions — Their Parameters and Measurement*, edited by L. Levi, Raven Press, New York, copyright 1974.)

proximately 15 min *after* completion of a session whereas the last four treatments were given at the usual presession pretreatment time of approximately 40 min. Figure 5 indicates that even though the animals had received two prior administrations of the compound outside of the test situation, they still showed a marked decrease in unpunished responding and only a minimal increase in punished responding when the first presession treatment was given. It still took three presession treatments before the anticonflict effect reached its maximum.

In a related study we have recently observed a similar phenomenon in a drug-naive conflict-trained squirrel monkey. In this case, a concurrent schedule of reinforcement was used similar to the one described by Cook and Catania (6). Monkeys worked to obtain food pellets by pressing either one of two levers positioned on each side of a food cup in the test chamber. The right lever was associated with a VI 6 min schedule of food delivery; responses on this lever were never punished. The left lever was associated simultaneously with a VI 1.5 min schedule of food delivery and a variable ratio (VR) 24 response schedule of footshock presentation. The shock level was adjusted for each monkey so that the initially higher level of responding on the left lever was suppressed until it was somewhat less than the unpunished right lever response rate. For reference purposes, Fig. 6 illustrates dose response curves for diazepam and chlordiazepoxide in monkeys with a history of prior exposure to benzodiazepines in this procedure. It may be seen that diazepam was more potent (minimally effective dose = 0.31 mg/kg orally) than chlordiazepoxide (0.62 mg/kg by mouth) and produced a larger magnitude of effect over the range of doses depicted. The effects of the 5 mg/kg diazepam dose on punished and unpunished response rates in drug-experienced animals (Fig. 6) can be contrasted with the effects of the same dose in a similarly trained drug-naive monkey (Fig. 7). In this monkey,

both punished and unpunished responding were markedly decreased after the first diazepam treatment. Although the second treatment was not administered until 2 wk later, and the third administration until 1 wk after the second, the pattern of the animal's response to the treatments changed characteristically. We are continuing to study this interesting phenomenon in order to determine to what extent metabolic and drug-behavior interaction factors participate in these effects.

The results of one such study are presented in Fig. 8. Two groups of rats were trained on the multiple VI30″ FR10 schedule. Only one group, however, was exposed to the footshock punishment condition during the FR periods. In Fig. 8 it can be seen that both groups

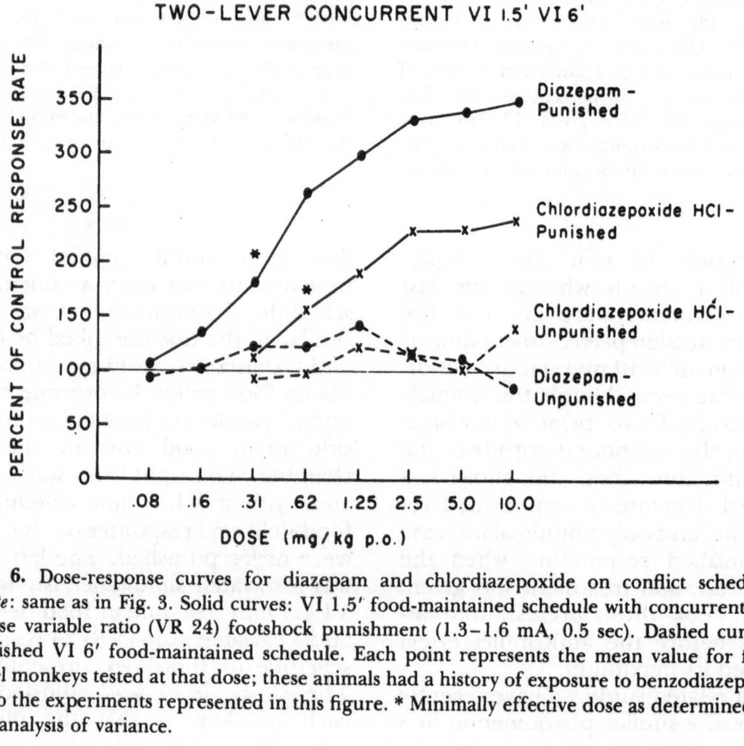

Figure 6. Dose-response curves for diazepam and chlordiazepoxide on conflict schedule. *Ordinate*: same as in Fig. 3. Solid curves: VI 1.5′ food-maintained schedule with concurrent 24-response variable ratio (VR 24) footshock punishment (1.3–1.6 mA, 0.5 sec). Dashed curves: unpunished VI 6′ food-maintained schedule. Each point represents the mean value for four squirrel monkeys tested at that dose; these animals had a history of exposure to benzodiazepines prior to the experiments represented in this figure. * Minimally effective dose as determined by group analysis of variance.

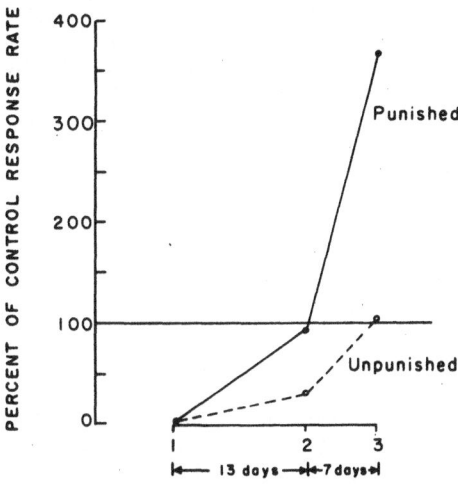

TWO-LEVER CONC. VI VI PUNISHMENT

SEQUENCE OF TREATMENTS

Figure 7. Effect of 5 mg/kg diazepam orally in a conflict trained drug-naive squirrel monkey. Experimental details as in Fig. 6. *Ordinate*: same as Fig. 3. *Abscissa*: monkey received diazepam treatment #1 and a marked reduction in punished and unpunished responding was observed. Unpunished rate was again decreased after treatment #2–13 days later, and punished rate was unchanged. After treatment #3, clear anticonflict effect was seen and unpunished responding was no longer affected.

had equivalent response rates during the VI segments. In contrast, the punished rats' FR rates were suppressed whereas the unpunished rats' FR rates were higher than the VI rates. Since other authors have considered the initial depression of VI rates in conflict-trained drug naive rats to reflect the sedative properties of benzodiazepines (4, 34), we wanted to determine whether such an effect would occur in both groups of our drug-naive animals. Figure 8 shows an interesting interaction between the schedule of reinforcement and the drug. In the punished group, the first treatment with chlordiazepoxide decreased VI response rate, with no change in FR response rate. However, in the unpunished rats, similar treatment did *not* produce a decrese in VI or FR rates. This finding indicates that additional work will be required

to define the meaning of the VI decrease in drug-naive punished subjects. For the present, caution is required when one attempts to label this as a reflection of a drug's sedative action.

Having stated this caution, we would like to set it aside temporarily in order to discuss the next topic. Wise, Berger and Stein (52) have proposed that a serotonergic mechanism is involved in mediating the suppressant effects of punishment on behavior. Using operant conditioning methods, we and others have studied this hypothesis and have used parachlorophenylalanine (PCPA), which depletes brain serotonin, as a research tool. We would like to point out that these studies have not consistently clarified serotonin's role in these phenomena. However, because of the high current interest in the biochemical mechanisms underlying behavior, we feel they deserve mention.

With respect to the two primary behavioral actions of benzodiazepines in the multiple schedule conflict model, Wise, Berger and Stein (52) have proposed that the depression of unpunished responding is associated with a drug induced decrease in norepinephrine turnover, and the conflict attenuation effect with a decrease in serotonin turnover. One of the findings on which they based their hypothesis was the reported ability of PCPA to attenuate conflict behavior in the Geller version of the conflict procedure (19, 42, 53). Figure 9 shows two experiments we conducted (10) with PCPA under different treatment regimes (300 mg/kg orally once, or 100 mg/kg orally daily for 3 days) in which

PCPA did not produce any pronounced changes in either punished or unpunished (not shown) response rates. Blakely and Parker (2) have reported similar results. We are currently attempting to determine whether the PCPA effect is specific to the Geller punishment schedule, since the negative findings were obtained in schedules that used a reduced shock density compared to the Geller schedule.

Although we did not observe any pronounced effect of PCPA by itself, we were surprised to note that when chlordiazepoxide was administered at the time of maximal serotonin depletion, the anticonflict response to chlordiazepoxide was completely blocked. A week later when the serotonin levels were returning to normal (28, 31) chlordiazepoxide again began to have its typical

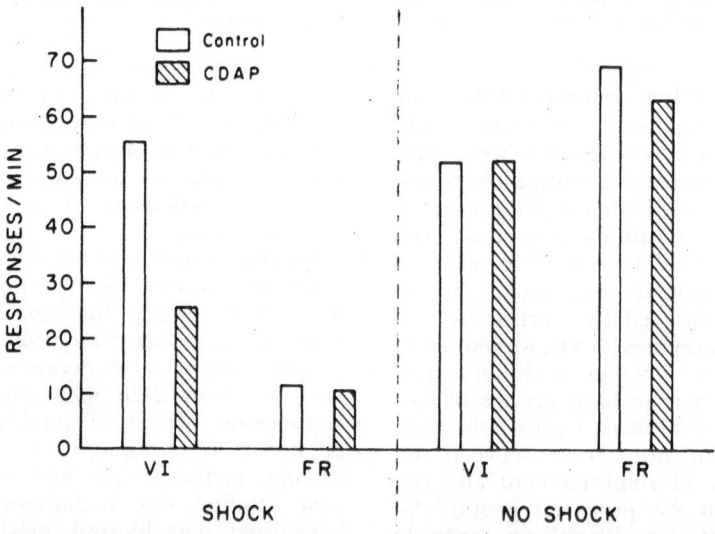

Figure 8. Comparison of first administration of chlordiazepoxide to drug-naive rats trained on multiple schedules with and without punishment. Open bars: mean ±SE of last three sessions prior to drug session. Striped bars: 10 mg/kg dose, first treatment. Left half: five rats were trained on multiple VI 30″ FR 10 schedule with footshock punishment at completion of fixed ratios. Right half: five rats were trained on same schedule without punishment. Only the punished rats showed a reduction in VI responding ($P < 0.005$) after drug administration.

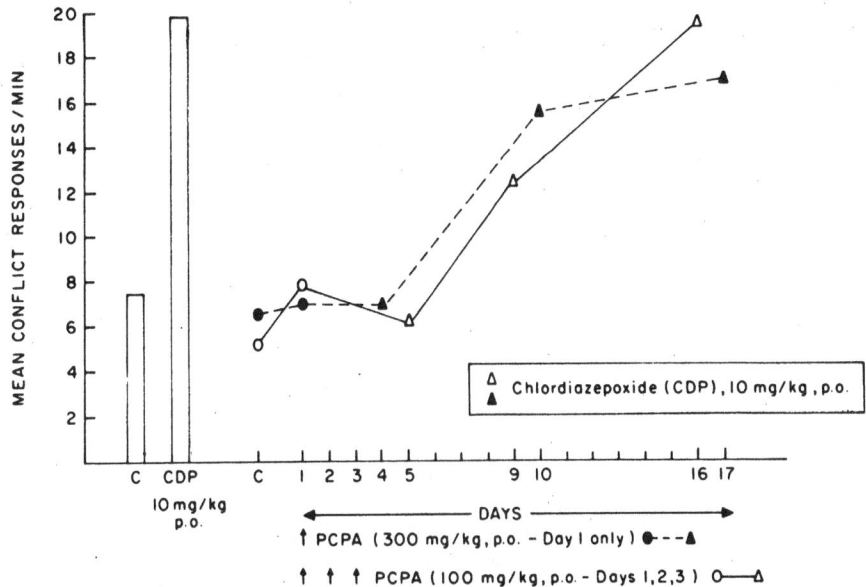

Figure 9. Effects of chlordiazepoxide on punished responding in multiple VI 30″ FR 10 conflict schedule in rats pretreated with parachlorophenylalanine (PCPA—free acid form). Four rats served as subjects. *Ordinate*: mean FR (punished) responses per minute. *Abscissa*: approximately one week prior to PCPA administration, the typical punishment-attenuating effect of chlordiazepoxide hydrochloride was measured and is indicated by the bar graphs (c = control). The four subjects received three consecutive daily doses of PCPA (suspended in 5% acacia) approximately 40 min prior to testing; two of the rats received a booster dose 5 days later. Chlordiazepoxide was given 2, 7, and 15 days following each rat's last PCPA dose (open symbols, solid curve). Subsequently, the experiment was replicated in the same rats with a single PCPA injection (300 mg/kg). Chlordiazepoxide was given 2, 9, and 15 days post-PCPA administration (filled symbols, dashed curve). In both PCPA experiments, the response to chlordiazepoxide was completely blocked 2 days after the last PCPA treatment and had returned to pre-PCPA levels by the 15th day.

effect, and by 15–16 days after PCPA the full effect was restored. (In a separate group of rats, we measured brain serotonin levels and found that our PCPA regimen produced effective serotonin depletion.) Although these results suggest that a serotonergic system may be important in these phenomena, this concept must still be regarded as speculative. Certainly, the effects of PCPA may involve substances other than serotonin. In addition, there are examples in which there seemed to be additivity between the effects of chlordiazepoxide and PCPA (26). We

have also reported such experiments in the multiple VI FR conflict procedure (10). Figure 10 illustrates still another observation made in rats performing on a Sidman avoidance schedule. Like high doses of chlordiazepoxide, PCPA by itself disrupted avoidance responding. In PCPA treated animals, when chlordiazepoxide was given at the point of assumed maximal serotonin decrease, its avoidance inhibiting effect was enhanced, perhaps by simple addition.

We would now like to comment on one last behavioral procedure

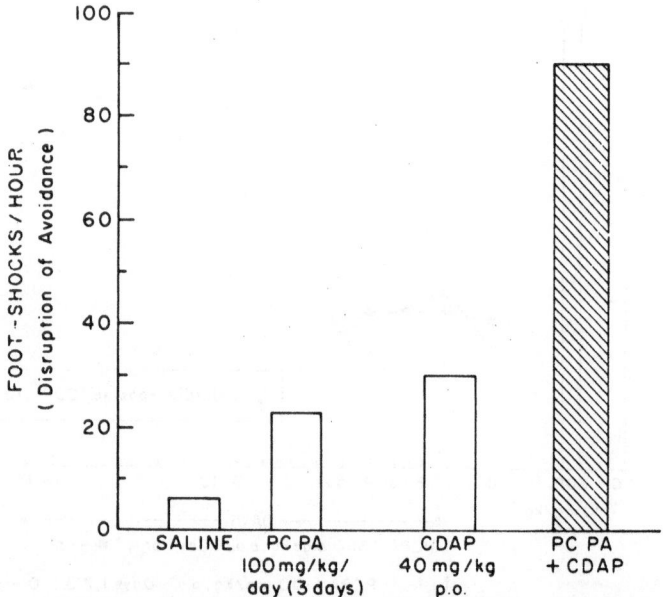

Figure 10. Effect of chlordiazepoxide on Sidman avoidance performance in rats pretreated with parachlorphenylalanine (PCPA—free acid form). Four rats served as subjects. *Ordinate*: footshocks per hour. *Abscissa*: saline bar represents control rates for shocks received. Both chlordiazepoxide and PCPA had avoidance-disrupting effects. There was at least an additive effect when the two drugs were given together.

that has been proposed for over 30 years as a model for human anxiety and fear and that has frequently been employed in pharmacological investigations. This is called the conditioned emotional response procedure (CER). There are many similarities to the conflict-punishment design: a main feature common to both methods is the use of shock in the presence of a discriminative stimulus to suppress ongoing responding for a rewarding substance, e.g., food. The major distinction be-between the two methods, however, is that the delivery of shock is response-contingent in the conflict-punishment procedure—the animal can avoid the shock by refraining from working for the positive reinforcement—but is not response-contingent in the CER design—the animal is shocked

whether it works or stops working for the food. Shock-produced suppression has been postulated to be due to anxiety or fear.

A review by Millenson and Leslie (37) illustrates that the conditioned emotional response procedure has been relatively insensitive, and inconsistent with respect to any positive drug effects that have been observed. Furthermore, the method seems to be quite nonspecific in that it does not seem to distinguish well among drug classes that are recognized as being different from one another clinically (e.g., major versus minor tranquilizers). It is quite understandable, therefore, that conflict methods have been accepted as being more reliable, consistent and discriminating tests (30). Millenson and Leslie (37), however, have recently tried to suggest dimensions or variables that would make some sense out of the seemingly inconsistent CER results. They have suggested that the treatment regimen, chronic versus acute dosing, helps to clarify the pattern of results. Phenothiazines, reserpine and meprobamate have been observed to reduce the amount of response suppression only when these agents were administered either chronically or at ataxic doses acutely; acute subataxic doses were without effect. Benzodiazepines, on the other hand, were active only after acute, but not chronic administration. The number of studies done to date, however, is too small to allow a conclusion yet that Millenson and Leslie's hypothesis regarding acute versus chronic dosing is correct. If correct, we might speculate that the pattern of results described in the last few sentences suggests that the CER paradigm is possibly most sensitive to major and minor tranquilizers at those times when their sedative effects are strongest.

The CER paradigm is a good example for the final point we wish to make. This procedure has a long history as an animal model for human "fear and anxiety" (37). Investigators have suggested that pharmacological alteration of behavior that has been controlled by an aversive reinforcer like footshock is due to an effect on "conditioned fear" processes assumed to be involved in such patterns of behavior. However, the available experimental evidence does not allow a conclusion that psychopharmacological agents act directly on a common unitary process of "fear." When one compares the conditioned avoidance procedure, the punishment (conflict) procedure, and the conditioned emotional response model, all of which might be assumed to involve "conditioned fear," it becomes apparent that, dependent on the experimental

design and the behavioral patterns developed in such experiments, specific drugs will modify some behavioral patterns maintained by aversive footshock and not other behaviors also dependent on the same type of aversive footshock. Experimental designs and factors such as schedules of reinforcement seem to be prepotent over the reinforcement itself in regard to the effects of certain psychopharmacological agents. Nevertheless, it is clear that schedules of reinforcement have played a valuable role in allowing empirical extrapolations to be made to humans from animal models.

The authors gratefully acknowledge the research contributions of Edward Boff, Fred S. Grodsky and John W. Sullivan to these studies, and the helpful comments of Arnold B. Davidson on the manuscript.

REFERENCES

1. BARTHOLINI, G., W. HAEFELY, M. JALFRE, H. H. KELLER AND A. PLETSCHER. Br. J. Pharmacol. 46: 736, 1972.
2. BLAKELY. T. A., AND L. F. PARKER. Pharmacol. Biochem. Behav. 1:609, 1973.
3. BUSER, P., L. COOK, C. GIURGEA, E. JACOBSEN, O. S. RAY, M. RICHELLE, A. P. SILVERMAN, L. STEIN AND Z. VOTAVA. In: The Neuroleptics. Mod. Probl. Pharmacopsychiatry, Vol. 5, edited by D. P. Bobon, P. A. J. Janssen and J. Bobon. Basel: Karger, 1970, p. 85–108.
4. CANNIZZARO, G., S. NIGITO, P. M. PROVENZANO AND T. VITIKOVA. Psychopharmacologia 26: 173, 1972.
5. CARLSSON, A., AND M. LINDQVIST. Acta Pharmacol. Toxicol. 20: 140, 1963.
6. COOK, L., AND A. C. CATANIA. Federation Proc. 23: 818, 1964.
7. COOK, L., AND A. B. DAVIDSON. In: The Benzodiazepines, edited by S. Garattini, E. Mussini and L. O. Randall. New York: Raven, 1973, p. 327–345.
8. COOK, L., AND R. T. KELLEHER. In: Neuro-Psychopharmacology, Vol. 2, edited by E. Rothlin. Amsterdam: Elsevier, 1961, p. 77–92.
9. COOK, L., AND R. T. KELLEHER. Ann. N.Y. Acad. Sci. 96: 315, 1962.

10. COOK, L., AND J. SEPINWALL. In: *Emotions —Their Parameters and Measurement*, edited by L. Levi. New York: Raven, 1975.

11. COOK, L., AND E. WEIDLEY. *Ann. N.Y. Acad. Sci.* 66: 740, 1957.

12. CORRODI, H., K. FUXE AND T. HÖKFELT. *Life Sci.* 6: 767, 1967.

13. CROW, T. J., AND C. GILLBE. *Nature New Biol.* 245: 27, 1973.

14. DAVIDSON, A. B., AND L. COOK. *Psychopharmacologia* 15: 159, 1969.

15. DAVIES, J. A., B. JACKSON AND P. H. REDFERN. *Neuropharmacology* 12: 735, 1973.

16. DeMAIO, D. *Arzneim.-Forsch.* 22: 919, 1972.

17. DEWS, P. B. *Naunyn-Schmiedebergs Arch. Exp. Pathol. Pharmakol.* 248: 296, 1964.

18. FOREE, D. D., F. H. MORETZ AND D. E. McMILLAN. *J. Exp. Anal. Behav.* 20: 291, 1973.

19. GELLER, I., AND K. BLUM. *Eur. J. Pharmacol.* 9: 319, 1970.

20. GELLER, I., AND J. SEIFTER. *Psychopharmacologia* 1: 482, 1960.

21. GOODMAN, L. S., AND A. GILMAN. *The Pharmacological Basis of Therapeutics*, 4th ed. New York: Macmillan, 1970, p. 155–165.

22. GROSS, H., AND E. LANGNER. *Int. Pharmacopsychiatry* 4: 220, 1970.

23. HANSON. H. M., J. J. WITOSLAWSKI AND E. H. CAMPBELL. *J. Exp. Anal. Behav.* 10: 565, 1967.

24. HANSON, H. M., C. A. STONE AND J. J. WITOSLAWSKI. *J. Pharmacol. Exp. Ther.* 173: 117, 1970.

25. IONESCO, R., S. U. NICA, L. OPROIU, A. NITURAD AND B. TUDORACHE. *Pharmakopsychiatr. Neuro-Psychopharmakol.* 6: 294, 1973.

26. JALFRE, M., M. A. MONACHON AND W. HAEFELY. In: *Proc. First Canadian International Symposium on Sleep*, edited by D. J. McClure. Montreal: Hoffmann-La Roche, 1972.

27. JANSSEN, P. A. J., C. J. E. NIEMEGEERS AND K. H. L. SCHELLEKENS. *Arzneim.-Forsch.* 15: 104, 1965.

28. JEQUIER, E., W. LOVENBERG AND A. SJOERDSMA. *Mol. Pharmacol.* 3: 247, 1967.

29. KELLEHER, R. T., W. C. RIDDLE AND L. COOK. *J. Exp. Anal. Behav.* 5: 3, 1962.

30. KELLEHER, R. T., AND W. H. MORSE. *Ergeb. Physiol. Biol. Chem. Exp. Pharmakol.* 60: 1, 1968.

31. KOE, B. K., AND A. WEISSMAN. *J. Pharmacol. Exp. Ther.* 154: 499, 1966.

32. LICHTENSTEIN (STITZER), M. Paper presented at 45th Annual Meeting, Eastern Psychological Association, Philadelphia, Pa., April 18–20, 1974.

33. MARGULES, D. L., AND L. STEIN. In: *Neuropsychopharmacology*, edited by H. Brill. Amsterdam: Excerpta Medica Foundation, 1967, p. 108–120.

34. MARGULES, D. L., AND L. STEIN. *Psychopharmacologia* 13: 74, 1968.

35. MATTHYSSE, S. *Federation Proc.* 32: 200, 1973.

36. McMILLAN, D. E. *J. Exp. Anal. Behav.* 19: 133, 1973.

37. MILLENSON, J. R., AND J. LESLIE. *Neuropharmacology* 13: 1, 1974.

38. MORPURGO, C., AND W. THEOBALD. *Psychopharmacologia* 6: 178, 1964.

39. NIEMEGEERS, C. J. E., F. J. VERBRUGGEN AND P. A. J. JANSSEN. *Psychopharmacologia* 16: 161, 1969.

40. NIEMEGEERS, C. J. E., F. J. VERBRUGGEN AND P. A. JANSSEN. *Psychopharmacologia* 16: 155, 1969.

41. PLETSCHER, A., AND M. DaPRADA. In: *Neuro-Psychopharmacology*, edited by H. Brill. Amsterdam: Excerpta Medica Foundation, 1967, p. 304–311.

42. ROBICHAUD, R. C., AND K. L. SLEDGE. *Life Sci.* 8: 965, 1969.

43. RUIZ, M., AND J. M. MONTI. *Eur. J. Pharmacol.* 20: 93, 1972.

44. SCHECKEL, C. L., AND E. BOFF. *Psychopharmacologia* 5: 198, 1964.

45. SEPINWALL, J., F. S. GRODSKY, J. W. SULLIVAN AND L. COOK. *Psychopharmacologia* 31: 375, 1973.

46. SIDMAN, M. *Science* 118: 157, 1953.

47. SNYDER, S. H., S. P. BANERJEE, H. I. YAMAMURA AND D. GREENBERG. *Science* 184: 1243, 1974.

48. STILLE, G., H. LAUENER AND E. EICHENBERGER. *Farmaco, Ed. Prat.* 26: 603, 1971.

49. THOMPSON, D. M., AND P. B. CORR. *J. Exp. Anal. Behav.* 21: 151, 1974.

50. UEKI, S., N. OGAWA, S. WATANABE, Y. GOMITA, Y. ARAKI, M. FUJIWARA, C. KAMEI, K. SHIMOMURA, M. INOUE, R. OHISHI, N. IBII AND K. TANAKA. *Nippon Yakurigaku Zasshi.* 69: 85, 1973.

51. VAN ROSSUM, J. M. In: *Neuro-Psychopharmacology*, edited by H. Brill. Amsterdam: Excerpta Medica Foundation, 1967, p. 321–329.

52. WISE, C. D., B. D. BERGER AND L. STEIN. *Science* 177: 180, 1972.

53. WISE, C. D., B. D. BERGER AND L. STEIN. *Biol. Psychiatry* 6: 3, 1973.

54. WÜTTKE, W., AND R. T. KELLEHER. *J. Pharmacol. Exp. Ther.* 172: 397, 1970.

Discrete trial analysis of drug action[1]

GEORGE A. HEISE

Department of Psychology, Indiana University
Bloomington, Indiana 47405

ABSTRACT

Discrete trial procedures permit exact control or description of the time of occurrence of stimuli, the probability of response occurrence, and the patterning of responses. They also make possible the experimental manipulation of the composition of the stimuli controlling behavior. The use of discrete trial procedures is illustrated here in an examination of the effects of scopolamine, a representative cholinergic blocker, on several aspects of behavior: *Memory.* Response alternation experiments, in which the spacing of discrete trials varies within the experimental session, show that, whereas accuracy of responding is consistently poorer under drug, the decline of accuracy with time since last trial is similar for drugged and nondrugged animals. Thus the drug does not affect memory "storage." *Inhibition.* Experiments in which discrete trials are presented in pairs, such that the correct response on Trial 2 of the pair is contingent upon Trial 1 events, show how the "disinhibiting" effect of scopolamine (as indicated by enhanced responding on "no go" trials) is augmented by increasing the time gap between Trial 1 and Trial 2, or by minimizing controlling stimuli on Trial 1. *Discrimination.* A variety of experiments suggest that scopolamine decreases the "detectability" of stimuli. Detectability effects, along with disinhibition observed under certain specific conditions, constitute the principal behavioral actions of scopolamine observed with discrete trial procedures.—HEISE, G. A. Discrete trial analysis of drug action. *Federation Proc.* 34: 1898–1903, 1975.

Since the pioneer studies of Dews (4), the sophisticated analysis of the behavioral action of drugs has been closely identified with so-called "free operant" procedures, in which *rate* of performing a simple response like lever pressing or key pecking is the principal dependent variable. There are good reasons besides historical primacy for this free operant emphasis. The continuous monitoring of rate of responding by means of cumulative records provides a direct and dramatic demonstration of systematic relationships between drug

[1] Supported by Grant MH 14658 from the National Institute of Mental Health.

treatment and behavioral output. Rate of response can directly reflect the effects of drug-induced changes on motivational or emotional variables. Finally, there is convincing evidence that the effects of drugs often depend not on the events maintaining the behavior (i.e., the motivational state or the type of reinforcer) but on the characteristics of the behavior itself and, specifically, on the rate of response (Kelleher and Morse, 10).

Nevertheless, free operant techniques that rely on rate measures are not optimal for measuring *all* behavioral effects of drugs. Sometimes it is necessary to control precisely the conditions of stimulus presentation—to specify precisely when a stimulus is presented and systematically to vary its characteristics. It may be desirable to determine the probability of a response, or to determine the probability that, if a response does occur, that the response is correct. Discrete trial procedures meet these needs. In these procedures a stimulus is presented at a time controlled by the experimenter and a single response to this stimulus is required for reward. The measure of responding in discrete trial procedures is a probability—the proportion of the stimulus occasions on which the required response occurs—rather than a rate.

The remainder of this paper examines the behavioral actions of scopolamine, as a representative cholinergic blocker. The behavioral effects that have been attributed to scopolamine are precisely those that call for a discrete trials analysis. Scopolamine has been alleged to be an "amnesic" drug that disrupts memory "storage" (9): this claim can be evaluated by specifying exactly the time of stimulus input and measuring the changes in performance that occur as a consequence of a time gap

between stimulus and response. Scopolamine increases the probability of occurrence of responses ordinarily suppressed as a consequence of nonreward (2): the generality and magnitude of this effect of the drug can be evaluated by measuring the probability of responding on "no go" (nonrewarded) trials. Scopolamine may also affect stimulus discrimination: such effects can be measured in experiments where the animal's ability to respond differentially is measured as the characteristics of trial stimuli are systematically varied.

The control technology in the discrete trial experiments to be described was similar to that used in free operant experiments. The experimental enclosures contained one or two response levers and were enclosed in a sound-deadening chamber. The stimuli that signaled the discrete trials—a pure tone, noise, or illumination of a panel light—were controlled automatically by relay circuitry or by computer, as was the programming of reinforcement (a drop of sugar-water) and the recording and tabulation of responses. As in free operant experiments, the experimental animal was not handled or subjected to other uncontrolled stimuli during the experimental session; a "pretrial delay" interposed between a response during the intertrial interval and the onset of the next trial effectively limited most lever responses to the period of trial presentation.

MEMORY

The essential feature of a "memory" experiment is the occurrence of a time gap between stimulus presentation and the occasion for performance of the response associated with that stimulus. When measuring the "amnesic" effects of a drug in such an experiment it is not sufficient merely to measure changes in performance produced by the drug: this confounds possible drug effects on input–output processes, i.e., on discrimination, with possible drug effects on specifically time-dependent processes, "storage." Drug effects on discrimination are indicated by changes in responding to stimuli present in the external environment at the time that the response occurs ("stimulus present" effects); these changes, although undeniably important and a readily measured consequence of scopolamine administration (see below), are not what are usually implied by drug effects on memory. A drug effect on memory implies, rather, an effect on storage, that is, on time-dependent processes. Effects on storage are measured under "stimulus absent" conditions, by determining changes in response probability as a function of the time gap between stimulus presentation and the associated response.

Obviously it would be desirable to determine these time-response functions by testing the same animals repeatedly at different stimulus–response time gaps, rather than to piece the functions together by assembling data from different animals at different delays. Testing the same animal repeatedly under comparable conditions in a memory experiment would appear at first to be impossible, since each successive test trial could alter what is learned and remembered. Delayed response and delayed matching situations provide a resolution of this apparent dilemma. In these procedures the animal is thoroughly trained on the behavior appropriate for all trials; e.g., in delayed spatial alternation (a form of delayed response) it is first trained always to perform the alternative response to that performed on the last trial. In addition, each trial poses a "unique" memory task (15): the animal must

perform that *particular* response (right or left response in the delayed alternation example) which is appropriate on that particular trial. The "appropriate" response on a trial is determined by the events (stimuli) that occurred on the immediately preceding trial (or trials) and were "remembered" over the intertrial interval. Since each trial poses a new or "unique" memory task the animal can be tested repeatedly—with different intertrial intervals and under different dosage conditions—with minimal cumulative effects from learning on preceding trials.

This, then, was the strategy that was used to determine the effects of scopolamine on memory storage in the rat. A "variable intertrial interval spatial alternation" task was employed (7). Rats were first thoroughly trained alternately to press the left and right levers on discrete trials, with the time between successive trials being varied. Each trial was a "unique" memory task, as defined above: in order to respond correctly on a trial, the animal had to "remember" what happened on the immediately preceding trial. Drug effects on memory storage were then determined by comparing the functions relating intertrial interval to accuracy of alternation performance measured under control and drug conditions.

Scopolamine effects on alternation performance were studied in two groups of rats—a "short" group for which the intertrial intervals, presented in a modified random order, were 2.5, 5, 10, 20, and 40 sec, and a "long" group for which the intertrial intervals were 10, 20, 40, 80, and 160 sec. Trials were signaled by the combined illumination of a panel light located midway between the two levers and the sounding of a 1,000 Hz tone; maximum trial duration was 5 sec. A single correct lever response during a trial (i.e., a response on the lever not pressed on the preceding trial) produced a drop of sugar-water reinforcement and terminated the trial. A single press on the incorrect lever terminated the trial without reinforcement; "correction" trials were presented following all incorrect trial responses.

An experimental session consisted of 100 reinforced trials—approximately 20 at each of the five different intertrial intervals. Thus the accuracy of responding for five different delays between stimulus and response was measured from an individual animal in a single session. Performance was scored in terms of the proportion of initial trials at each intertrial interval on which, if the animal responded, it responded correctly. ("Initial" trials were trials that immediately followed trials on which the animal responded correctly.)

The accuracy-by-interval functions obtained with the short intertrial interval (2.5–40 sec) group on control runs and following administration of various doses of scopolamine are presented in Fig. 1. Figure 1 shows that accuracy decreased regularly with increasing intertrial interval, and that scopolamine reduced accuracy in proportion to dose. Statistical analyses showed, also, that the various control and drug accuracy-by-interval curves were parallel. The differences between drugged and control performance were as great for the 2.5 sec intertrial interval, the shortest delay tested, as for much longer time intervals.

The results show that scopolamine did not alter memory storage; the *rate* of decline in accuracy was unaffected by drug treatment. Instead, the drug apparently affected discrimination processes: decrements as large as those observed would presumably have been obtained even if there had been no time gap at all

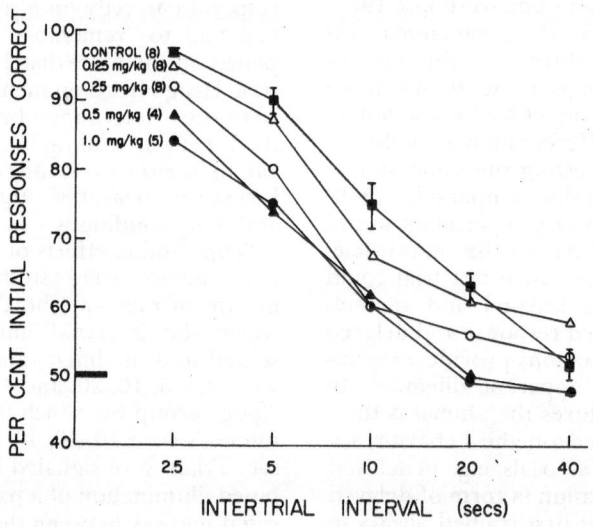

Figure 1. Effect of scopolamine on accuracy of variable intertrial interval spatial alternation performance by the 2.5–40 sec group (no. = 8). Numbers in parenthesis indicate number of animals contributing data; where less than 8, some rats did not respond under drug. Brackets indicate ±1 standard error for control performance. The 0.125 and 1.0 mg/kg doses were each given once and the 0.25 and 0.5 mg/kg doses were each given twice to each rat.

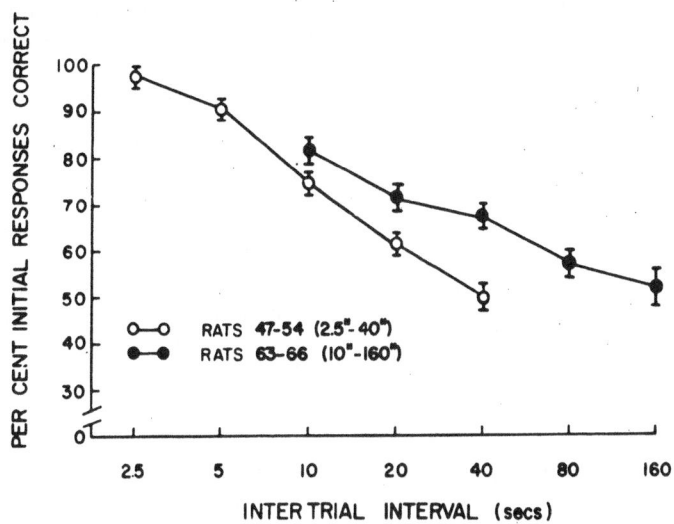

Figure 2. Accuracy of base line control performance of variable interval spatial alternation performance of 2.5–40 sec group (no. = 8) and 10–160 sec group (no. = 4). Brackets indicate ±1 standard error.

underlying physiological or biochemical "memory decay" process. The form of the curves is not invariant: the accuracy of performance at a given delay can be significantly changed by environmental manipulations that would not be expected to alter a memory decay process.

It will be recalled that, in addition between stimulus and response. In other words, the drug impaired the animal's ability to respond appropriately to stimulus inputs, but did not alter retention of whatever the animal did learn on a trial.

Similar findings were obtained by Glick and Jarvik (5), who studied the effect of scopolamine on variable interval delayed matching performance in monkeys. These authors also concluded that scopolamine did not affect storage.

A caveat must accompany the interpretation of these results. The accuracy-by-interval curves in Fig. 1 are not direct behavioral reflections of an to the group of rats trained on the

2.5–40 sec variable interval schedule of trials that produced the scopolamine results already described, a second group of rats was trained on a variable interval schedule of trials ranging from 10 to 160 secs. Figure 2 compares the baseline (nondrug) accuracy-by-interval curves obtained for the 2.5–40 and 10–160 sec groups. The animals in the 10–160 sec group responded more accurately than the 2.5–40 sec animals on 10, 20, and 40 secs —the three intervals common to both groups; all three of these differences in performance between the two groups are statistically significant. These differences in performance were not due merely to fortuitous selection of "superior" rats in the 10–160 sec group. Krank, Christoph and Leinwand (11) have obtained analogous results by training the *same* rats on two different variable interval schedules on alternate experimental sessions.

Thus the accuracy of performance at a particular delay in the variable interval spatial alternation experiment depends not only on the length of the delay but also on the "spectrum" of delays presented during the experimental session. D'Amato (3) contends that the decrease in accuracy after a delay in "unique" memory experiments of this type does not reflect the decay over time of "memory storage." It reflects, rather, the animal's ability on a trial n to discriminate between the immediately preceding trial (trial $(n-1)$) and the trial that precedes trial n-1 (i.e., trial n-2). Accuracy of performance of this discrimination will therefore depend on the duration and sequence of the intertrial intervals and trial events that must be discriminated. The decrease in accuracy produced by scopolamine would accordingly be interpreted as impairment in the animal's ability to make this discrimination.

INCREASED RESPONDING

Carlton, in a seminal review (2), postulated that a principal effect of scopolamine was to increase responding ordinarily suppressed due to nonreward. Although experiments cited by Carlton (2), Heise, Laughlin and Keller (8), and other authors definitely establish that scopolamine does increase responding on nonrewarded trials, it is not clear whether the drug *selectively* increases nonrewarded trial-responding, or whether, instead, the drug impairs discrimination between the occasions for responding and nonresponding. A "selective" increase in nonrewarded responding would constitute evidence that the drug "disinhibits" behavior. An impairment in discrimination (as will be discussed more extensively below) could indicate either disinhibition of responding or a change in sensitivity to stimuli.

If scopolamine selectively increases responding on nonrewarded trials,

then the increase in responding on the "no go" trials of a discrete trial "go/no go" stimulus-present discrimination should be greater than the decrease in responding on the "go" trials of this discrimination. (In a go/no go discrimination, responses made in the presence of the no go or negative stimulus are not rewarded.) The top portion of Fig. 3 shows typical results from a discrimination experiment, in which the go and no go stimuli were, respectively, 1,000 and 3,000 Hz tones. The drug evidently did *not* selectively increase responding on the no go trials: the small increase in no go trial responding that did occur under the drug was accompanied by an equivalent decrease in responding on the go trials. Thus the modest increase in no go responding observed in this experiment may plausibly be ascribed to a drug effect on discrimination.

In go/no go discrimination experiments like the one just described, possible selective effects of a drug on no go trials are confounded with effects on discrimination. A more definitive evaluation of drug effects on discrimination is obtained from a go/go discrimination experiment, in which no go trials do not occur. The bottom portion of Fig. 3 shows the effects of scopolamine on a go/go discrimination in which rats were rewarded for pressing the right lever on trials with a 1,000 Hz tone and for pressing the left lever on trials with a 3,000 Hz tone. The drug impaired the go/go discrimination between the two tones to about the same extent as it had impaired the go/no go discrimination between these tones.

Experiments have also been carried out in which the same animals performed both go/no go and go/go discriminations during the same experimental sessions. (Halgren, Milar: personal communications.) In these latter experiments, also, the small

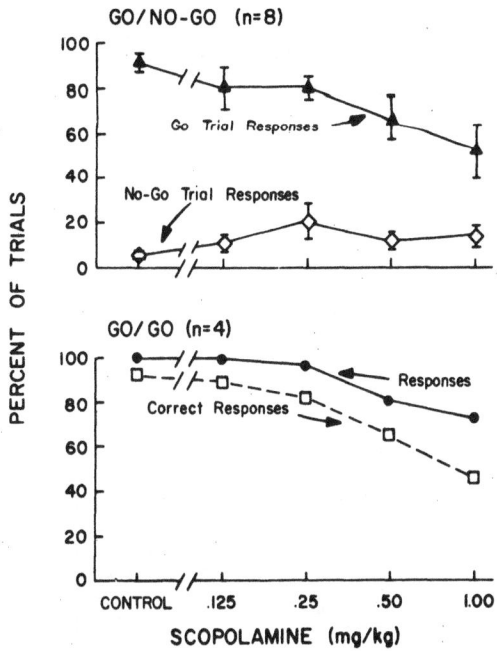

Figure 3. Dose-response curves for effects of scopolamine on two types of discrimination between 1,000 and 3,000 Hz tones. *Top*: Go/no go discrimination curves showing the mean pecentage of go and no go trials on which the animals responded. *Bottom*: Go/go discrimination curves showing the mean percentage of all trials on which the animals responded, and the mean percentage of all trials on which the animals responded correctly. (From Bradley: *Methods in Brain Research*. Copyright 1975 John Wiley & Sons Ltd.)

increases in responding on no go trials of the go/no go discrimination produced by scopolamine were paralleled by decrements in performance on the go/go discrimination.

Nevertheless, dramatic increases in non-rewarded responding have been reported in some experiments (e.g., Hearst, (6)). In contrast to the stimulus-present go/no go discriminations experiment described above, those experiments in which increased responding did occur were characterized by *1*) a delay or "gap" between the time of presentation of the controlling stimulus and the occasion for the response, and *2*) minimal or reduced stimulus control of responding. These factors were varied systematically in a series of

experiments designed to delineate more precisely the conditions under which scopolamine increases responding on no go trials.

The essential features of the "go/no go two-trial" experimental paradigm employed are diagrammed in Fig. 4. The experiment incorporated features of the "memory" and the go/no go discrimination experiments previously described. Trials were presented in pairs, and 40 pairs of trials comprised an experimental session. As indicated in the figure, Trial 1 of each pair was a "stimulus present" trial providing differential stimuli (different tones, differential reinforcement, feedback from differential responding, or some combination of these); Trial 2, which followed

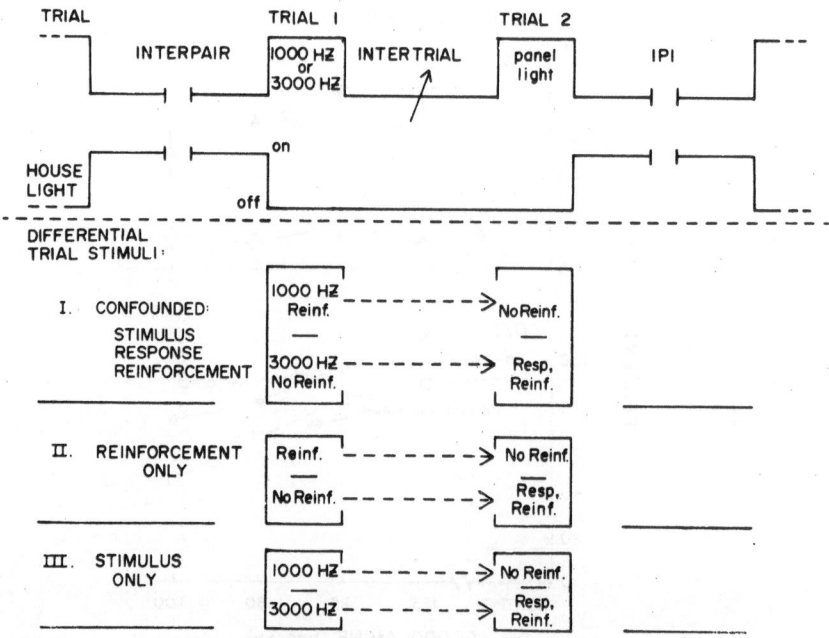

Figure 4. Schematic representation of the go/no go two-trial experimental design. *Top*: Trial pair. Trial 1 and Trial 2 of the pair were separated by a variable duration intertrial interval; each pair was separated from the preceding and following pairs by a fixed duration (40 sec) interpair interval (IPI). The house light in the experimental chamber was "off" from the onset of Trial 1 to the termination of Trial 2. *Bottom*: Experimental design, showing the Trial 1 conditions for each of the three groups (confounded, reinforcement only, and stimulus only), and the two different Trial 1 stimuli presented to each group. (Further explanation in text).

Trial 1 after a time gap, was a "stimulus absent" trial. Trial 2 was always signalled by the same nondifferential environmental stimulus, illumination of a panel light. Go or no go reponding on Trial 1 was controlled by the differential environmental stimuli present on the trial, whereas go or no go responding on Trial 2 was controlled by the stimuli or events that took place on Trial 1.

As indicated in Fig. 4, effects of scopolamine on no go responding were measured as a function of Trial 1 composition and of the length of the intertrial interval between Trial 1 and Trial 2. Different groups of rats were trained on each of three Trial 1

conditions, which differed in effectiveness in controlling behavior.

The "confounded" condition was so labeled because any or all of the different stimulus events on Trial 1 could have controlled Trial 2 performance. Trial 1 was a go/no go discrimination between 1,000 and 3,000 Hz tones; the rats were rewarded for responding to one of the tones. If the conditions under which responses were rewarded were presented on Trial 1, then responses on Trial 2 were not rewarded, and vice versa.

For "reinforcement only," as the label implies, the presence or absence of reward on Trial 1 was the differential cue that controlled performance on Trial 2. The same stimulus (a 1,000 Hz tone) was always the signal for Trial 1 in "reinforcement only"; the animal nearly always responded on Trial 1 but responding on only some of the trials was rewarded. As in the confounded condition, if the response on Trial 1 was rewarded, the response on Trial 2 was not, and vice versa.

Finally, for the "stimulus only" condition, either a 1,000 or 3,000 Hz tone was presented on Trial 1 but responses on Trial 1 were never rewarded (and the animals did not respond). When one of the tones was presented on Trial 1, a response on Trial 2 was rewarded, and when the other tone was presented on Trial 1, Trial 2 responses were not rewarded.

The effects of scopolamine dose and of time since presentation of Trial 1 on the performance of no go trials in these experiments are presented in Fig. 5. Fig. 5 shows both Trial 1 and Trial 2 performance for the confounded condition, and only Trial 2 performance for the reinforcement only and stimulus only conditions.

In the confounded condition, scopolamine did not increase responding on Trial 1 no go trials, nor did Trial 2 responding increase substantially under the drug even when the delay between the Trial 1 controlling stimulus and the Trial 2 response was 20 sec, the longest delay tested. For the reinforcement only condition, in contrast, the drug dramatically increased responding on Trial 2 no go trials at the 20 sec delay.

Trial 1 control of Trial 2 responding was weakest for the stimulus only condition. Intuitively, this would be expected because the animal merely "listened" on Trial 1. Objectively, the weakness in Trial 1 control was indicated by the relatively greater difficulty in training the animals for

the stimulus only condition, and by the comparatively high level of baseline no go responding that obtained under control conditions (see Fig. 5). With stimulus only a 5-sec delay was sufficient for a marked drug effect on no go responding.[2] Increases in no go responding were not observed with this brief delay for the other two conditions.

In contrast, scopolamine effects on go trial responding were relatively slight, comparable in magnitude to those observed in the go trials of the simple go/no go discrimination experiment described above (see Fig. 3).

Thus the present results indicate that scopolamine may selectively increase responding on nonreinforced trials, but this increase was demonstrated only under conditions of diminished stimulus control of the response. Specific conditions that produce this diminished control are the combined occurrence of a time gap between stimulus and response and a reduction in the usual complexity of the controlling stimulus.

SUMMARY AND CONCLUSIONS

Discrete trial and free operant procedures together make possible a reasonably comprehensive account of the effects of scopolamine on behavior in terms of the interaction between temporal factors and discrimination processes.

Scopolamine did not alter memory storage, a time-dependent process. The importance of temporal factors in determining the qualitative and quantitative effects of scopolamine on behavior is well established, however. Laties and Weiss (12), working in the

---

[2] It was not possible to obtain time-response or dose-response functions from the "stimulus only" animals because repeated drugging altered delayed "no go" trial performance under drug. These effects are currently being investigated in our laboratory.

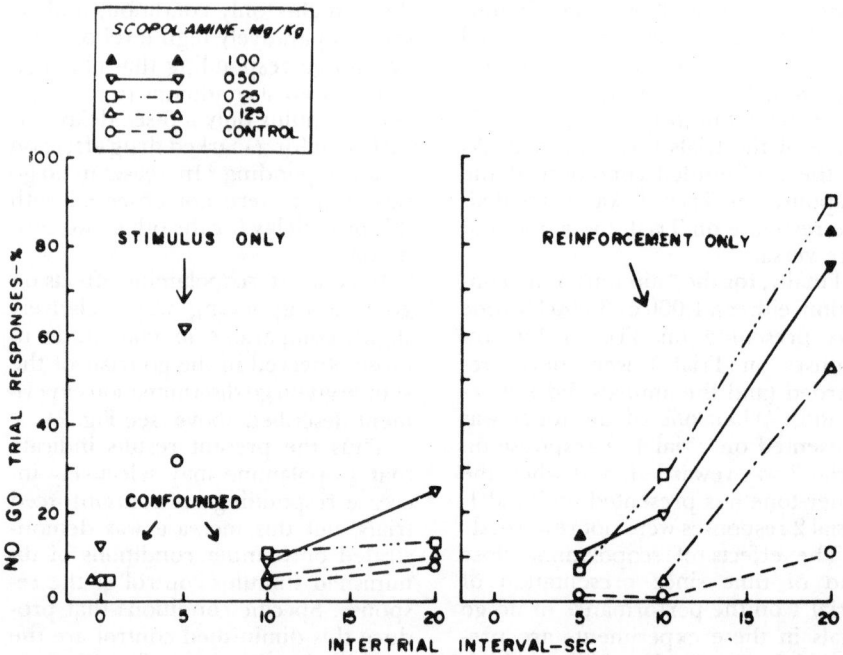

Figure 5. Effects of scopolamine and intertrial interval duration on no go trial responding in the go/no go two-trial experiment. *Left panel*: Trial 1 (intertrial interval (ITI) = 0) and Trial 2 (ITI = 10 and 20 sec) results for confounded group, and Trial 2 results (ITI = 5 sec) for the stimulus only group. Note relatively high control no go responding (responses occurred on 32.5% of no go trials) for stimulus only group. *Right panel*: Trial 2 no go trial responding for reinforcement only group. (Further explanation in text).

free operant tradition, demonstrated with pigeons that scopolamine impaired "internally controlled" (stimulus absent) responding much more than "externally controlled" (stimulus present) responding. The present results from the two-trial experiment confirm this finding and show, in addition, that scopolamine selectively increases no go responding if the controlling stimulus is attenuated and there is sufficient delay between the controlling stimulus and its associated response.

The present analysis has repeatedly called attention to scopolamine effects on discrimination. As previously discussed, the drug effects observed in the variable interval spatial alter-

nation experiment were probably due to a decrement in discrimination rather than memory storage. Scopolamine impaired both go/go and go/no go discrimination performance when stimulus presentation and response occurrence were concurrent and, as mentioned above, produced additional effects on no go responding when a delay was interposed between stimulus and response.

These drug effects on "discrimination" obviously require further analysis. Warburton (14) recognized that decrements in discrimination performance could indicate either a drug effect on stimulus input or a "disinhibition" of response output (or, presumably, a combination of the two), and carried out theory of signal detection experiments designed to separate the two effects. He identified stimulus input effects with effects on "stimulus sensitivity" or "signal-to-noise ratio" in the theory of signal detection, analysis, and "disinhibitory" effects on response output with effects on "response criterion" or "bias" in the theory of signal detection analysis. The theory of signal detection analysis was applied both to go/no go stimulus present discrimination experiments (14, 15) and to a "stimulus absent" experiment (1) in which rats responded on a 15-sec differential reinforcement of low rates schedule (that is, the rat was rewarded when it responded 15 sec or more after its previous response).

Warburton's results were clear: scopolamine reduced stimulus sensitivity, but in neither the stimulus present nor the stimulus absent theory of signal detection experiments did it shift response bias. Warburton attributed the changes in sensitivity to a drug-induced impairment in "attention" mechanisms and denied, on the basis of his results, that the drug disinhibits responding.

The Warburton analysis does not, of course, establish that there are *no* circumstances in which scopolamine disinhibits responding. Nevertheless, if a wide range of theory of signal detection experiments should, like the Brown and Warburton stimulus-absent theory of signal detection experiment (1), find no scopolamine effects on response bias, then we must conclude that scopolamine does not disinhibit responding. We suspect, however, that stimulus control in the Brown and Warburton experiment was functionally analogous to that of the two-trial stimulus-absent "confounded" experiment described above, in which selective increases in no go responding were not observed. We predict that stimulus-absent theory of sig-

nal detection experiments in which stimulus control is analogous to that in the two-trial stimulus only and reinforcement only experiments (and in which marked increases in no go responding did occur) will show that scopolamine does indeed induce response disinhibition.

Robert Conner was an indispensable collaborator on the original research reported in this paper.

REFERENCES

1. BROWN, K., AND D. M. WARBURTON. Attenuation of stimulus sensitivity by scopolamine. *Psychonomic Sci.* 22: 297, 1971.
2. CARLTON, P. L. Cholinergic mechanisms in the control of behavior by the brain. *Psychol. Rev.* 70: 19, 1963.
3. D'AMATO, M. R. Delayed matching and short-term memory in monkeys. In: *The Psychology of Learning and Motivation: Advances in Research and Theory, Vol. 7,* edited by G. H. Bower. New York: Academic, 1973, p. 227.
4. DEWS, P. B. Studies on Behavior. I. Differential Sensitivity to Pentobarbital of Pecking Performance in Pigeons Depending on the Schedule of Reward. *J. Pharmac. Exp. Ther.* 113: 393, 1955.
5. GLICK, S. D., AND M. E. JARVIK. Differential effects of amphetamine and scopolamine on matching performance of monkeys with lateral frontal lesions. *J. Comp. Physiol. Psychol.* 73: 307, 1970.
6. HEARST, E. Effects of scopolamine on discriminated responding in the rat. *J. Pharmac. Exp. Ther.* 126: 349, 1959.
7. HEISE, G. A., AND R. CONNER. Scopolamine effects on time dependent processes ("memory storage") in the rat measured in a variable interval spatial alternation task. Presented at Annual Meeting of the Midwestern Psychological Association, May 10–12, 1973, Chicago, Illinois.
8. HEISE, G. A., N. LAUGHLIN AND E. KELLER. A behavioral and pharmacological analysis of reinforcement withdrawal. *Psychopharmacologia* 16: 345, 1970.
9. INNES, I. R., AND M. NICKERSON. Drugs inhibiting the action of acetylcholine on structures innervated by postganglionic parasympathetic nerves (antimuscarinic or atropinic drugs). In: *The Pharmacological Basis of Therapeutics,* edited by

L. S. Goodman and A. Gilman. New York: Macmillan, 1965, 3rd ed, p. 521.

10. KELLEHER, R. T., AND W. H. MORSE. Determinants of the specificity of behavioral effects of drugs. *Ergeb. Physiol. Biol. Chem. Exp. Pharmakol.* 60: 1, 1968.

11. KRANK, M., G. CHRISTOPH AND S. LEINWAND. Proactive influences on spatial alternation in the rat. Presented at the Annual Meeting of the Midwestern Psychological Association, Chicago, Illinois, May 2–4, 1974.

12. LATIES, V. G., AND B. WEISS. Influence of drugs on behavior controlled by internal and external stimuli. *J. Pharmac. Exp. Ther.* 152: 388, 1966.

13. WARBURTON, D. M. The cholinergic control of internal inhibition. In: *Inhibition and Learning*, edited by R. Boakes and M. S. Halliday. London: Academic, 1972, p. 431.

14. WARBURTON, D. M., AND K. BROWN. Scopolamine-induced attenuation of stimulus sensitivity. *Nature* 230: 126, 1971.

15. WEISKRANTZ, L. Memory. In: *Analysis of Behavioral Change*, edited by L. Weiskrantz. New York: Harper and Row, 1968, p. 158.

Index